STUDENT RESOURCE MANUAL

to accompany

Elementary & Intermediate
Algebra
a combined course

by Charles P. McKeague

DR. JOHN GARLOW

Tarrant County Junior College

LORI PALMER

Utah Valley State College

BROOKS/COLE

THOMSON LEARNING

Australia • Canada • Mexico • Singapore • Spain
United Kingdom • United States

**For more information about our products,
contact us at:
Thomson Learning Academic Resource Center
1-800-423-0563**

**For permission to use material from this text,
contact us by:
Phone: 1-800-730-2214
Fax: 1-800-731-2215
Web: www.thomsonrights.com**

Asia
Thomson Learning
60 Albert Complex, #15-01
Alpert Complex
Singapore 189969

Australia
Nelson Thomson Learning
102 Dodds Street
South Street
South Melbourne, Victoria 3205
Australia

Canada
Nelson Thomson Learning
1120 Birchmount Road
Toronto, Ontario M1K 5G4
Canada

Europe/Middle East/South Africa
Thomson Learning
Berkshire House
168-173 High Holborn
London WC1 V7AA
United Kingdom

Latin America
Thomson Learning
Seneca, 53
Colonia Polanco
11560 Mexico D.F.
Mexico

Spain
Paraninfo Thomson Learning
Calle/Magallanes, 25
28015 Madrid, Spain

PREFACE

The first part of this Student Resource Manual contains complete solutions to every odd-numbered problem in the problem sets, chapter review sets, and the cumulative review sets for <u>Elementary and Intermediate Algebra: A Combined Course</u> by Charles McKeague. In addition it contains solutions to every problem in the chapter tests. If there are any errors, I apologize and take the responsibility for them.

The second part of this manual contains 50 real-world application problems by Utah Valley State College's Lori Palmer and is designed to show that algebra concepts arise in real life. To research these applications, Lori has conducted over 20 engaging interviews with professionals in such fields as aviation, optics, banking, and environmental sciences. Answers to the problems are presented at the back of the manual.

This manual should be used as a reference after attempting the solutions on your own. The importance of your initial effort can not be overstated. The quality of your learning experience in mathematics is directly related to the effort you apply in working the problems assigned to you.

I would like to thank Jay Campbell of Saunders College Publishing for his editorial assistance and for giving me the opportunity for authoring this manual and Charles P. McKeague for writing the textbook and carefully revising it to make the writing of the solutions go more smoothly. I also want to thank my wife Suzanne Garlow for the keyboarding, proof reading, and final formatting of the manual.

John Garlow
Tarrant County College-Southeast Campus
Arlington, Texas

TABLE OF CONTENTS

PREFACE iv

CHAPTER 1
The Basics 1

CHAPTER 2
Linear Equations and Inequalities 14

CHAPTER 3
Graphing and Linear Systems 55

CHAPTER 4
Exponents and Polynomials 87

CHAPTER 5
Factoring 111

CHAPTER 6
Rational Expressions 132

CHAPTER 7
Transitions 163

CHAPTER 8
Equations and Inequalities in Two Variables 208

CHAPTER 9
Rational Exponents and Roots 238

CHAPTER 10
Quadratic Functions 275

CHAPTER 11
Exponential and Logarithmic Functions 321

CHAPTER 12
Sequences and Series 351

CHAPTER 13
Conic Sections 379

SOLUTIONS TO SELECTED PROBLEMS

CHAPTER 1
The Basics

1.1 The Basics

1. $x + 5 = 14$

3. $5y < 30$

5. $3y \le y + 6$

7. $\frac{x}{3} = x + 2$

9. $3^2 = 3 \cdot 3 = 9$

11. $7^2 = 7 \cdot 7 = 49$

13. $2^3 = 2 \cdot 2 \cdot 2 = 8$

15. $4^3 = 4 \cdot 4 \cdot 4 = 64$

17. $2^4 = 2 \cdot 2 \cdot 2 \cdot 2 = 16$

19. $10^2 = 10 \cdot 10 = 100$

21. $11^2 = 11 \cdot 11 = 121$

23. $2 \cdot 3 + 5 = 6 + 5 = 11$

25. $2(3 + 5) = 2(8) = 16$

27. $5 + 2 \cdot 6 = 5 + 12 = 17$

29. $(5 + 2) \cdot 6 = 7 \cdot 6 = 42$

31. $5 \cdot 4 + 5 \cdot 2 = 20 + 10 = 30$

33. $5(4 + 2) = 5(6) = 30$

35. $8 + 2(5 + 3) = 8 + 2(8) = 8 + 16 = 24$

37. $(8 + 2)(5 + 3) = 10(8) = 80$

39. $20 + 2(8 - 5) + 1 = 20 + 2(3) + 1 = 20 + 6 + 1 = 27$

41. $5 + 2(3 \cdot 4 - 1) + 8 = 5 + 2(12 - 1) + 8 = 5 + 2(11) + 8 = 5 + 22 + 8 = 35$

43. $8 + 10 \div 2 = 8 + 5 = 13$

45. $4 + 8 \div 4 - 2 = 4 + 2 - 2 = 4$

47. $3 + 12 \div 3 + 6 \cdot 5 = 3 + 4 + 30 = 37$

49. $3 \cdot 8 + 10 \div 2 + 4 \cdot 2 = 24 + 5 + 8 = 37$

51. $(5 + 3)(5 - 3) = (8)(2) = 16$

53. $5^2 - 3^2 = 5 \cdot 5 - 3 \cdot 3 = 25 - 9 = 16$

55. $(4 + 5)^2 = 9^2 = 9 \cdot 9 = 81$

57. $4^2 + 5^2 = 4 \cdot 4 + 5 \cdot 5 = 16 + 25 = 41$

59. $3 \cdot 10^2 + 4 \cdot 10 + 5 = 300 + 40 + 5 = 345$

61. $2 \cdot 10^3 + 3 \cdot 10^2 + 4 \cdot 10 + 5 = 2000 + 300 + 40 + 5 = 2345$

63. $10 - 2(4 \cdot 5 - 16) = 10 - 2(20 - 16) = 10 - 2(4) = 10 - 8 = 2$

65. $4[7 + 3(2 \cdot 9 - 8)] = 4[7 + 3(18 - 8)] = 4[7 + 3(10)] = 4(7 + 30) = 4(37) = 148$

67. $5(7 - 3) + 8(6 - 4) = 5(4) + 8(2) = 20 + 16 = 36$

69. $3(4 \cdot 5 - 12) + 6(7 \cdot 6 - 40) = 3(20 - 12) + 6(42 - 40) = 3(8) + 6(2) = 24 + 12 = 36$

71. $3^4 + 4^2 \div 2^3 - 5^2 = 81 + 16 \div 8 - 25 = 81 + 2 - 25 = 58$

73. $5^2 + 3^4 \div 9^2 + 6^2 = 25 + 81 \div 81 + 36 = 25 + 1 + 36 = 62$

75. The next number is 5.

77. The next number is 10.

79. The next number is $5^2 = 25$.

81. Since $2 + 2 = 4$ and $2 + 4 = 6$ the next number is $4 + 6 = 10$.

83. (a) $6(100) = 600$ mg   (b) $2(45) + 3(47) = 231$ mg

85. See the table in the back of the textbook.

87. $2(10 + 4) = 2(14) = 28$ shares.

89. $3 \cdot 50 - 14 = 150 - 14 = 136$ dollars

91. $5 \cdot 2 = 10$ cookies in the package.

93. $210 \cdot 2 = 420$ calories.

95. $7 \cdot 32 = 224$ chips in the bag.

97. $80 + 15 = 95$ grams.

1.2 Real Numbers

1. See the number line in the back of the textbook.

3. See the number line in the back of the textbook.

5. See the number line in the back of the textbook.

7. See the number line in the back of the textbook.

9. $\frac{3}{4} = \frac{3 \cdot 6}{4 \cdot 6} = \frac{18}{24}$

11. $\frac{1}{2} = \frac{1 \cdot 12}{2 \cdot 12} = \frac{12}{24}$

13. $\frac{5}{8} = \frac{5 \cdot 3}{8 \cdot 3} = \frac{15}{24}$

15. $\frac{3}{5} = \frac{3 \cdot 12}{5 \cdot 12} = \frac{36}{60}$

17. $\frac{11}{30} = \frac{11 \cdot 2}{30 \cdot 2} = \frac{22}{60}$

19. opposite -10, reciprocal $\frac{1}{10}$, absolute value 10.

21. opposite $-\frac{3}{4}$, reciprocal $\frac{4}{3}$, absolute value $\frac{3}{4}$.

23. opposite $-\frac{11}{2}$, reciprocal $\frac{2}{11}$, absolute value $\frac{11}{2}$.

25. opposite 3, reciprocal $-\frac{1}{3}$, absolute value 3.

27. opposite $\frac{2}{5}$, reciprocal $-\frac{5}{2}$, absolute value $\frac{2}{5}$.

29. opposite $-x$, reciprocal $\frac{1}{x}$, absolute value $|x|$ (The distance between x and 0 on the number line.)

31. $-5 < -3$

33. $-3 > -7$

35. Since $|-4| = 4$ and $-|-4| = -4$, $|-4| > -|-4|$

37. Since $-|-7| = -7$, $7 > -|-7|$

39. $-\frac{3}{4} < -\frac{1}{4}$

41. $-\frac{3}{2} < -\frac{3}{4}$

43. $|8 - 2| = |6| = 6$

45. $|5 \cdot 2^3 - 2 \cdot 3^2| = |5 \cdot 8 - 2 \cdot 9| = |40 - 18| = |22| = 22$

47. $|7 - 2| - |4 - 2| = |5| - |2| = 5 - 2 = 3$

49. $10 - |7 - 2(5 - 3)| = 10 - |7 - 2(2)| = 10 - |7 - 4| = 10 - |3| = 10 - 3 = 7$

51. $15 - |8 - 2(3 \cdot 4 - 9)| - 10 = 15 - |8 - 2(12 - 9)| - 10$

$$= 15 - |8 - 2(3)| - 10$$
$$= 15 - |8 - 6| - 10$$
$$= 15 - |2| - 10$$
$$= 15 - 2 - 10$$
$$= 3$$

53. $\frac{2}{3} \cdot \frac{4}{5} = \frac{2 \cdot 4}{3 \cdot 5} = \frac{8}{15}$

55. $\frac{1}{2}(3) = \frac{1}{2} \cdot \frac{3}{1} = \frac{3}{2}$

57. $\frac{1}{4}(5) = \frac{1}{4} \cdot \frac{5}{1} = \frac{1 \cdot 5}{4 \cdot 1} = \frac{5}{4}$

59. $\frac{4}{3} \cdot \frac{3}{4} = \frac{12}{12} = 1$

61. $6\left(\frac{1}{6}\right) = \frac{6}{1} \cdot \frac{1}{6} = \frac{6 \cdot 1}{1 \cdot 6} = \frac{6}{6} = 1$

63. $3 \cdot \frac{1}{3} = \frac{3}{1} \cdot \frac{1}{3} = \frac{3}{3} = 1$

65. $\left(\frac{3}{4}\right)^2 = \frac{3}{4} \cdot \frac{3}{4} = \frac{3 \cdot 3}{4 \cdot 4} = \frac{9}{16}$

67. $\left(\frac{2}{3}\right)^3 = \frac{2}{3} \cdot \frac{2}{3} \cdot \frac{2}{3} = \frac{8}{27}$

69. $\left(\frac{1}{10}\right)^4 = \frac{1}{10} \cdot \frac{1}{10} \cdot \frac{1}{10} \cdot \frac{1}{10} = \frac{1 \cdot 1 \cdot 1 \cdot 1}{10 \cdot 10 \cdot 10 \cdot 10} = \frac{1}{10,000}$

71. The next number is $\frac{1}{9}$.

73. The next number is $\frac{1}{5^2} = \frac{1}{25}$.

75. The perimeter is $4(1 \text{ in.}) = 4$ in., and the area is $(1 \text{ in.})^2 = 1 \text{ in.}^2$.

77. The perimeter is $2(1.5 \text{ inches}) + 2(0.75 \text{ inches}) = 3.0$ inches $+1.50$ inches $= 4.5$ inches
    and the area is $(1.5 \text{ inches})(0.75 \text{ inches}) = 1.125 \text{ inches}^2$.

79. The perimeter is $2.75 \text{ cm} + 4 \text{ cm} + 3.5 \text{ cm} = 10.25$ cm, and the area is $\frac{1}{2}(4 \text{ cm})(2.5 \text{ cm}) = 5.0 \text{ cm}^2$.

81. A loss of 8 yds corresponds to -8 on the number line. The total yds gained corresponds to -2 yds.

83. The temperature can be represented as $-64°$. The new (warmer) temperature corresponds to $-54°$.

85. $-15°$

87. See the table in the back of the textbook.

89. His position corresponds to -100 feet. His new (deeper) position corresponds to -105 feet.

91. The area is given by: $8\frac{1}{2} \cdot 11 = \frac{17}{2} \cdot \frac{11}{1} = \frac{187}{2} = 93.5 \text{ in.}^2$.

93. The calories consumed would be: $2(544) + (299) = 1387$ calories

95. The calories consumed by the 180 lb person would be $3(653) = 1,959$ calories, while the calories consumed
    by the 120 lb person would be $3(435) = 1,305$ calories. Thus the 180 lb person consumed $1959 - 1305 = 654$
    more calories.

1.3 Addition of Real Numbers

1. $3 + 5 = 8$, $3 + (-5) = -2$, $-3 + 5 = 2$, $(-3) + (-5) = -8$

3. $15 + 20 = 35$, $15 + (-20) = -5$, $-15 + 20 = -5$, $(-15) + (-20) = -35$

5. $6 + (-3) = 3$

7. $13 + (-20) = -7$

9. $18 + (-32) = -14$

11. $-6 + 3 = -3$

13. $-30 + 5 = -25$

15. $-6 + (-6) = -12$

17. $-9 + (-10) = -19$

19. $-10 + (-15) = -25$

21. $5 + (-6) + (-7) = 5 + (-13) = -8$

23. $-7 + 8 + (-5) = -12 + 8 = -4$

25. $5 + [6 + (-2)] + (-3) = 5 + 4 + (-3) = 9 + (-3) = 6$

27. $[6 + (-2)] + [3 + (-1)] = 4 + 2 = 6$

29. $20 + (-6) + [3 + (-9)] = 20 + (-6) + (-6) = 20 + (-12) = 8$

31. $-3 + (-2) + [5 + (-4)] = -3 + (-2) + 1 = -5 + 1 = -4$

33. $(-9 + 2) + [5 + (-8)] + (-4) = -7 + (-3) + (-4) + -14$

35. $[-6 + (-4)] + [7 + (-5)] + (-9) = -10 + 2 + (-9) = -19 + 2 = -17$

37. $(-6 + 9) + (-5) + (-4 + 3) + 7 = 3 + (-5) + (-1) + 7 = 10 + (-6) = 4$

39. $-5 + 2(-3 + 7) = -5 + 2(4) = -5 + 8 = 3$

41. $9 + 3(-8 + 10) = 9 + 3(2) = 9 + 6 = 15$

43. $-10 + 2(-6 + 8) + (-2) = -10 + 2(2) + (-2) = -10 + 4 + (-2) = -12 + 4 = -8$

45. $2(-4 + 7) + 3(-6 + 8) = 2(3) + 3(2) = 6 + 6 = 12$

47. The pattern is to add 5, so the next two terms are $18 + 5 = 23 + 5 = 28$.

49. The pattern is to add 5, so the next two terms are $25 + 5 = 30$ and $30 + 5 = 35$.

51. The pattern is to add -5, so the next two terms are $5 + (-5) = 0$ and $0 + (-5) = -5$.

53. The pattern is to add -6, so the next two terms are $-6 + (-6) = -12$ and $-12 + (-6) = -18$.

55. The pattern is to add -4, so the next two terms are $0 + (-4) = -4$ and $-4 + (-4) = -8$.

57. Yes, since each successive odd number is 2 added to the previous one.

59. $5 + 9 = 14$ 61. $[-7 + (-5)] + 4 = -12 + 4 = -8$

63. $[-2 + (-3)] + 10 = -5 + 10 = 5$ 65. The number is 3, since $-8 + 3 = -5$

67. The number is -3, since $-6 + (-3) = -9$ 69. $-12° + 4° = -8°$.

71. $\$10 + (-\$6) + (-\$8) = \$10 + (-\$14) = -\$4$.   73. The new balance is $-\$30 + \$40 = \$10$.

75. The sequence of her wages is: $6.50, $6.75, $7.00, $7.25, $7.50, $7.75, $8.00.

77. (a) See the table in the back of the textbook.

   (b) $36,036, + 42,862 = $78,898$

1.4 Subtraction of Real Numbers

1. $5 - 8 = 5 + (-8) = -3$ 3. $3 - 9 = 3 + (-9) = -6$

5. $5 - 5 = 5 + (-5) = 0$ 7. $-8 - 2 = -8 + (-2) = -10$

9. $-4 - 12 = -4 + (-12) = -16$ 11. $-6 - 6 = -6 + (-6) = -12$

13. $-8 - (-1) = -8 + 1 = -7$ 15. $15 - (-20) = 15 + 20 = 35$

17. $-4 - (-4) = -4 + 4 = 0$ 19. $3 - 2 - 5 = 3 + (-2) + (-5) = 3 + (-7) = -4$

21. $9 - 2 - 3 = 9 + (-2) + (-3) = 9 + (-5) = 4$ 23. $-6 - 8 - 10 = -6 + (-8) + (-10) = -24$

25. $-22 + 4 - 10 = -22 + 4 + (-10) = -32 + 4 = -28$ 27. $10 - (-20) - 5 = 10 + 20 + (-5) = 30 + (-5) = 25$

29. $8 - (2 - 3) - 5 = 8 - (-1) - 5 = 8 + 1 + (-5) = 9 + (-5) = 4$

31. $7 - (3 - 9) - 6 = 7 - (-6) - 6 = 7 + 6 + (-6) = 13 + (-6) = 7$

33. $5 - (-8 - 6) - 2 = 5 - (-14) - 2 = 5 + 14 + (-2) = 19 + (-2) = 17$

35. $-(5 - 7) - (2 - 8) = -(-2) - (-6) = 2 + 6 = 8$ 37. $-(3 - 10) - (6 - 3) = -(-7) - 3 = 7 + (-3) = 4$

39. $16 - [(4 - 5) - 1] = 16 - (-1 - 1) = 16 - (-2) = 16 + 2 = 18$

41. $5 - [(2 - 3) - 4] = 5 - (-1 - 4) = 5 - (-5) = 5 + 5 = 10$

43. $21 - [-(3 - 4) - 2] - 5 = 21 - [-(-1) - 2] - 5 = 21 - (1 - 2) - 5 = 21 - (-1) - 5 = 21 + 1 + (-5) = 22 + (-5) = 17$

45. $2 \cdot 8 - 3 \cdot 5 = 16 - 15 = 16 + (-15) = 1$ 47. $3 \cdot 5 - 2 \cdot 7 = 15 - 14 = 15 + (-14) = 1$

49. $5 \cdot 9 - 2 \cdot 3 - 6 \cdot 2 = 45 - 6 - 12 = 45 + (-6) + (-12) = 45 + (-18) = 27$

51. $3 \cdot 8 - 2 \cdot 4 - 6 \cdot 7 = 24 - 8 - 42 = 24 + (-8) + (-42) = 24 + (-50) = -26$

53. $2 \cdot 3^2 - 5 \cdot 2^2 = 2 \cdot 9 - 5 \cdot 4 = 18 - 20 = 18 + (-20) = -2$

55. $4 \cdot 3^3 - 5 \cdot 2^3 = 4 \cdot 27 - 5 \cdot 8 = 108 - 40 = 108 + (-40) = 68$

57. $-7 - 4 = -7 + (-4) = -11$ 59. $12 - (-8) = 12 + 8 = 20$

61. $-5 - (-7) = -5 + 7 = 2$ 63. $[4 + (-5)] - 17 = -1 - 17 = -1 + (-17) = -18$

65. $8 - 5 = 8 + (-5) = 3$ 67. $-8 - 5 = -8 + (-5) = -13$

69. $8 - (-5) = 8 + 5 = 13$ 71. The number is 10, since $8 - 10 = 8 + (-10) = -2$.

73. The number is -2, since $8 - (-2) = 8 + 2 = 10$.   75. $\$1,500 - \$730$

77. $-\$35 + \$15 - \$20 = -\$35 + (-\$20) + \$15 = -\$55 + \$15 = -\$40$

79. $\$98 - \$65 - \$53 = \$98 + (-\$65) + (-\$53) = \$98 + (-\$118) = -\$20$

81. The sequence of values is \$4500, \$3950, \$3400, \$2850, and \$2300.   This is an arithmetic sequence, since

   $-\$550$ is added to each value to obtain the new value.

83. The difference is 1000 feet -231 feet = 769 feet.

85. He is 439 feet from the starting line.   87. 2 seconds have gone by.

89. (a) See the table in the back of the textbook.

   (b) $205 - 121 = 84$ million tons

91. (a) See the table in the back of the textbook.

   (b) $33 - 28 = 5$ cents per minute

93. The angles add to 90°, so $x = 90° - 55° = 35°$.   95. The angles add to 180°, so $x = 180° - 120° = 60°$.

1.5 Properties of Real Numbers

1. Commutative property of addition

3. Multiplicative inverse property

5. Commutative property of addition

7. Distributive property

9. Commutative and associative properties of addition

11. Commutative and associative properties of addition

13. Commutative property of addition

15. Commutative and associative properties of multiplication

17. Commutative property of multiplication

19. Additive inverse property

21. $3(x+2) = 3x+6$

23. $9(a+b) = 9a+9b$

25. $3(0) = 0$

27. $3 + (-3) = 0$

29. $10(1) = 10$

31. $4 + (2+x) = (4+2) + x = 6 + x$

33. $(x+2) + 7 = x + (2+7) = x + 9$

35. $3(5x) = (3 \cdot 5)x = 15x$

37. $9(6y) = (9 \cdot 6)y = 54y$

39. $\frac{1}{2}(3a) = (\frac{1}{2} \cdot 3)a = \frac{3}{2}a$

41. $\frac{1}{3}(3x) = (\frac{1}{3} \cdot 3)x = 1x = x$

43. $\frac{1}{2}(2y) = (\frac{1}{2} \cdot 2)y = 1y = y$

45. $\frac{3}{4}(\frac{4}{3}x) = (\frac{3}{4} \cdot \frac{4}{3})x = 1x = x$

47. $\frac{6}{5}(\frac{5}{6}a) = (\frac{6}{5} \cdot \frac{5}{6})a = 1a = a$

49. $8(x+2) = 8 \cdot x + 8 \cdot 2 = 8x + 16$

51. $8(x-2) = 8 \cdot x - 8 \cdot 2 = 8x - 16$

53. $4(y+1) = 4 \cdot y + 4 \cdot 1 = 4y + 4$

55. $3(6x+5) = 3 \cdot 6x + 3 \cdot 5 = 18x + 15$

57. $2(3a+7) = 2 \cdot 3a + 2 \cdot 7 = 6a + 14$

59. $9(6y-8) = 9 \cdot 6y - 9 \cdot 8 = 54y - 72$

61. $\frac{1}{2}(3x-6) = \frac{1}{2}(3x) - \frac{1}{2}(6) = \frac{3}{2}x - 3$

63. $\frac{1}{3}(3x+6) = \frac{1}{3} \cdot 3x + \frac{1}{3} \cdot 6 = x + 2$

65. $3(x+y) = 3x + 3y$

67. $8(a-b) = 8a - 8b$

69. $6(2x+3y) = 6(2x) + 6(3y) = 12x + 18y$

71. $4(3a-2b) = 4 \cdot 3a - 4 \cdot 2b = 12a - 8b$

73. $\frac{1}{2}(6x+4y) = \frac{1}{2}(6x) + \frac{1}{2}(4y) = 3x + 2y$

75. $4(a+4) + 9 = 4a + 16 + 9 = 4a + 25$

77. $2(3x+5) + 2 = 6x + 10 + 2 = 6x + 12$

79. $7(2x+4) + 10 = 14x + 28 + 10 = 14x + 38$

81. No. The man cannot reverse the order of putting on his socks and putting put on his shoes.

83. No. The skydiver must jump out of the plane before pulling the rip cord.

85. Division is not a commutative operation. For example, $8 \div 4 = 2$ while $4 \div 8 = \frac{1}{2}$.

87. $12(2400 - 480) = 12(1920) = \$23,040$

   $12 \cdot 2400 - 12 \cdot 480 = 28,800 - 5,760 = \$23,040$

89. $P = 2w + 2l = 2(w + l)$

1.6 Multiplication of Real Numbers

1. $7(-6) = -42$

3. $-7(3) = -21$

5. $-8(2) = -16$

7. $-3(-1) = 3$

9. $-11(-11) = 121$

11. $-3(2)(-1) = 6$

13. $-3(-4)(-5) = -60$

15. $-2(-4)(-3)(-1) = 24$

17. $(-7)^2 = (-7)(-7) = 49$

19. $(-3)^3 = (-3)(-3)(-3) = -27$

21. $-2(2-5) = -2(-3) = 6$

23. $-5(8-10) = -5(-2) = 10$

25. $(4-7)(6-9) = (-3)(-3) = 9$

27. $(-3-2)(-5-4) = (-5)(-9) = 45$

29. $-3(-6)+4(-1) = 18+(-4) = 14$

31. $2(3)-3(-4)+4(-5) = 6+12+(-20) = 18+(-20) = -2$

33. $4(-3)^2 + 5(-6)^2 = 4(9)+5(36) = 36+180 = 216$

35. $7(-2)^3 - 2(-3)^3 = 7(-8)-2(-27) = -56+54 = -2$

37. $6-4(8-2) = 6-4(6) = 6-24 = 6+(-24) = -18$

39. $9-4(3-8) = 9-4(-5) = 9+20 = 29$

41. $-4(3-8)-6(2-5) = -4(-5)-6(-3) = 20+18 = 38$

43. $7-2[-6-4(-3)] = 7-2(-6+12) = 7-2(6) = 7-12 = 7+(-12) = -5$

45. $7-3[2(-4-4)-3(-1-1)] = 7-3[2(-8)-3(-2)] = 7-3(-16+6) = 7-3(-10) = 7+30 = 37$

47. $8-6[-2(-3-1)+4(-2-3)] = 8-6[-2(-4)+4(-5)] = 8-6[8+(-20)] = 8-6(-12) = 8+72 = 80$

49. $-\frac{2}{3} \cdot \frac{5}{7} = -\frac{2 \cdot 5}{3 \cdot 7} = -\frac{10}{21}$

51. $-8(\frac{1}{2}) = -\frac{8}{1} \cdot \frac{1}{2} = -\frac{8}{2} = -4$

53. $-\frac{3}{4}(-\frac{4}{3}) = -\frac{3}{4} \cdot (-\frac{4}{3}) = \frac{12}{12} = 1$

55. $(-\frac{3}{4})^2 = (-\frac{3}{4})(-\frac{3}{4}) = \frac{9}{16}$

57. $(-\frac{2}{3})^3 = (-\frac{2}{3})(-\frac{2}{3})(-\frac{2}{3}) = -\frac{8}{27}$

59. $-2(4x) = (-2 \cdot 4)x = -8x$

61. $-7(-6x) = [-7(-6)]x = 42x$

63. $-\frac{1}{3}(-3x) = [-\frac{1}{3} \cdot (-3)]x = 1x = x$

65. $-4(-\frac{1}{4}x) = [-4(-\frac{1}{4})]x = 1x = x$

67. $-4(a+2) = -4a+(-4)(2) = -4a-8$

69. $-\frac{1}{2}(3x-6) = -\frac{3}{2}x - \frac{1}{2}(-6) = -\frac{3}{2}x+3$

71. $-3(2x-5)-7 = -6x+15-7 = -6x+8$

73. $-5(3x+4)-10 = -15x-20-10 = -15x-30$

75. $3(-10)+5 = -30+5 = -25$

77. $2(-4x) = -8x$

79. $-9 \cdot 2 - 8 = -18+(-8) = -26$

81. The pattern is to multiply by 2, so the next number is $4 \cdot 2 = 8$.

83. The pattern is to multiply by -2, so the next number is $40 \cdot (-2) = -80$.

85. The pattern is to multiply by $\frac{1}{2}$, so the next number is $\frac{1}{4} \cdot \frac{1}{2} = \frac{1}{8}$.

87. The pattern is to multiply by $\frac{1}{2}$, so the next number is $2 \cdot \frac{1}{2} = 1$.

89. The pattern is to multiply by -2, so the next number is $12 \cdot (-2) = -24$.

91. The amount lost is: $20(\$3) = \$60$.

93. The temperature is: $25° - 4(6°) = 25° - 24° = 1°$

95. $500, $1000, $2000, $4000, $8000, and $16,000. Yes, this is a geometric sequence, since each value is 2 times the preceding value.

97. $2(630) - 3(265) = +465$ calories (gain)

1.7 Division of Real Numbers

1. $\frac{8}{-4} = -2$

3. $\frac{-48}{16} = -3$

5. $\frac{-7}{21} = -\frac{1}{3}$

7. $\frac{-39}{-13} = 3$

9. $\frac{-6}{-42} = \frac{1}{7}$

11. $\frac{0}{-32} = 0$

13. $-3 + 12 = 9$

15. $-3 - 12 = -3 + (-12) = -15$

17. $-3(12) = -36$

19. $-3 \div 12 = \frac{-3}{12} = -\frac{1}{4}$

21. $\frac{4}{5} \div \frac{3}{4} = \frac{4}{5} \cdot \frac{4}{3} = \frac{16}{15}$

23. $-\frac{5}{6} \div \left(-\frac{5}{8}\right) = -\frac{5}{6} \cdot \left(-\frac{8}{5}\right) = \frac{40}{30} = \frac{4}{3}$

25. $\frac{10}{13} \div \left(-\frac{5}{4}\right) = \frac{10}{13} \cdot \left(-\frac{4}{5}\right) = -\frac{40}{65} = -\frac{8}{13}$

27. $-\frac{5}{6} \div \frac{5}{6} = -\frac{5}{6} \cdot \frac{6}{5} = -\frac{30}{30} = -1$

29. $-\frac{3}{4} \div \left(-\frac{3}{4}\right) = -\frac{3}{4} \cdot \left(-\frac{4}{3}\right) = \frac{12}{12} = 1$

31. $\frac{3(-2)}{-10} = \frac{-6}{-10} = \frac{3}{5}$

33. $\frac{-5(-5)}{-15} = \frac{25}{-15} = -\frac{5}{3}$

35. $\frac{-8(-7)}{-28} = \frac{56}{-28} = -2$

37. $\frac{27}{4-13} = \frac{27}{-9} = -3$

39. $\frac{20-6}{5-5} = \frac{14}{0} = $ undefined

41. $\frac{-3+9}{2(5)-10} = \frac{6}{10-10} = \frac{6}{0} = $ undefined

43. $\frac{15(-5)-25}{2(-10)} = \frac{-75-25}{-20} = \frac{-100}{-20} = 5$

45. $\frac{27-2(-4)}{-3(5)} = \frac{27+8}{-15} = \frac{35}{-15} = -\frac{7}{3}$

47. $\frac{12-6(-2)}{12(-2)} = \frac{12+12}{-24} = \frac{24}{-24} = -1$

49. $\frac{5^2-2^2}{-5+2} = \frac{25-4}{-3} = \frac{21}{-3} = -7$

51. $\frac{8^2-2^2}{8^2+2^2} = \frac{64-4}{64+4} = \frac{60}{68} = \frac{15}{17}$

53. $\frac{(5+3)^2}{-5^2-3^2} = \frac{8^2}{-25-9} = \frac{64}{-34} = -\frac{32}{17}$

55. $\frac{(8-4)^2}{8^2-4^2} = \frac{4^2}{64-16} = \frac{16}{48} = \frac{1}{3}$

57. $\frac{-4 \cdot 3^2 - 5 \cdot 2^2}{-8(7)} = \frac{-4 \cdot 9 - 5 \cdot 4}{-56} = \frac{-36-20}{-56} = \frac{-56}{-56} = 1$

59. $\frac{3 \cdot 10^2 + 4 \cdot 10 + 5}{345} = \frac{300+40+5}{345} = \frac{345}{345} = 1$

61. $\frac{7-[(2-3)-4]}{-1-2-3} = \frac{7-(-1-4)}{-6} = \frac{7-(-5)}{-6} = \frac{7+5}{-6} = \frac{12}{-6} = -2$

63. $\frac{6(-4)-2(5-8)}{-6-3-5} = \frac{-24-2(-3)}{-14} = \frac{-24+6}{-14} = \frac{-18}{-14} = \frac{9}{7}$

65. $\frac{3(-5-3)+4(7-9)}{5(-2)+3(-4)} = \frac{3(-8)+4(-2)}{-10+(-12)} = \frac{-24+(-8)}{-22} = \frac{-32}{-22} = \frac{16}{11}$

67. $\frac{|3-9|}{3-9} = \frac{|-6|}{-6} = \frac{6}{-6} = -1$

69. The quotient is $\frac{-12}{-4} = 3$.

71. The number is -10, since $\frac{-10}{-5} = 2$.

73. The number is -3, since $\frac{27}{-3} = -9$.

75. The expression is: $\frac{-20}{4} - 3 = -5 - 3 = -8$

77. Each person would lose: $\frac{13600-15000}{4} = \frac{-1400}{4} = -350 = \$350$ loss per person

79. The change per hour is: $\frac{61°-75°}{4} = \frac{-14°}{4} = -3.5$ Drops 3.5° F. each hour.

1.8 Subsets of the Real Numbers

1. $A \cup B = \{0, 1, 2, 3, 4, 5, 6\}$

3. $A \cap C = \varnothing$

5. $A \cup (B \cap C) = \{0, 1, 2, 3, 4, 5, 6\}$

7. $\{x | x \in A \text{ and } x < 4\} = \{0, 2\}$.

9. $\{x | x \in A \text{ and } x \notin B\} = \{0, 6\}$.

11. $\{x | x \in A \text{ or } x \in C\} = \{0, 1, 2, 3, 4, 5, 6, 7\}$.

13. The whole numbers are: 0, 1

15. The rational numbers are: $-3, -2.5, 0, 1, \frac{3}{2}$

17. The real numbers are: $\{-3, -2.5, 0, 1, \frac{3}{2}, \sqrt{15}\}$

19. The integers are: $-10, -8, -2, 9$

21. The irrational numbers are: π

23. True

25. False

27. False

29. True

31. This number is composite: $48 = 6 \cdot 8 = (2 \cdot 3) \cdot (2 \cdot 2 \cdot 2) = 2^4 \cdot 3$

33. This number is prime.

35. This number is composite: $1023 = 3 \cdot 341 = 3 \cdot 11 \cdot 31$

37. $144 = 12 \cdot 12 = (3 \cdot 4) \cdot (3 \cdot 4) = (3 \cdot 2 \cdot 2) \cdot (3 \cdot 2 \cdot 2) = 2^4 \cdot 3^2$

39. $38 = 2 \cdot 19$

41. $105 = 5 \cdot 21 = 5 \cdot (3 \cdot 7) = 3 \cdot 5 \cdot 7$

43. $180 = 10 \cdot 18 = (2 \cdot 5) \cdot (3 \cdot 2 \cdot 3) = 2^2 \cdot 3^2 \cdot 5$

45. $385 = 5 \cdot 77 = 5 \cdot (7 \cdot 11) = 5 \cdot 7 \cdot 11$

47. $121 = 11 \cdot 11 = 11^2$

49. $420 = 10 \cdot 42 = (2 \cdot 5) \cdot (7 \cdot 6) = (2 \cdot 5) \cdot (7 \cdot 2 \cdot 3) = 2^2 \cdot 3 \cdot 5 \cdot 7$

51. $620 = 10 \cdot 62 = (2 \cdot 5) \cdot (2 \cdot 31) = 2^2 \cdot 5 \cdot 31$

53. $\frac{105}{165} = \frac{3 \cdot 5 \cdot 7}{3 \cdot 5 \cdot 11} = \frac{7}{11}$

55. $\frac{525}{735} = \frac{3 \cdot 5 \cdot 5 \cdot 7}{3 \cdot 5 \cdot 7 \cdot 7} = \frac{5}{7}$

57. $\frac{385}{455} = \frac{5 \cdot 7 \cdot 11}{5 \cdot 7 \cdot 13} = \frac{11}{13}$

59. $\frac{322}{345} = \frac{2 \cdot 7 \cdot 23}{3 \cdot 5 \cdot 23} = \frac{2 \cdot 7}{3 \cdot 5} = \frac{14}{15}$

61. $\frac{205}{369} = \frac{5 \cdot 41}{3 \cdot 3 \cdot 41} = \frac{5}{3 \cdot 3} = \frac{5}{9}$

63. $\frac{215}{344} = \frac{5 \cdot 43}{2 \cdot 2 \cdot 2 \cdot 43} = \frac{5}{2 \cdot 2 \cdot 2} = \frac{5}{8}$

65. $6^3 = (2 \cdot 3)^3 = 2^3 \cdot 3^3$

67. $9^4 \cdot 16^2 = (3 \cdot 3)^4 \cdot (2 \cdot 2 \cdot 2 \cdot 2)^2 = 2^8 \cdot 3^8$

69. $3 \cdot 8 + 3 \cdot 7 + 3 \cdot 5 = 24 + 21 + 15 = 60 = 6 \cdot 10 = (2 \cdot 3) \cdot (2 \cdot 5) = 2^2 \cdot 3 \cdot 5$

71. They are not a subset of the irrational numbers.

73. 8, 21, 34

1.9 Addition and Subtraction with Fractions

1. $\frac{3}{6} + \frac{1}{6} = \frac{4}{6} = \frac{2}{3}$

3. $\frac{3}{8} - \frac{5}{8} = -\frac{2}{8} = -\frac{1}{4}$

5. $-\frac{1}{4} + \frac{3}{4} = \frac{2}{4} = \frac{1}{2}$

7. $\frac{x}{3} - \frac{1}{3} = \frac{x-1}{3}$

9. $\frac{1}{4} + \frac{2}{4} + \frac{3}{4} = \frac{6}{4} = \frac{3}{2}$

11. $\frac{x+7}{2} - \frac{1}{2} = \frac{x+7-1}{2} = \frac{x+6}{2}$

13. $\frac{1}{10} - \frac{3}{10} - \frac{4}{10} = -\frac{6}{10} = -\frac{3}{5}$

15. $\frac{1}{a} + \frac{4}{a} + \frac{5}{a} = \frac{10}{a}$

17. $\frac{1}{8} + \frac{3}{4} = \frac{1}{8} + \frac{3 \cdot 2}{4 \cdot 2} = \frac{1}{8} + \frac{6}{8} = \frac{7}{8}$

19. $\frac{3}{10} - \frac{1}{5} = \frac{3}{10} - \frac{1 \cdot 2}{5 \cdot 2} = \frac{3}{10} - \frac{2}{10} = \frac{1}{10}$

21. $\frac{4}{9} + \frac{1}{3} = \frac{4}{9} + \frac{1 \cdot 3}{3 \cdot 3} = \frac{4}{9} + \frac{3}{9} = \frac{7}{9}$

23. $2 + \frac{1}{3} = \frac{2 \cdot 3}{1 \cdot 3} + \frac{1}{3} = \frac{6}{3} + \frac{1}{3} = \frac{7}{3}$

25. $-\frac{3}{4} + 1 = -\frac{3}{4} + \frac{1 \cdot 4}{1 \cdot 4} = -\frac{3}{4} + \frac{4}{4} = \frac{1}{4}$

27. $\frac{1}{2} + \frac{2}{3} = \frac{1 \cdot 3}{2 \cdot 3} + \frac{2 \cdot 2}{3 \cdot 2} = \frac{3}{6} + \frac{4}{6} = \frac{7}{6}$

29. $\frac{5}{12} - \left(-\frac{3}{8}\right) = \frac{5}{12} + \frac{3}{8} = \frac{5 \cdot 2}{12 \cdot 2} + \frac{3 \cdot 3}{8 \cdot 3} = \frac{10}{24} + \frac{9}{24} = \frac{19}{24}$

31. $-\frac{1}{20} + \frac{8}{30} = -\frac{1 \cdot 3}{20 \cdot 3} + \frac{8 \cdot 2}{30 \cdot 2} = -\frac{3}{60} + \frac{16}{60} = \frac{13}{60}$

33. First factor the denominators to find the LCM:

$30 = 2 \cdot 3 \cdot 5$

$42 = 2 \cdot 3 \cdot 7$

$LCM = 2 \cdot 3 \cdot 5 \cdot 7 = 210$

Combining the fractions: $\frac{17}{30} + \frac{11}{42} = \frac{17 \cdot 7}{30 \cdot 7} + \frac{11 \cdot 5}{42 \cdot 5} = \frac{119}{210} + \frac{55}{210} = \frac{174}{210} = \frac{2 \cdot 3 \cdot 29}{2 \cdot 3 \cdot 5 \cdot 7} = \frac{29}{5 \cdot 7} = \frac{29}{35}$

35. First factor the denominators to find the LCM:

$84 = 2 \cdot 2 \cdot 3 \cdot 7$

$90 = 2 \cdot 3 \cdot 3 \cdot 5$

$LCM = 2 \cdot 2 \cdot 3 \cdot 3 \cdot 5 \cdot 7 = 1260$

Combining the fractions: $\frac{25}{84} + \frac{41}{90} = \frac{25 \cdot 15}{84 \cdot 15} + \frac{41 \cdot 14}{90 \cdot 14} = \frac{375}{1260} + \frac{574}{1260} = \frac{949}{1260}$

37. First factor the denominators to find the LCM:

$126 = 2 \cdot 3 \cdot 3 \cdot 7$

$180 = 2 \cdot 2 \cdot 3 \cdot 3 \cdot 5$

$LCM = 2 \cdot 2 \cdot 3 \cdot 3 \cdot 5 \cdot 7 = 1260$

Combining the fractions:

$\frac{13}{126} - \frac{13}{180} = \frac{13 \cdot 10}{126 \cdot 10} - \frac{13 \cdot 7}{180 \cdot 7} = \frac{130}{1260} - \frac{91}{1260} = \frac{39}{1260} = \frac{3 \cdot 13}{2 \cdot 2 \cdot 3 \cdot 3 \cdot 5 \cdot 7} = \frac{13}{2 \cdot 2 \cdot 3 \cdot 5 \cdot 7} = \frac{13}{420}$

39. Combining the fractions: $\frac{3}{4} + \frac{1}{8} + \frac{5}{6} = \frac{3 \cdot 6}{4 \cdot 6} + \frac{1 \cdot 3}{8 \cdot 3} + \frac{5 \cdot 4}{6 \cdot 4} = \frac{18}{24} + \frac{3}{24} + \frac{20}{24} = \frac{41}{24}$

41. Combining the fractions: $\frac{1}{2} + \frac{1}{3} + \frac{1}{4} + \frac{1}{6} = \frac{1 \cdot 6}{2 \cdot 6} + \frac{1 \cdot 4}{3 \cdot 4} + \frac{1 \cdot 3}{4 \cdot 3} + \frac{1 \cdot 2}{6 \cdot 2} = \frac{6}{12} + \frac{4}{12} + \frac{3}{12} + \frac{2}{12} = \frac{15}{12} = \frac{5}{4}$

43. The sum is given by: $\frac{3}{7} + 2 + \frac{1}{9} = \frac{3 \cdot 9}{7 \cdot 9} + \frac{2 \cdot 63}{1 \cdot 63} + \frac{1 \cdot 7}{9 \cdot 7} = \frac{27}{63} + \frac{126}{63} + \frac{7}{63} = \frac{160}{63}$

45. The difference is given by: $\frac{7}{8} - \frac{1}{4} = \frac{7}{8} - \frac{1 \cdot 2}{4 \cdot 2} = \frac{7}{8} - \frac{2}{8} = \frac{5}{8}$

47. The pattern is to add $-\frac{1}{3}$, so the fourth term is: $-\frac{1}{3} + \left(-\frac{1}{3}\right) = -\frac{2}{3}$

49. The pattern is to add $\frac{2}{3}$, so the fourth term is: $\frac{5}{3} + \frac{2}{3} = \frac{7}{3}$

51. The pattern is to multiply by $\frac{1}{5}$, so the fourth term is: $\frac{1}{25} \cdot \frac{1}{5} = \frac{1}{125}$

Chapter 1 Review

1. $-7 + (-10) = -17$

3. $(-3 + 12) + 5 = 9 + 5 = 14$

5. $9 - (-3) = 9 + 3 = 12$

7. $(-3)(-7) - 6 = 21 - 6 = 15$

9. $2(-8 \cdot 3x) = 2(-24x) = -48x$

11. $(-40/8) - 7 = -5 - 7 = -5 + (-7) = -12$

13-17 See the graph in the back of the textbook.

19. $|12| = 12$

21. $\left|-\frac{4}{5}\right| = \frac{4}{5}$

23. $|-1.8| = 1.8$

25. The opposite is -6, and the reciprocal is $\frac{1}{6}$.

27. The opposite is 9, and the reciprocal is $-\frac{1}{9}$.

29. $\left(\frac{2}{5}\right)\left(\frac{3}{7}\right) = \frac{6}{35}$

31. $\left(-\frac{4}{5}\right)\left(\frac{25}{16}\right) = -\frac{4 \cdot 25}{5 \cdot 16} = -\frac{100}{80} = -\frac{5}{4}$

33. $-18 + (-20) = -38$

35. $(-5) + (-10) + (-7) = -15 + (-7) = -22$

37. $(-21) + 40 + (-23) + 5 = 19 + (-23) + 5 = -4 + 5 = 1$

39. $14 - (-8) = 14 + 8 = 22$

41. $4 - 9 - 15 = 4 + (-9) + (-15) = -5 + (-15) = -20$

43. $5 - (-10 - 2) - 3 = 5 - [-10 + (-2)] - 3 = 5 - (-12) - 3 = 5 + 12 + (-3) = 17 + (-3) = 14$

45. $20 - [-(10 - 3) - 8] - 7 = 20 - [-(7) - 8] - 7 = 20 - [-7 + (-8)] - 7 = 20 - [-15] - 7 = 35 + (-7) = 28$

47. $4(-3) = -12$

49. $(-1)(-3)(-1)(-4) = 3(-1)(-4) = (-3)(-4) = 12$

51. $\frac{-9}{36} = -\frac{1}{4}$

53. $4 \cdot 5 + 3 = 20 + 3 = 23$

55. $2^3 - 4 \cdot 3^2 + 5^2 = 8 - 4 \cdot 9 + 25 = 8 - 36 + 25 = 8 + (-36) + 25 = -28 + 25 = -3$

57. $20 + 8 \div 4 + 2 \cdot 5 = 20 + 2 + 10 = 22 + 10 = 32$

59. $-4(-5) + 10 = 20 + 10 = 30$

61. $3(4 - 7)^2 - 5(3 - 8)^2 = 3[4 + (-7)]^2 - 5[3 + (-8)]^2$
$$= 3(-3)^2 - 5(-5)^2$$
$$= 3(9) - 5(25)$$
$$= 27 - 125$$
$$= 27 + (-125)$$
$$= -98$$

63. $\frac{4(-3)}{-6} = \frac{-12}{-6} = \frac{12}{6} = 2$

65. $\frac{15-10}{6-6} = \frac{5}{0} =$ undefined

67. $\frac{2(-7)+(-11)(-4)}{7-(-3)} = \frac{-14+44}{10} = \frac{30}{10} = 3$

69. $8(1) = 8$ multiplicative identity

71. $5 + (-5) = 0$ additive inverse

73. $8 + 0 = 8$ additive identity

75. $5(w-6) = 5w - 30$ distributive property

77. $4(7a) = (4 \cdot 7)a = 28a$

79. $\frac{4}{5}(\frac{5}{4}y) = (\frac{4}{5} \cdot \frac{5}{4})y = \frac{20}{20}y = 1y = y$

81. $3(2a-4) = 3(2a) - 3(4) = 6a - 12$

83. $(-1/2)(3x-6) = (-1/2)(3x) - (-1/2)(6)$

$= (-1/2)(3x/1) - (-1/2)(6/1)$

$= (-3x/2) - (-6/2)$

$= (-3/2x) + 3$

85. 0, 5 Remember: Counting numbers and the number 0 are integers.

87. 0, 5, -3   Remember: Whole numbers and the opposites of all the counting numbers

89. $840 = 84 \cdot 10$

$= 7 \cdot 12 \cdot 2 \cdot 5$

$= 7 \cdot 3 \cdot 4 \cdot 2 \cdot 5$

$= 7 \cdot 3 \cdot 2 \cdot 2 \cdot 2 \cdot 5$

$= 2^3 \cdot 3 \cdot 5 \cdot 7$

91. $\frac{9}{70} + \frac{11}{84} = \frac{9 \cdot 6}{70 \cdot 6} + \frac{11 \cdot 5}{84 \cdot 5}$ $70 = 2 \cdot 5 \cdot 7$ LCD $= 2^2 \cdot 3 \cdot 5 \cdot 7 = 420$

$= \frac{54}{420} + \frac{55}{420}$ $84 = 2^2 \cdot 3 \cdot 7$

$= \frac{109}{420}$

93. $10, -30, 90, -270, 810, \ldots$ For each new number multiply the previous number by -3.

95. $4, 6, 8, 10, 12, \ldots$ For each new number add 2 to the previous number.

97. $1, -1/2, 1/4, -1/8, 1/16, \ldots$ For each new number multiply the previous number by $-1/2$.

Chapter 1 Test

1. $x + 3 = 8$

2. $5y = 15$

3. $5^2 + 3(9 - 7) + 3^2 = 5^2 + 3(2) + 3^2 = 25 + 6 + 9 = 40$

4. $10 - 6 \div 3 + 2^3 = 10 - 6 \div 3 + 8 = 10 - 2 + 8 = 18 - 2 = 16$

5. Opposite 4,   Reciprocal $-\frac{1}{4}$,   Absolute value $|-4| = 4$

6. Opposite $-\frac{3}{4}$,   Reciprocal $\frac{4}{3}$,   Absolute value $\left|\frac{3}{4}\right| = \frac{3}{4}$

7. $3 + (-7) = -4$

8. $|-9 + (-6)| + |-3 + 5| = |-15| + |2| = 15 + 2 = 17$

9. $-4 - 8 = -4 + (-8) = -12$

10. $9 - (7 - 2) - 4 = 9 - 5 - 4 = 9 + (-5) + (-4) = 9 + (-9) = 0$

11. c. (associative property of addition)

12. e. (distributive property)

13. d. (associative property of multiplication)

14. a. (commutative property of addition)

15. $-3(7) = -21$ 16. $-4(8)(-2) = 64$

17. $8\left(-\frac{1}{4}\right) = \frac{8}{1}\left(-\frac{1}{4}\right) = -\frac{8}{4} = -2$ 18. $\left(-\frac{2}{3}\right)^3 = \left(-\frac{2}{3}\right)\left(-\frac{2}{3}\right)\left(-\frac{2}{3}\right) = -\frac{8}{27}$

19. $-3(-4) - 8 = 12 - 8 = 4$

20. $5(-6)^2 - 3(-2)^3 = 5(36) - 3(-8) = 180 + 24 = 204$

21. $7 - 3(2 - 8) = 7 - 3(-6) = 7 + 18 = 25$

22. $4 - 2[-3(-1 + 5) + 4(-3)] = 4 - 2[-3(4) + 4(-3)] = 4 - 2(-12 - 12) = 4 - 2(-24) = 4 + 48 = 52$

23. $\frac{4(-5) - 2(7)}{-10 - 7} = \frac{-20 - 14}{-17} = \frac{-34}{-17} = 2$

24. $\frac{2(-3 - 1) + 4(-5 + 2)}{-3(2) - 4} = \frac{2(-4) + 4(-3)}{-6 - 4} = \frac{-8 - 12}{-10} = \frac{-20}{-10} = 2$

25. $3 + (5 + 2x) = (3 + 5) + 2x = 8 + 2x$ 26. $-2(-5x) = [-2(-5)]x = 10x$

27. $2(3x + 5) = 2(3x) + 2(5) = 6x + 10$ 28. $(-1/2)(4x - 2) = (-1/2)(4x) + (-1/2)(-2) = -2x + 1$

29. The integers are 1 and -8.   30. The rational numbers are 1, 1.5, 3/4, -8.

31. The irrational numbers are $\sqrt{2}$.   32. The real numbers are 1, 1.5, $\sqrt{2}$, 3/4, and -8.

33. $592 = 4 \cdot 148 = (2 \cdot 2) \cdot (4 \cdot 37) = (2 \cdot 2) \cdot (2 \cdot 2 \cdot 37) = 2^4 \cdot 37$

34. $1340 = 10 \cdot 134 = (2 \cdot 5) \cdot (2 \cdot 67) = 2^2 \cdot 5 \cdot 67$

35. First factor the denominators to find the LCM:

$$15 = 3 \cdot 5$$

$$42 = 2 \cdot 3 \cdot 7$$

LCM$= 2 \cdot 3 \cdot 5 \cdot 7 = 210$

Combining the fractions: $\frac{5}{15} + \frac{11}{42} = \frac{5 \cdot 14}{15 \cdot 14} + \frac{11 \cdot 5}{42 \cdot 5} = \frac{70}{210} + \frac{55}{210} = \frac{125}{210} = \frac{5 \cdot 25}{5 \cdot 42} = \frac{25}{42}$

36. Combining the fractions: $\frac{5}{x} + \frac{3}{2} = \frac{5 \cdot 2}{x \cdot 2} + \frac{3 \cdot x}{2 \cdot x} = \frac{10}{2x} + \frac{3x}{2x} = \frac{10+3x}{2x}$

37. $8 + (-3) = 5$ 38. $-24 - 2 = -24 + (-2) = -26$

39. $-5(-4) = 20$ 40. $\frac{-24}{-2} = 12$

41. The pattern is to add 5, so the next term is $7 + 5 = 12$.

42. The pattern is to multiply by $-\frac{1}{2}$ so the next term is $-1\left(-\frac{1}{2}\right) = \frac{1}{2}$.

CHAPTER 2

Linear Equations and Inequalities

2.1 Simplifying Expressions

1. $3x - 6x = (3 - 6)x = -3x$

3. $-2a + a = (-2 + 1)a = -a$

5. $7x + 3x + 2x = (7 + 3 + 2)x = 12x$

7. $3a - 2a + 5a = (3 - 2 + 5)a = 6a$

9. $4x - 3 + 2x = 4x + 2x - 3 = 6x - 3$

11. $3a + 4a + 5 = 7a + 5$

13. $2x - 3 + 3x - 2 = 2x + 3x - 3 - 2 = 5x - 5$

15. $3a - 1 + a + 3 = 3a + a - 1 + 3 = 4a + 2$

17. $-4x + 8 - 5x - 10 = -4x - 5x + 8 - 10 = -9x - 2$

19. $7a + 3 + 2a + 3a = 7a + 2a + 3a + 3 = 12a + 3$

21. $5(2x - 1) + 4 = 10x - 5 + 4 = 10x - 1$

23. $7(3y + 2) - 8 = 21y + 14 - 8 = 21y + 6$

25. $-3(2x - 1) + 5 = -6x + 3 + 5 = -6x + 8$

27. $5 - 2(a + 1) = 5 - 2a - 2 = -2a - 2 + 5 = -2a + 3$

29. $6 - 4(x - 5) = 6 - 4x + 20 = -4x + 20 + 6 = -4x + 26$

31. $-9 - 4(2 - y) + 1 = -9 - 8 + 4y + 1 = 4y + 1 - 9 - 8 = 4y - 16$

33. $-6 + 2(2 - 3x) + 1 = -6 + 4 - 6x + 1 = -6x - 6 + 4 + 1 = -6x - 1$

35. $(4x - 7) - (2x + 5) = 4x - 7 - 2x - 5 = 4x - 2x - 7 - 5 = 2x - 12$

37. $8(2a + 4) - (6a - 1) = 16a + 32 - 6a + 1 = 16a - 6a + 32 + 1 = 10a + 33$

39. $3(x - 2) + (x - 3) = 3x - 6 + x - 3 = 3x + x - 6 - 3 = 4x - 9$

41. $4(2y - 8) - (y + 7) = 8y - 32 - y - 7 = 8y - y - 32 - 7 = 7y - 39$

43. $-9(2x + 1) - (x + 5) = -18x - 9 - x - 5 = -18x - x - 9 - 5 = -19x - 14$

45. When $x = 2$: $3x - 1 = 3(2) - 1 = 6 - 1 = 5$

47. When $x = 2$: $-2x - 5 = -2(2) - 5 = -4 - 5 = -9$

49. When $x = 2$: $x^2 - 8x + 16 = (2)^2 - 8(2) + 16 = 4 - 16 + 16 = 4$

51. When $x = 2$: $(x - 4)^2 = (2 - 4)^2 = (-2)^2 = 4$

53. When $x = -5$: $7x - 4 - x - 3 = 7(-5) - 4 - (-5) - 3 = -35 - 4 + 5 - 3 = -42 + 5 = -37$

   Now simplifying the expression: $7x - 4 - x - 3 = 7x - x - 4 - 3 = 6x - 7$

   When $x = -5$: $6x - 7 = 6(-5) - 7 = -30 - 7 = -37$

Note that the two values are the same.

55. When $x = -5$: $5(2x+1) + 4 = 5[2(-5)+1] + 4 = 5(-10+1) + 4 = 5(-9) + 4 = -45 + 4 = -41$

Now simplifying the expression: $5(2x+1) + 4 = 10x + 5 + 4 = 10x + 9$

When $x = -5$: $10x + 9 = 10(-5) + 9 = -50 + 9 = -41$

57. When $x = -3$ and $y = 5$, $x^2 - 2xy + y^2 = (-3)^2 - 2(-3)(5) + (5)^2 = 9 + 30 + 25 = 64$

59. When $x = -3$ and $y = 5$: $(x-y)^2 = (-3-5)^2 = (-8)^2 = 64$

61. When $x = -3$ and $y = 5$, $x^2 + 6xy + 9y^2 = (-3)^2 + 6(-3)(5) + 9(5)^2 = 9 - 90 + 225 = 144$

63. When $x = -3$ and $y = 5$: $(x+3y)^2 = [-3 + 3(5)]^2 = (-3+15)^2 = (12)^2 = 144$

65. When $x = \frac{1}{2}$, $12x - 3 = 12(\frac{1}{2}) - 3 = 6 - 3 = 3$

67. When $x = \frac{1}{4}$: $12x - 3 = 12(\frac{1}{4}) - 3 = 3 - 3 = 0$

69. When $x = \frac{3}{2}$, $12x - 3 = 12(\frac{3}{2}) - 3 = 18 - 3 = 15$

71. When $x = \frac{3}{4}$: $12x - 3 = 12(\frac{3}{4}) - 3 = 9 - 3 = 6$

73. $2n + 3$ when $n = 1$, $2(1) + 3 = 5$

$2n + 3$ when $n = 2$, $2(2) + 3 = 7$

$2n + 3$ when $n = 3$, $2(3) + 3 = 9$

$2n + 3$ when $n = 4$, $2(4) + 3 = 11$

75. $n^2 + 1$ when $n = 1$, $(1)^2 + 1 = 2$

$n^2 + 1$ when $n = 2$, $(2)^2 + 1 = 5$

$n^2 + 1$ when $n = 3$, $(3)^2 + 1 = 10$

$n^2 + 1$ when $n = 4$, $(4)^2 + 1 = 17$

77. See the tables in the back of the textbook.

79. $\quad\quad T = -0.0035A + 70$

(a) When $A = 8,000$ feet

$\quad\quad T = -0.0035(8,000) + 70$

$\quad\quad\quad = 42°\,F$

(b) When $A = 12,000$ feet

$\quad\quad T = -00.0035(12,000) + 70$

$\quad\quad\quad = 28°\,F$

(c) when $A = 24,000$ feet

$\quad\quad T = -0.0035(24,000) + 70$

$\quad\quad\quad = -14°\,F$

81. $\quad\quad C = 35 + 0.25t$

(a) When $t = 10$

$\quad\quad C = 35 + 0.25(10)$

$\quad\quad\quad = \$37.50$

(b) When $t = 20$

$\quad\quad C = 35 + 0.25(20)$

$\quad\quad\quad = \$40.00$

(c) When $t = 30$

$\quad\quad C = 35 + 0.25(30)$

$\quad\quad\quad = \$42.50$

83. $\quad\quad T = G - 0.21G - 0.08G$

$\quad\quad\quad = 0.71G$

When $G = 1,250$

$\quad\quad T = 0.71(1,250)$

$\quad\quad\quad = \$887.50$

85. $x - 5$

When $x = -2$: $x - 5 = (-2) - 5 = -7$

87. $2(x + 10)$

When $x = -2$: $2(x + 10) = 2(-2 + 10)$

$\quad\quad\quad\quad\quad\quad\quad\quad = 2(8) = 16$

89. $\dfrac{10}{x}$

When $x = -2$: $\dfrac{10}{x} = \dfrac{10}{-2} = -5$

91. $\left[3x + (-2)\right] - 5 = 3x - 2 - 5 = 3x - 7$

When $x = -2$: $3x - 7 = 3(-2) - 7$

$\quad\quad\quad\quad\quad\quad\quad = -6 - 7 = -13$

93. $-3 - \dfrac{1}{2} = \dfrac{-3}{1} - \dfrac{1}{2} = \dfrac{-3\cdot 2}{1\cdot 2} - \dfrac{1}{2} = \dfrac{-6}{2} - \dfrac{1}{2} = -\dfrac{7}{2}$

95. $\dfrac{4}{5} + \dfrac{1}{10} + \dfrac{3}{8} = \dfrac{4\cdot 8}{5\cdot 8} + \dfrac{1\cdot 4}{10\cdot 4} + \dfrac{3\cdot 5}{8\cdot 5} = \dfrac{32}{40} + \dfrac{4}{40} + \dfrac{15}{40} = \dfrac{51}{40}$

2.2 Addition Property of Equality

1.
$$x - 3 = 8$$
$$x - 3 + 3 = 8 + 3$$
$$x = 11$$

3.
$$x + 2 = 6$$
$$x + 2 + (-2) = 6 + (-2)$$
$$x = 4$$

5.
$$a + \frac{1}{2} = -\frac{1}{4}$$
$$a + \frac{1}{2} + \left(-\frac{1}{2}\right) = -\frac{1}{4} + \left(-\frac{1}{2}\right)$$
$$a = -\frac{1}{4} + \left(-\frac{2}{4}\right)$$
$$a = -\frac{3}{4}$$

7.
$$x + 2.3 = -3.5$$
$$x + 2.3 + (-2.3) = -3.5 + (-2.3)$$
$$x = -5.8$$

9.
$$y + 11 = -6$$
$$y + 11 + (-11) = -6 + (-11)$$
$$y = -17$$

11.
$$x - \frac{5}{8} = -\frac{3}{4}$$
$$x - \frac{5}{8} + \frac{5}{8} = -\frac{3}{4} + \frac{5}{8}$$
$$x = -\frac{6}{8} + \frac{5}{8}$$
$$x = -\frac{1}{8}$$

13.
$$m - 6 = -10$$
$$m - 6 + 6 = -10 + 6$$
$$m = -4$$

15.
$$6.9 + x = 3.3$$
$$-6.9 + 6.9 + x = -6.9 + 3.3$$
$$x = -3.6$$

17.
$$5 = a + 4$$
$$5 + (-4) = a + 4 + (-4)$$
$$a = 1$$

19.
$$-\frac{5}{9} = x - \frac{2}{5}$$
$$-\frac{5}{9} + \frac{2}{5} = x - \frac{2}{5} + \frac{2}{5}$$
$$-\frac{25}{45} + \frac{18}{45} = x$$
$$x = -\frac{7}{45}$$

21. $4x + 2 - 3x = 4 + 1$
$x + 2 = 5$
$x + 2 + (2) = 5 + (-2)$
$x = 3$

23. $8a - \dfrac{1}{2} - 7a = \dfrac{3}{4} + \dfrac{1}{8}$
$a - \dfrac{1}{2} = \dfrac{6}{8} + \dfrac{1}{8}$
$a - \dfrac{1}{2} = \dfrac{7}{8}$
$a - \dfrac{1}{2} + \dfrac{1}{2} = \dfrac{7}{8} + \dfrac{1}{2}$
$a = \dfrac{7}{8} + \dfrac{4}{8}$
$a = \dfrac{11}{8}$

25. $-3 - 4x + 5x = 18$
$-3 + x = 18$
$3 - 3 + x = 3 + 18$
$x = 21$

27. $-11x + 2 + 10x + 2x = 9$
$x + 2 = 9$
$x + 2 + (-2) = 9 + \{-2\}$
$x = 7$

29. $-2.5 + 4.8 = 8x - 1.2 - 7x$
$2.3 = x - 1.2$
$2.3 + 1.2 = x - 1.2 + 1.2$
$x = 3.5$

31. $2y - 10 + 3y - 4y = 18 - 6$
$y - 10 = 12$
$y - 10 + 10 = 12 + 10$
$y = 22$

33. $15 - 21 = 8x + 3x - 10x$
$x = -6$

35. $24 - 3 + 8a - 5a - 2a = 21$
$21 + a = 21$
$-21 + 21 + a = -21 + 21$
$a = 0$

37. $2(x + 3) - x = 4$
$2x + 6 - x = 4$
$x + 6 = 4$
$x + 6 + (-6) = 4 + (-6)$
$x = -2$

39. $-3(x - 4) + 4x = 3 - 7$
$-3x + 12 + 4x = -4$
$x + 12 = -4$
$x + 12 + (-12) = -4 + (-12)$
$x = -16$

41. $5(2a + 1) - 9a = 8 - 6$
$10a + 5 - 9a = 2$
$a + 5 = 2$
$a + 5 + (-5) = 2 + (-5)$
$a = -3$

43. $-(x + 3) + 2x - 1 = 6$
$-x - 3 + 2x - 1 = 6$
$x - 4 = 6$
$x - 4 + 4 = 6 + 4$
$x = 10$

45. $4y - 3(y - 6) + 2 = 8$

$\quad\quad 4y - 3y + 18 + 2 = 8$

$\quad\quad\quad\quad\quad y + 20 = 8$

$\quad\quad y + 20 + (-20) = 8 + (-20)$

$\quad\quad\quad\quad\quad\quad\quad y = -12$

47. $2(3x + 1) - 5(x + 2) = 1 - 10$

$\quad\quad 6x + 2 - 5x - 10 = -9$

$\quad\quad\quad\quad\quad\quad x - 8 = -9$

$\quad\quad\quad\quad x - 8 + 8 = -9 + 8$

$\quad\quad\quad\quad\quad\quad\quad x = -1$

49. $-3(2m - 9) + 7(m - 4) = 12 - 9$

$\quad\quad -6m + 27 + 7m - 28 = 3$

$\quad\quad\quad\quad\quad\quad\quad m - 1 = 3$

$\quad\quad\quad\quad\quad m - 1 + 1 = 3 + 1$

$\quad\quad\quad\quad\quad\quad\quad\quad m = 4$

51. $\quad\quad\quad\quad 4x = 3x + 2$

$\quad\quad 4x + (-3x) = 3x + (-3x) + 2$

$\quad\quad\quad\quad\quad\quad x = 2$

53. $\quad\quad\quad\quad 8a = 7a - 5$

$\quad\quad 8a + (-7a) = 7a + (-7a) - 5$

$\quad\quad\quad\quad\quad\quad a = -5$

55. $\quad\quad\quad\quad 2x = 3x + 1$

$\quad\quad (-2x) + 2x = (-2x) + 3x + 1$

$\quad\quad\quad\quad\quad\quad 0 = x + 1$

$\quad\quad 0 + (-1) = x + 1 + (-1)$

$\quad\quad\quad\quad\quad\quad x = -1$

57. $\quad\quad\quad\quad 3y + 4 = 2y + 1$

$\quad\quad 3y + (-2y) + 4 = 2y + (-2y) + 1$

$\quad\quad y + 4 + (-4) = 1 + (-4)$

$\quad\quad\quad\quad\quad\quad y = -3$

59. $\quad\quad\quad\quad 2m - 3 = m + 5$

$\quad\quad 2m + (-m) - 3 = m + (-m) + 5$

$\quad\quad\quad\quad\quad\quad m - 3 = 5$

$\quad\quad\quad\quad m - 3 + 3 = 5 + 3$

$\quad\quad\quad\quad\quad\quad m = 8$

61. $\quad\quad\quad\quad 4x - 7 = 5x + 1$

$\quad\quad 4x + (-4x) - 7 = 5x + (-4x) + 1$

$\quad\quad\quad\quad\quad -7 = x + 1$

$\quad\quad -7 + (-1) = x + 1 + (-1)$

$\quad\quad\quad\quad\quad\quad x = -8$

63. $\quad\quad\quad 5x - \dfrac{2}{3} = 4x + \dfrac{4}{3}$

$\quad\quad 5x + (-4x) - \dfrac{2}{3} = 4x + (-4x) + \dfrac{4}{3}$

$\quad\quad\quad\quad\quad x - \dfrac{2}{3} = \dfrac{4}{3}$

$\quad\quad x - \dfrac{2}{3} + \dfrac{2}{3} = \dfrac{4}{3} + \dfrac{2}{3}$

$\quad\quad\quad\quad\quad\quad x = \dfrac{6}{3} = 2$

65.
$$8a - 7.1 = 7a + 3.9$$
$$8a + (-7a) - 7.1 = 7a + (-7a) + 3.9$$
$$a - 7.1 = 3.9$$
$$a - 7.1 + 7.1 = 3.9 + 7.1$$
$$a = 11$$

67. $x - 2 + 5 = 6$
$$x + 3 = 6$$
$$x = 3$$
$$y + 6 + 1 = 6$$
$$y + 7 = 6$$
$$y = -1$$
$$4 + 2 + z = 6$$
$$6 + z = 6$$
$$z = 0$$
$$x = 3, y = -1, z = 0$$

69. $T + R + A = 100$

(a) $T = 88, A = 6$
$$88 + R + 6 = 100$$
$$R + 94 = 100$$
$$R = 6\%$$

(b) $T = 0, A = 95$
$$0 + R + 95 = 100$$
$$R = 5\%$$

(c) $T = 0, R = 98$
$$0 + 98 + A = 100$$
$$A = 2\%$$

(d) $T = 0, A = 25$
$$0 + R + 25 = 100$$
$$R = 75\%$$

71. $x + 55 + 55 = 180$
$$x + 110 = 180$$
$$x = 70°$$

73. $x - 3 = 8$
$$x = 11$$

75.
$$2y + 3 = y + 5$$
$$2y + (-y) + 3 = y + (-y) + 5$$
$$y + 3 = 5$$
$$y = 2$$

77. $3(6x) = (3 \cdot 6)x = 18x$

79. $\dfrac{1}{5}(5x) = \left(\dfrac{1}{5} \cdot 5\right)x = 1x = x$

81. $8\left(\dfrac{1}{8}y\right) = \left(8 \cdot \dfrac{1}{8}\right)y = 1y = y$

83. $-2\left(-\dfrac{1}{2}x\right) = \left[-2 \cdot \left(-\dfrac{1}{2}\right)\right]x = 1x = x$

85. $-\dfrac{4}{3}\left(-\dfrac{3}{4}a\right)=\left[-\dfrac{4}{3}\cdot\left(-\dfrac{3}{4}\right)\right]a=1a=a$

2.3 Multiplication Property of Equality

1. $5x=10$

$\dfrac{1}{5}(5x)=\dfrac{1}{5}(10)$

$x=2$

3. $7a=28$

$\dfrac{1}{7}(7a)=\dfrac{1}{7}(28)$

$a=4$

5. $-8x=4$

$-\dfrac{1}{8}(-8x)=-\dfrac{1}{8}(4)$

$x=-\dfrac{1}{2}$

7. $8m=-16$

$\dfrac{1}{8}(8m)=\dfrac{1}{8}(-16)$

$m=-2$

9. $-3x=-9$

$-\dfrac{1}{3}(-3x)=-\dfrac{1}{3}(-9)$

$x=3$

11. $-7y=-28$

$-\dfrac{1}{7}(-7y)=-\dfrac{1}{7}(-28)$

$y=4$

13. $2x=0$

$\dfrac{1}{2}(2x)=\dfrac{1}{2}(0)$

$x=0$

15. $-5x=0$

$-\dfrac{1}{5}(-5x)=-\dfrac{1}{5}(0)$

$x=0$

17. $\dfrac{x}{3}=2$

$3\left(\dfrac{x}{3}\right)=3(2)$

$x=6$

19. $-\dfrac{m}{5}=10$

$-5\left(-\dfrac{m}{5}\right)=-5(10)$

$m=-50$

21. $-\dfrac{x}{2}=-\dfrac{3}{4}$

$-2\left(-\dfrac{x}{2}\right)=-2\left(-\dfrac{3}{4}\right)$

$x=\dfrac{3}{2}$

23. $\dfrac{2}{3}a=8$

$\dfrac{3}{2}\left(\dfrac{2}{3}a\right)=\dfrac{3}{2}(8)$

$a=12$

25.
$$-\frac{3}{5}x = \frac{9}{5}$$
$$-\frac{5}{3}\left(-\frac{3}{5}x\right) = -\frac{5}{3}\left(\frac{9}{5}\right)$$
$$x = -3$$

27.
$$-\frac{5}{8}y = -20$$
$$-\frac{8}{5}\left(-\frac{5}{8}y\right) = -\frac{8}{5}(-20)$$
$$y = 32$$

29. $-4x - 2x + 3x = 24$
$$-3x = 24$$
$$-\frac{1}{3}(-3x) = -\frac{1}{3}(24)$$
$$x = -8$$

31. $4x + 8x - 2x = 15 - 10$
$$10x = 5$$
$$\frac{1}{10}(10x) = \frac{1}{10}(5)$$
$$x = \frac{1}{2}$$

33.
$$-3 - 5 = 3x + 5x - 10x$$
$$-\frac{1}{2}(-8) = -\frac{1}{2}(-2x)$$
$$x = 4$$

35. $18 - 13 = \frac{1}{2}a + \frac{3}{4}a - \frac{5}{8}a$
$$8(5) = 8\left(\frac{1}{2}a + \frac{3}{4}a - \frac{5}{8}a\right)$$
$$40 = 4a + 6a - 5a$$
$$40 = 5a$$
$$\frac{1}{5}(40) = \frac{1}{5}(5a)$$
$$a = 8$$

37.
$$-x = 4$$
$$-1(-x) = -1(4)$$
$$x = -4$$

39.
$$-x = -4$$
$$-1(-x) = -1(-4)$$
$$x = 4$$

41.
$$15 = -a$$
$$-1(15) = -1(-a)$$
$$a = -15$$

43.
$$-y = \frac{1}{2}$$
$$-1(-y) = -1\left(\frac{1}{2}\right)$$
$$y = -\frac{1}{2}$$

45.
$$3x - 2 = 7$$
$$3x - 2 + 2 = 7 + 2$$
$$3x = 9$$
$$\frac{1}{3}(3x) = \frac{1}{3}(9)$$
$$x = 3$$

47.
$$2a + 1 = 3$$
$$2a + 1 + (-1) = 3 + (-1)$$
$$\frac{1}{2}(2a) = \frac{1}{2}(2)$$
$$a = 1$$

49.　$\dfrac{1}{8}+\dfrac{1}{2}x=\dfrac{1}{4}$

$8\left(\dfrac{1}{8}+\dfrac{1}{2}x\right)=8\left(\dfrac{1}{4}\right)$

$1+4x=2$

$(-1)+1+4x=(-1)+2$

$4x=1$

$\dfrac{1}{4}(4x)=\dfrac{1}{4}(1)$

$x=\dfrac{1}{4}$

51.　$6x=2x-12$

$6x+(-2x)=2x+(-2x)-12$

$4x=-12$

$\dfrac{1}{4}(4x)=\dfrac{1}{4}(-12)$

$x=-3$

53.　$2y=-4y+18$

$2y+4y=-4y+4y+18$

$6y=18$

$\dfrac{1}{6}(6y)=\dfrac{1}{6}(18)$

$y=3$

55.　$-7x=-3x-8$

$-7x+3x=-3x+3x-8$

$-4x=-8$

$-\dfrac{1}{4}(-4x)=-\dfrac{1}{4}(-8)$

$x=2$

57.　$8x+4=2x-5$

$8x+(-2x)+4=2x+(-2x)-5$

$6x+4=-5$

$6x+4+(-4)=-5+(-4)$

$6x=-9$

$\dfrac{1}{6}(6x)=\dfrac{1}{6}(-9)$

$x=-\dfrac{3}{2}$

59.　$x+\dfrac{1}{2}=\dfrac{1}{4}x-\dfrac{5}{8}$

$8\left(x+\dfrac{1}{2}\right)=8\left(\dfrac{1}{4}x-\dfrac{5}{8}\right)$

$8x+4=2x-5$

$8x+(-2x)+4=2x+(-2x)-5$

$6x+4=-5$

$6x+4+(-4)=-5+(-4)$

$6x=-9$

$\dfrac{1}{6}(6x)=\dfrac{1}{6}(-9)$

$x=-\dfrac{3}{2}$

61.　$6m-3=m+2$

$6m+(-m)-3=m+(-m)+2$

$5m-3=2$

$5m-3+3=2+3$

$5m=5$

$\dfrac{1}{5}(5m)=\dfrac{1}{5}(5)$

$m=1$

63.
$$\frac{1}{2}m - \frac{1}{4} = \frac{1}{12}m + \frac{1}{6}$$
$$12\left(\frac{1}{2}m - \frac{1}{4}\right) = 12\left(\frac{1}{12}m + \frac{1}{6}\right)$$
$$6m - 3 = m + 2$$
$$6m + (-m) - 3 = m + (-m) + 2$$
$$5m - 3 = 2$$
$$5m - 3 + 3 = 2 + 3$$
$$5m = 5$$
$$\frac{1}{5}(5m) = \frac{1}{5}(5)$$
$$m = 1$$

65.
$$9y + 2 = 6y - 4$$
$$9y + (-6y) + 2 = 6y + (-6y) - 4$$
$$3y + 2 = -4$$
$$3y + 2 + (-2) = -4 + (-2)$$
$$3y = -6$$
$$\frac{1}{3}(3y) = \frac{1}{3}(-6)$$
$$y = -2$$

67.
$$\frac{3}{2}y + \frac{1}{3} = y - \frac{2}{3}$$
$$6\left(\frac{3}{2}y + \frac{1}{3}\right) = 6\left(y - \frac{2}{3}\right)$$
$$9y + 2 = 6y - 4$$
$$9y + (-6y) + 2 = 6y + (-6y) - 4$$
$$3y + 2 = -4$$
$$3y + 2 + (-2) = -4 + (-2)$$
$$3y = -6$$
$$\frac{1}{3}(3y) = \frac{1}{3}(-6)$$
$$y = -2$$

69.
$$7.5x = 1500$$
$$\frac{1}{7.5}(7.5x) = \frac{1}{7.5}(1500)$$
$$x = 200$$

71.
$$1 + 3(2) + 3x = 13$$
$$1 + 6 + 3x = 13$$
$$7 + 3x = 13$$
$$7 + (-7) + 3x = 13 + (-7)$$
$$3x = 6$$
$$\frac{1}{3}(3x) = \frac{1}{3}(6)$$
$$x = 2$$

73.　(a) $AB = 12$, $AY = 15$, and $AC = 20$
$$\frac{AX}{12} = \frac{15}{20}$$
$$12\left(\frac{AX}{12}\right) = 12\left(\frac{15}{20}\right)$$
$$AX = 9$$

(b) $XY = 8$, $BC = 10$, and $AX = 12$
$$\frac{12}{AB} = \frac{8}{10}$$
$$AB\left(\frac{12}{AB}\right) = AB\left(\frac{8}{10}\right)$$
$$12 = \frac{8}{10}AB$$
$$\frac{10}{8}(12) = \frac{10}{8}\left(\frac{8}{10}AB\right)$$
$$15 = AB$$

73.

Continued

(c) $YC = 6$, $XB = 4$, and $AX = 8$

$$\frac{8}{8+4} = \frac{AY}{AY+6}$$

$$12(AY+6)\left(\frac{8}{12}\right) = 12(AY+6)\left(\frac{AY}{AY+6}\right)$$

$$8(AY+6) = 12AY$$

$$8AY+48 = 12AY$$

$$8AY+(-8AY)+48 = 12AY+(-8AY)$$

$$48 = 4AY$$

$$\frac{1}{4}(48) = \frac{1}{4}(4AY)$$

$$12 = AY$$

$$AC = AY+YC = 12+6 = 18$$

75.

$$3x+2 = 19$$

$$3x+2+(-2) = 19+(-2)$$

$$3x = 17$$

$$\frac{1}{3}(3x) = \frac{1}{3}(17)$$

$$x = \frac{17}{3}$$

77.

$$2(x+10) = 40$$

$$\frac{1}{2}(2)(x+10) = \frac{1}{2}(40)$$

$$x+10 = 20$$

$$x+10+(-10) = 20+(-10)$$

$$x = 10$$

79. Using the distributive property and combining like terms: $5(2x-8)-3 = 10x-40-3 = 10x-43$

81. Using the distributive property and combining like terms:

$-2(3x+5)+3(x-1) = -6x-10+3x-3 = -6x+3x-10-3 = -3x-13$

83. Using the distributive property and combining like terms: $7-3(2y+1) = 7-6y-3 = -6y+7-3 = -6y+4$

85. Using the distributive property and combining like terms: $4x-(9x-3)+4 = 4x-9x+3+4 = -5x+7$

2.4 Solving Linear Equations

1. $2(x+3)=12$

$2x+6=12$

$2x+6+(-6)=12+(-6)$

$2x=6$

$\frac{1}{2}(2x)=\frac{1}{2}(6)$

$x=3$

3. $6(x-1)=-18$

$6x-6=-18$

$6x-6+6=-18+6$

$6x=-12$

$\frac{1}{6}(6x)=\frac{1}{6}(-12)$

$x=-2$

5. $2(4a+1)=-6$

$8a+2=-6$

$8a+2+(-2)=-6+(-2)$

$8a=-8$

$\frac{1}{8}(8a)=\frac{1}{8}(-8)$

$a=-1$

7. $14=2(5x-3)$

$14=10x-6$

$14+6=10x-6+6$

$20=10x$

$\frac{1}{10}(20)=\frac{1}{10}(10x)$

$x=2$

9. $-2(3y+5)=14$

$-6y-10=14$

$-6y-10+10=14+10$

$-6y=24$

$-\frac{1}{6}(-6y)=-\frac{1}{6}(24)$

$y=-4$

11. $-5(2a+4)=0$

$-10a-20=0$

$-10a-20+20=0+20$

$-10a=20$

$-\frac{1}{10}(-10a)=-\frac{1}{10}(20)$

$a=-2$

13. $1=\frac{1}{2}(4x+2)$

$1=2x+1$

$1+(-1)=2x+1+(-1)$

$0=2x$

$\frac{1}{2}(0)=\frac{1}{2}(2x)$

$x=0$

15. $3(t-4)+5=-4$

$3t-12+5=-4$

$3t-7=-4$

$3t-7+7=-4+7$

$3t=3$

$\frac{1}{3}(3t)=\frac{1}{3}(3)$

$t=1$

17. $4(2y+1)-7=1$

$\quad 8y+4-7=1$

$\quad\quad 8y-3=1$

$\quad 8y-3+3=1+3$

$\quad\quad\quad 8y=4$

$\quad \dfrac{1}{8}(8y)=\dfrac{1}{8}(4)$

$\quad\quad\quad y=\dfrac{1}{2}$

19. $\quad \dfrac{1}{2}(x-3)=\dfrac{1}{4}(x+1)$

$\quad\quad \dfrac{1}{2}x-\dfrac{3}{2}=\dfrac{1}{4}x+\dfrac{1}{4}$

$\quad 4\left(\dfrac{1}{2}x-\dfrac{3}{2}\right)=4\left(\dfrac{1}{4}x+\dfrac{1}{4}\right)$

$\quad\quad 2x-6=x+1$

$\quad 2x+(-x)-6=x+(-x)+1$

$\quad\quad\quad x-6=1$

$\quad\quad x-6+6=1+6$

$\quad\quad\quad\quad x=7$

21. $\quad -0.7(2x-7)=0.3(11-4x)$

$\quad\quad -1.4x+4.9=3.3-1.2x$

$\quad -1.4x+1.2x+4.9=3.3-1.2x+1.2x$

$\quad\quad\quad -0.2x+4.9=3.3$

$\quad -0.2x+4.9+(-4.9)=3.3+(-4.9)$

$\quad\quad\quad -0.2x=-1.6$

$\quad\quad\quad \dfrac{-0.2x}{-0.2}=\dfrac{-1.6}{-0.2}$

$\quad\quad\quad\quad x=8$

23. $\quad -2(3y+1)=3(1-6y)-9$

$\quad\quad -6y-2=3-18y-9$

$\quad\quad -6y-2=-18y-6$

$\quad -6y+18y-2=-18y+18y-6$

$\quad\quad\quad 12y-2=-6$

$\quad\quad 12y-2+2=-6+2$

$\quad\quad\quad 12y=-4$

$\quad\quad \dfrac{1}{12}(12y)=\dfrac{1}{12}(-4)$

$\quad\quad\quad y=-\dfrac{1}{3}$

25. $\dfrac{3}{4}(8x-4)+3=\dfrac{2}{5}(5x+10)-1$

$\quad\quad 6x-3+3=2x+4-1$

$\quad\quad\quad 6x=2x+3$

$\quad 6x+(-2x)=2x+(-2x)+3$

$\quad\quad\quad 4x=3$

$\quad\quad \dfrac{1}{4}(4x)=\dfrac{1}{4}(3)$

$\quad\quad\quad x=\dfrac{3}{4}$

27. $0.06x+0.08(100-x)=6.5$

$\quad 0.06x+8-0.08x=6.5$

$\quad\quad -0.02x+8=6.5$

$\quad -0.02x+8+(-8)=6.5+(-8)$

$\quad\quad\quad -0.02x=-1.5$

$\quad\quad \dfrac{-0.02x}{-0.02}=\dfrac{-1.5}{-0.02}$

$\quad\quad\quad x=75$

29. $6 - 5(2a - 3) = 1$
$6 - 10a + 15 = 1$
$-10a + 21 = 1$
$-10a + 21 + (-21) = 1 + (-21)$
$-10a = -20$
$-\dfrac{1}{10}(-10a) = -\dfrac{1}{10}(-20)$
$a = 2$

31. $0.2x - 0.5 = 0.5 - 0.2(2x - 13)$
$0.2x - 0.5 = 0.5 - 0.4x + 2.6$
$0.2x - 0.5 = -0.4x + 3.1$
$0.2x + 0.4x - 0.5 = -0.4x + 0.4x + 3.1$
$0.6x - 0.5 = 3.1$
$0.6x - 0.5 + 0.5 = 3.1 + 0.5$
$0.6x = 3.6$
$\dfrac{0.6x}{0.6} = \dfrac{3.6}{0.6}$
$x = 6$

33. $2(t - 3) + 3(t - 2) = 28$
$2t - 6 + 3t - 6 = 28$
$5t - 12 = 28$
$5t - 12 + 12 = 28 + 12$
$5t = 40$
$\dfrac{1}{5}(5t) = \dfrac{1}{5}(40)$
$t = 8$

35. $5(x - 2) - (3x + 4) = 3(6x - 8) + 10$
$5x - 10 - 3x - 4 = 18x - 24 + 10$
$2x - 14 = 18x - 14$
$2x + (-18x) - 14 = 18x + (-18x) - 14$
$-16x - 14 = -14$
$-16x - 14 + 14 = -14 + 14$
$-16x = 0$
$-\dfrac{1}{16}(-16x) = -\dfrac{1}{16}(0)$
$x = 0$

37. $2(5x - 3) - (2x - 4) = 5 - (6x + 1)$
$10x - 6 - 2x + 4 = 5 - 6x - 1$
$8x - 2 = -6x + 4$
$8x + 6x - 2 = -6x + 6x + 4$
$14x - 2 + 2 = 4 + 2$
$14x = 6$
$\dfrac{1}{14}(14x) = \dfrac{1}{14}(6)$
$x = \dfrac{3}{7}$

39. $-(3x + 1) - (4x - 7) = 4 - (3x + 2)$
$-3x - 1 - 4x + 7 = 4 - 3x - 2$
$-7x + 6 = -3x + 2$
$-7x + 3x + 6 = -3x + 3x + 2$
$-4x + 6 = 2$
$-4x + 6 + (-6) = 2 + (-6)$
$-4x = -4$
$-\dfrac{1}{4}(-4x) = -\dfrac{1}{4}(-4)$
$x = 1$

41. $\frac{1}{2}(3) = \frac{1}{2} \cdot \frac{3}{1} = \frac{3}{2}$

43. $\frac{2}{3}(6) = \frac{2}{3} \cdot \frac{6}{1} = \frac{12}{3} = 4$

45. $\frac{5}{9} \cdot \frac{9}{5} = \frac{45}{45} = 1$

47. $2(3x - 5) = 2 \cdot 3x - 2 \cdot 5 = 6x - 10$

49. $\frac{1}{2}(3x + 6) = \frac{1}{2} \cdot (3x) + \frac{1}{2} \cdot 6 = \frac{3}{2}x + 3$

51. $\frac{1}{3}(-3x + 6) = \frac{1}{3} \cdot (-3x) + \frac{1}{3} \cdot 6 = -x + 2$

2.5 Formulas

1. Substituting $P = 300$ and $w = 50$:
$$P = 2l + 2w$$
$$300 = 2l + 2(50)$$
$$300 = 2l + 100$$
$$200 = 2l$$
$$l = 100$$

3. Substituting $x = 3$:
$$2(3) + 3y = 6$$
$$6 + 3y = 6$$
$$6 + (-6) + 3y = 6 + (-6)$$
$$3y = 0$$
$$y = 0$$

5. Substituting $x = 0$:
$$2(0) + 3y = 6$$
$$0 + 3y = 6$$
$$3y = 6$$
$$y = 2$$

7. Substituting $y = 2$:
$$2x - 5(2) = 20$$
$$2x - 10 = 20$$
$$2x - 10 + 10 = 20 + 10$$
$$2x = 30$$
$$x = 15$$

9. Substituting $y = 0$:
$$2x - 5(0) = 20$$
$$2x - 0 = 20$$
$$2x = 20$$
$$x = 10$$

11. Substituting $y = 7$:
$$7 = 2x - 1$$
$$7 + 1 = 2x - 1 + 1$$
$$2x = 8$$
$$x = 4$$

13. Substituting $y = 3$:
$$3 = 2x - 1$$
$$3 + 1 = 2x - 1 + 1$$
$$2x = 4$$
$$x = 2$$

15. Solving for l:
$$lw = A$$
$$\frac{lw}{w} = \frac{A}{w}$$
$$l = \frac{A}{w}$$

17. Solving for r:
$$rt = d$$
$$\frac{rt}{t} = \frac{d}{t}$$
$$r = \frac{d}{t}$$

19. Solving for h:
$$lwh = V$$
$$\frac{lwh}{lw} = \frac{V}{lw}$$
$$h = \frac{V}{lw}$$

21. Solving for P:

$$PV = nRT$$

$$\frac{PV}{V} = \frac{nRT}{V}$$

$$P = \frac{nRT}{V}$$

23. Solving for a:

$$a + b + c = P$$

$$a + b + c - b - c = P - b - c$$

$$a = P - b - c$$

25. Solving for x:

$$x - 3y = -1$$

$$x - 3y + 3y = -1 + 3y$$

$$x = 3y - 1$$

27. Solving for y:

$$-3x + y = 6$$

$$-3x + 3x + y = 6 + 3x$$

$$y = 3x + 6$$

29. Solving for y:

$$2x + 3y = 6$$

$$-2x + 2x + 3y = -2x + 6$$

$$3y = -2x + 6$$

$$\frac{1}{3}(3y) = \frac{1}{3}(-2x + 6)$$

$$y = -\frac{2}{3}x + 2$$

31. Solving for y:

$$6x + 3y = 12$$

$$-6x + 6x + 3y = -6x + 12$$

$$3y = -6x + 12$$

$$\frac{1}{3}(3y) = \frac{1}{3}(-6x + 12)$$

$$y = -2x + 4$$

33. Solving for y:

$$5x - 2y = 3$$

$$-5x + 5x - 2y = -5x + 3$$

$$-2y = -5x + 3$$

$$-\frac{1}{2}(-2y) = -\frac{1}{2}(-5x + 3)$$

$$y = \frac{5}{2}x - \frac{3}{2}$$

35. Solving for w:

$$2l + 2w = P$$

$$2l - 2l + 2w = P - 2l$$

$$2w = P - 2l$$

$$\frac{2w}{2} = \frac{P - 2l}{2}$$

$$w = \frac{P - 2l}{2}$$

37. Solving for v:

$$vt + 16t^2 = h$$

$$vt + 16t^2 - 16t^2 = h - 16t^2$$

$$vt = h - 16t^2$$

$$\frac{vt}{t} = \frac{h - 16t}{t}$$

$$v = \frac{h - 16t^2}{t}$$

39. Solving for h:

$$\pi r^2 + 2\pi rh = A$$

$$\pi r^2 - \pi r^2 + 2\pi rh = A - \pi r^2$$

$$2\pi rh = A - \pi r^2$$

$$\frac{2\pi rh}{2\pi r} = \frac{A - \pi r^2}{2\pi r}$$

$$h = \frac{A - \pi r^2}{2\pi r}$$

41. $$\frac{x}{2}+\frac{y}{3}=1$$
$$-\frac{x}{2}+\frac{x}{2}+\frac{y}{3}=-\frac{x}{2}+1$$
$$\frac{y}{3}=-\frac{x}{2}+1$$
$$3\left(\frac{y}{3}\right)=3\left(-\frac{x}{2}+1\right)$$
$$y=-\frac{3}{2}x+3$$

43. $$\frac{x}{7}-\frac{y}{3}=1$$
$$-\frac{x}{7}+\frac{x}{7}-\frac{y}{3}=-\frac{x}{7}+1$$
$$-\frac{y}{3}=-\frac{x}{7}+1$$
$$-3\left(-\frac{y}{3}\right)=-3\left(-\frac{x}{7}+1\right)$$
$$y=\frac{3}{7}x-3$$

45. $$-\frac{1}{4}x+\frac{1}{8}y=1$$
$$-\frac{1}{4}x+\frac{1}{4}x+\frac{1}{8}y=1+\frac{1}{4}x$$
$$\frac{1}{8}y=\frac{1}{4}x+1$$
$$8\left(\frac{1}{8}y\right)=8\left(\frac{1}{4}x+1\right)$$
$$y=2x+8$$

47. The complement of $30°$ is $90°-30°=60°,$ and the supplement is $180°-30°=150°.$

49. The complement of $45°$ is $90°-45°=45°,$ and the supplement is $180°-45°=135°.$

51. $x=0.25\cdot40$
$x=10$
The number 10 is 25% of 40.

53. $x=0.12\cdot2000$
$x=240$
The number 240 is 12% of 2000.

55. $$x\cdot28=7$$
$$8x=7$$
$$\frac{1}{28}(28x)=\frac{1}{28}(7)$$
$$x=0.25=25\%$$
The number 7 is 25% of 28.

57. $$x\cdot40=14$$
$$40x=14$$
$$\frac{1}{40}(40x)=\frac{1}{40}(14)$$
$$x=0.35=35\%$$
The number 14 is 35% of 40.

59. $0.50\cdot x=32$
$$\frac{0.50x}{0.50}=\frac{32}{0.50}$$
$$x=64$$
The number 32 is 50% of 64.

61. $0.12\cdot x=240$
$$\frac{0.12x}{0.12}=\frac{240}{0.12}$$
$$x=2000$$
The number 240 is 12% of 2000.

63. (a) 62.8% of T is $68,840,000$

$$0.628T = 68,840,000$$

(b) $\dfrac{0.628T}{0.628} = \dfrac{68,840,000}{0.628}$

$$T = 109,617,834 \text{ people}$$

65. (a) 12 is what percent of 20?

$$12 = x \cdot 20$$

$$x = \frac{12}{20} = 0.6 = 60\% \text{ silver}$$

(b) $100\% - 60\% = 40\%$ copper

67. Substituting $F = 212$:

$$C = \frac{5}{9}(212 - 32)$$

$$= \frac{5}{9}(180)$$

$$= 100^\circ C$$

Yes

69. Substituting $F = 68$:

$$C = \frac{5}{9}(68 - 32)$$

$$= \frac{5}{9}(36)$$

$$= 20^\circ$$

Yes

71.

$$\frac{9}{5}C + 32 = F$$

$$\frac{9}{5}C + 32 - 32 = F - 32$$

$$\frac{9}{5}C = F - 32$$

$$\frac{5}{9}\left(\frac{9}{5}C\right) = \frac{5}{9}(F - 32)$$

$$C = \frac{5}{9}(F - 32)$$

73. Find what percent of 150 is 90:

$$x \cdot 150 = 90$$

$$\frac{1}{150}(150x) = \frac{1}{150}(90)$$

$$x = 0.60 = 60\%$$

60% of the calories in one serving of vanilla ice cream are fat calories.

75. Find what percent of 98 is 26:

$$x \cdot 98 = 26$$

$$\frac{1}{98}(98x) = \frac{1}{98}(26)$$

$$x \approx 0.265 = 26.5\%$$

26.5% of one serving of frozen yogurt are carbohydrates.

77. $T = -0.0035A + 56$

See the table in the back of the textbook.

79. (a)   If $C = 44$ and $\pi = \dfrac{22}{7}$

$$2 \cdot \frac{22}{7} r = 44$$

$$\frac{44}{7} r = 44$$

$$\frac{7}{44}\left(\frac{44}{7} r\right) = \frac{7}{44}(44)$$

$$r = 7 \text{ meters}$$

(b)   If $C = 9.42$ and $\pi = 3.14$

$$2 \cdot (3.14) r = 9.42$$

$$6.28r = 9.42$$

$$\frac{6.28r}{6.28} = \frac{9.42}{6.28}$$

$$r = 1.5 \text{ inches}$$

81. The sum of 4 and 1 is 5.

83. The difference of 6 and 2 is 4.

85. An equivalent expression is: $2(6 + 3)$

87. An equivalent expression is: $2(5) + 3$

2.6 Applications

1. Let $x =$ the number.

$$x + 5 = 13$$
$$x = 8$$

The number is 8.

3. Let $x =$ the number.

$$2x + 4 = 14$$
$$2x = 10$$
$$x = 5$$

The number is 5.

5. Let $x =$ the number.

$$5(x + 7) = 30$$
$$5x + 35 = 30$$
$$5x = -5$$
$$x = -1$$

The number is -1.

7. Let x and $x + 2 =$ the two numbers.

$$x + x + 2 = 8$$
$$2x + 2 = 8$$
$$2x = 6$$
$$x = 3$$
$$x + 2 = 5$$

The two numbers are 3 and 5.

9. Let x and $3x-4 =$ the two numbers.

$$(x+3x-4)+5 = 25$$
$$4x+1 = 25$$
$$4x = 24$$
$$x = 6$$
$$3x-4 = 3(6)-4 = 14$$

The two numbers are 6 and 14.

11.

	Five Years Ago	Now
Fred	$x+4-5 = x-1$	$x+4$
Barney	$x-5$	x

$$x-1+x-5 = 48$$
$$2x-6 = 48$$
$$2x = 54$$
$$x = 27$$
$$x+4 = 31$$

Barney is 27 and Fred is 31.

13.

	Now	Three years from now
Jack	$2x$	$2x+3$
Lacy	x	$x+3$

$$2x+3+x+3 = 54$$
$$3x+6 = 54$$
$$3x = 48$$
$$x = 16$$
$$2x = 32$$

Lacy is 16 and Jack is 32.

15.

	Now	Two years from now
Pat	$x+20$	$x+20+2 = x+22$
Patrick	x	$x+2$

$$x+22 = 2(x+2)$$
$$x+22 = 2x+4$$
$$2 = x+4$$
$$x = 18$$
$$20 = 38$$

Patrick is 18 and Pat is 38.

17. Let $w =$ the width and $w+5 =$ the length

$$2w+2(w+5) = 34$$
$$2w+2w+10 = 34$$
$$4w+10 = 34$$
$$4x = 24$$
$$w = 6$$
$$w+5 = 11$$

The length is 11 inches and the
width is 6 inches.

19. Let s represent the side of the square.

$$4s = 48$$
$$s = 12$$

The length of one side is 12 meters.

21. Let $w =$ the width and $2w-3 =$ the length.

$$2w+2(2w-3) = 54$$
$$2w+4w-6 = 54$$
$$6w-6 = 54$$
$$6w = 60$$
$$w = 10$$
$$2w-3 = 2(10)-3 = 17$$

The length is 17 inches and the
width is 10 inches.

23.

	Nickels	Dimes
Number	x	$x+9$
Value (cents)	$5(x)$	$10(x+9)$

$$5(x)+10(x+9) = 210$$
$$x+10x+90 = 210$$
$$15x+90 = 210$$
$$15x = 120$$
$$x = 8$$
$$x+9 = 17$$

Sue has 8 nickels and 17 dimes

25.

	Dimes	Quarters
Number	x	$2x$
Value (cents)	$10(x)$	$25(2x)$

$$10(x) + 25(2x) = 900$$
$$10x + 50x = 900$$
$$60x = 900$$
$$x = 15$$
$$2x = 30$$

You have 15 dimes and 30 quarters.

27.

	Nickels	Dimes	Quarters
Number	x	$x+3$	$x+5$
Value (cents)	$5(x)$	$10(x+3)$	$25(x+5)$

$$5(x) + 10(x+3) + 25(x+5) = 435$$
$$5x + 10x + 30 + 25x + 125 = 435$$
$$40x + 155 = 435$$
$$40x = 280$$
$$x = 7$$
$$x + 3 = 10$$
$$x + 5 = 12$$

Katie has 7 nickels, 10 dimes, and 12 quarters.

29. 4 is less than 10

31. 9 is greater than or equal to -5

33. $12 < 20$

35. $-8 < -6$

37. $|8 - 3| - |5 - 2| = |5| - |3| = 5 - 3 = 2$

39. $15 - |9 - 3(7 - 5)| = 15 - |9 - 3(2)| = 15 - |9 - 6| = 15 - |3| = 15 - 3 = 12$

2.7 More Applications

1.

	Dollars Invested at 8%	Dollars Invested at 9%
Number of	x	$x + 2000$
Interest on	$0.08(x)$	$0.09(x + 2000)$

$$0.08(x) + 0.09(x + 2000) = 860$$
$$0.08x + 0.09x + 180 = 860$$
$$0.17x + 180 = 860$$
$$0.17x = 680$$
$$x = 4000$$
$$x + 2000 = 6000$$

You have \$4,000 invested at 8% and \$6,000 invested at 9%.

3.

	Dollars Invested at 10%	Dollars Invested at 12%
Number of	x	$x + 500$
Interest on	$0.10(x)$	$0.12(x + 500)$

$$0.10(x) + 0.12(x + 500) = 214$$
$$0.10x + 0.12x + 60 = 214$$
$$0.22x + 60 = 214$$
$$0.22x = 154$$
$$x = 700$$
$$x + 500 = 1200$$

Tyler has \$700 invested at 10% and \$1,200 invested at 12%.

5.

	Dollars Invested at 8%	Dollars Invested at 9%	Dollars Invested at 10%
Number of	x	$2x$	$3x$
Interest on	$0.08(x)$	$0.09(2x)$	$0.10(3x)$

$$0.08(x) + 0.09(2x) + 0.10(3x) = 280$$
$$0.08x + 0.18x + 0.30x = 280$$
$$0.56x = 280$$
$$x = 500$$
$$2x = 1000$$
$$3x = 1500$$

She has invested \$500 at 8%, \$1,000 at 9% and \$1,500 at 10%.

7. Let x represent the measure of the two equal angles, so $x + x = 2x$ represents the measure of the third angle. Since the sum of the three angles is $180°$, the equation is:

$$x + x + 2x = 180°$$
$$4x = 180°$$
$$x = 45°$$
$$x = 90°$$

The measures of the three angles are $45°$, $45°$, and $90°$.

9. Let $x =$ the largest angle. Then $\frac{1}{5}x =$ the smallest angle, and $2\left(\frac{1}{5}x\right) = \frac{2}{5}x =$ the third angle.

$$x + \frac{1}{5}x + \frac{2}{5}x = 180°$$
$$\frac{5}{5}x + \frac{1}{5}x + \frac{2}{5}x = 180°$$
$$\frac{8}{5}x = 180$$
$$x = 112.5°$$
$$\frac{1}{5}x = 22.5°$$
$$\frac{2}{5}x = 45°$$

The three angles are $22.5°, 45°, 112.5°$.

11. Let $x =$ the other acute angle, and $90°$ is the measure of the right angle. The sum of the three angles is $180°$.

$$x + 37° + 90° = 180°$$
$$x + 127° = 180°$$
$$x = 53°$$

The other two angles are $53°$ and $90°$.

13. Let $x =$ the total minutes for the call. Then $\$0.41$ is charged for the first minute, and $\$0.32$ is charged for the additional $x - 1$ minutes.

$$0.41(1) + 0.32(x - 1) = 5.21$$
$$0.41 + 0.32x - 0.32 = 5.21$$
$$0.32x + 0.09 = 5.21$$
$$0.32x = 5.12$$
$$x = 16$$

The call was 16 minutes long.

15. Let $x =$ the hours JoAnn worked that week. Then \$12/hour is paid for the first 35 hours and \$18/hour is paid for the additional $x - 35$ hours.

$$12(35) + 18(x - 35) = 492$$
$$420 + 18x - 630 = 492$$
$$18x - 210 = 492$$
$$18x = 702$$
$$x = 39$$

JoAnn worked 39 hours that week.

17. Let $x =$ the number of children's tickets Stacey sold, so $2x$ represents the number of adult tickets sold.

$$6.00(2x) + 4.50(x) = 115.50$$
$$12x + 4.5x = 115.5$$
$$16.5x = 115.5$$
$$x = 7$$
$$2x = 14$$

Stacey sold 7 children's tickets and 14 adult tickets.

19. For Jeff, the total time traveled is $\dfrac{425 \text{ miles}}{55 \text{ miles/hour}} \approx 7.72$ hours ≈ 463 minutes. Since he left at 11:00 AM, he will arrive at 6:43 PM. For Carla, the total time traveled is $\dfrac{425 \text{ miles}}{65 \text{ miles/hour}} \approx 6.54$ hours ≈ 392 minutes. Since she left at 1:00 PM, she will arrive at 7:32 PM. Thus Jeff will arrive in Lake Tahoe first.

21. Since $\dfrac{1}{5}$ mile $= 0.2$ mile, the taxi charge is \$1.25 for the first $\dfrac{1}{5}$ mile and \$0.25 per fifth mile for the remaining 7.3 miles. Since 7.3 miles $= \dfrac{7.3}{0.2} = 36.5$ fifths, the total charge is:

$1.25 + \$0.25(36.5) \approx \$10.38$.

23. The first $\dfrac{1}{5}$ mile is \$1.25, and the remaining $12.4 - 0.2 = 12.2$ miles will be charged at \$0.25 per fifth mile. Since 12.2 miles $= \dfrac{12.2}{0.2} = 61$ fifths, the total charge is:

$1.25 + \$0.25(61) = \$16.50$
Yes, the meter is working correctly.

25. If all 36 people are Elk's Lodge members (which would be the least amount), the cost of the lessons would be $\$3(36) = \$108$. Since half of the money is paid to Ike and Nancy, the least amount they could make is $\frac{1}{2}(\$108) = \$54$.

27. Yes. The total receipts were $160, which is possible if there were 10 Elk's members and 26 nonmembers. Computing the total receipts: $10(\$3) + 26(\$5) = \$30 + \$130 = \$160$.

29.   Let $x =$ salary in 1998
       $0.07x =$ salary increase
    $x + 0.07x = 85,200$
       $1.07x = 85,200$
          $x = 79,626$
          The salary in 1998 was $79,626

31. The pattern is to add -4, so the next number is:
    $$-4 + (-4) = -8$$

33. The pattern is to multiply by $-\frac{1}{2}$, so the next number is:
    $$-\frac{3}{2}\left(-\frac{1}{2}\right) = \frac{3}{4}$$

35. Each number is the square of a number, with alternating signs. For example,
    $1^2 = 1$, $-2^2 = -4$, $3^2 = 9$, and $-4^2 = -16$. Based on this pattern, the next number is:
    $5^2 = 25$.

37. The pattern is to add $-\frac{1}{2}$, so the next number is: $\frac{1}{2} + \left(-\frac{1}{2}\right) = 0$

2.8 Linear Inequalities

1. $x - 5 < 7$

     $x - 5 + 5 < 7 + 5$

          $x < 12$

See the graph in the back of the textbook.

3. $a - 4 \le 8$

     $a - 4 + 4 \le 8 + 4$

          $a \le 12$

See the graph in the back of the textbook.

5. $x - 4.3 > 8.7$

     $x - 4.3 + 4.3 > 8.7 + 4.3$

          $x > 13$

See the graph in the back of the textbook.

7. $y + 6 \ge 10$

     $y + 6 + (-6) \ge 10 + (-6)$

          $y \ge 4$

See the graph in the back of the textbook.

9. $2 < x - 7$

     $2 + 7 < x - 7 + 7$

          $9 < x$

          $x > 9$

See the graph in the back of the textbook.

11. $3x < 6$

     $\dfrac{1}{3}(3x) < \dfrac{1}{3}(6)$

          $x < 2$

See the graph in the back of the textbook.

13. $5a \le 25$

     $\dfrac{1}{5}(5a) \le \dfrac{1}{5}(25)$

          $a \le 5$

See the graph in the back of the textbook.

15. $\dfrac{x}{3} > 5$

     $3\left(\dfrac{x}{3}\right) > 3(5)$

          $x > 15$

See the graph in the back of the textbook.

17. $-2x > 6$

     $-\dfrac{1}{2}(-2x) < -\dfrac{1}{2}(6)$

          $x < -3$

See the graph in the back of the textbook.

19. $-3x \ge -18$

     $-\dfrac{1}{3}(-3x) \le -\dfrac{1}{3}(-18)$

          $x \le 6$

See the graph in the back of the textbook.

21. $-\dfrac{x}{5} \le 10$

     $-5\left(-\dfrac{x}{5}\right) \ge -5(10)$

          $x \ge -50$

See the graph in the back of the textbook.

23. $-\dfrac{2}{3}y > 4$

     $-\dfrac{3}{2}\left(-\dfrac{2}{3}y\right) < -\dfrac{3}{2}(4)$

          $y < -6$

See the graph in the back of the textbook.

25.
$$2x-3<9$$
$$2x-3+3<9+3$$
$$2x<12$$
$$\frac{1}{2}(2x)<\frac{1}{2}(12)$$
$$x<6$$

See the graph in the back of the textbook.

27.
$$-\frac{1}{5}y-\frac{1}{3}\le\frac{2}{3}$$
$$-\frac{1}{5}y-\frac{1}{3}+\frac{1}{3}\le\frac{2}{3}+\frac{1}{3}$$
$$-\frac{1}{5}y\le1$$
$$-5\left(-\frac{1}{5}y\right)\ge-5(1)$$
$$y\ge-5$$

See the graph in the back of the textbook.

29.
$$-4x+1>-11$$
$$-4x+1+(-1)>-11+(-1)$$
$$-4x>-12$$
$$-\frac{1}{4}(-4x)<-\frac{1}{4}(-12)$$
$$x<3$$

See the graph in the back of the textbook.

31.
$$\frac{2}{3}x-5\le7$$
$$\frac{2}{3}x-5+5\le7+5$$
$$\frac{2}{3}x\le12$$
$$\frac{3}{2}\left(\frac{2}{3}x\right)\le\frac{3}{2}(12)$$
$$x\le18$$

See the graph in the back of the textbook.

33.
$$-\frac{2}{5}a-3>5$$
$$-\frac{2}{5}a-3+3>5+3$$
$$-\frac{2}{5}a>8$$
$$-\frac{5}{2}\left(-\frac{2}{5}a\right)<-\frac{5}{2}(8)$$
$$a<-20$$

See the graph in the back of the textbook.

35.
$$5-\frac{3}{5}y>-10$$
$$-5+5-\frac{3}{5}y>-5+(-10)$$
$$-\frac{3}{5}y>-15$$
$$-\frac{5}{3}\left(-\frac{3}{5}y\right)<-\frac{5}{3}(-15)$$
$$y<25$$

See the graph in the back of the textbook.

37.
$$0.3(a+1)\le1.2$$
$$0.3a+0.3\le1.2$$
$$0.3a+0.3+(-0.3)\le1.2+(-0.3)$$
$$0.3a\le0.9$$
$$\frac{0.3a}{0.3}\le\frac{0.9}{0.3}$$
$$a\le3$$

See the graph in the back of the textbook.

39.
$$2(5-2x)\le-20$$
$$10-4x\le-20$$
$$-10+10-4x\le-10+(-20)$$
$$-4x\le-30$$
$$-\frac{1}{4}(-4x)\ge-\frac{1}{4}(-30)$$
$$x\ge\frac{15}{2}$$

See the graph in the back of the textbook.

41.
$$3x - 5 > 8x$$
$$-3x + 3x - 5 > -3x + 8x$$
$$-5 > 5x$$
$$\frac{1}{5}(-5) > \frac{1}{5}(5x)$$
$$-1 > x$$
$$x < -1$$

See the graph in the back of the textbook.

43.
$$\frac{1}{3}y - \frac{1}{2} \leq \frac{5}{6}y + \frac{1}{2}$$
$$6\left(\frac{1}{3}y - \frac{1}{2}\right) \leq 6\left(\frac{5}{6}y + \frac{1}{2}\right)$$
$$2y - 3 \leq 5y + 3$$
$$-5y + 2y - 3 \leq -5y + 5y + 3$$
$$-3y - 3 + 3 \leq 3 + 3$$
$$-3y \leq 6$$
$$-\frac{1}{3}(-3y) \geq -\frac{1}{3}(6)$$
$$y \geq -2$$

See the graph in the back of the textbook.

45.
$$-2.8x + 8.4 < -14x - 2.8$$
$$-2.8x + 14x + 8.4 < 14x - 14x - 2.8$$
$$11.2x + 8.4 < -2.8$$
$$11.2x + 8.4 - 8.4 < -2.8 - 8.4$$
$$11.2x < -11.2$$
$$\frac{11.2x}{11.2} < \frac{-11.2}{11.2}$$
$$x < -1$$

See the graph in the back of the textbook.

47.
$$3(m - 2) - 4 \geq 7m + 14$$
$$3m - 6 - 4 \geq 7m + 14$$
$$3m - 10 \geq 7m + 14$$
$$-7m + 3m - 10 \geq -7m + 7m + 14$$
$$-4m - 10 \geq 14$$
$$-4m - 10 + 10 \geq 14 + 10$$
$$-4m \geq 24$$
$$-\frac{1}{4}(-4m) \leq -\frac{1}{4}(24)$$
$$m \leq -6$$

See the graph in the back of the textbook.

49.
$$3 - 4(x - 2) \leq -5x + 6$$
$$3 - 4x + 8 \leq -5x + 6$$
$$-4x + 11 \leq -5x + 6$$
$$-4x + 5x + 11 \leq -5x + 5x + 6$$
$$x + 11 \leq 6$$
$$x + 11 + (-11) \leq 6 + (-11)$$
$$x \leq -5$$

51. $3x + 2y < 6$
$$2y < -3x + 6$$
$$y < -\frac{3}{2}x + 3$$

53. $2x - 5y > 10$
$$-5y > -2x + 10$$
$$y < \frac{2}{5}x - 2$$

55. $-3x + 7y \leq 21$
$$7y \leq 3x + 21$$
$$y \leq \frac{3}{7}x + 3$$

57. $2x - 4y \geq -4$

    $-4y \geq -2x - 4$

    $y \leq \dfrac{1}{2}x + 1$

59. $x < 3$

61. $x \geq 3$

63.   Let $x = $ first integer

    $x + 1 = $ next consecutive integer

    $x + x + 1 \geq 583$

    $2x + 1 \geq 583$

    $2x \geq 582$

    $x \geq 291$

65. Let x represent the number.

    $2x + 6 < 10$

    $2x < 4$

    $x < 2$

67. Let x represent the number.

    $4x > x - 8$

    $3x > -8$

    $x > -\dfrac{8}{3}$

69. Let $w = $ the width, $3w = $ the length. Using the formula for perimeter:

    $2(w) + 2(3w) \geq 48$

    $2w + 6w \geq 48$

    $8w \geq 48$

    $w \geq 6$

The width is at least 6 meters.

71. Let x, $x + 2$, and $x + 4 = $ the sides of the triangle.

    $x + (x + 2) + (x + 4) > 24$

    $3x + 6 > 24$

    $3x > 18$

    $x > 6$

The shortest side is an even number greater than 6 inches.

73. $t \geq 100$

75. Loss: $7.50x < 1500$

          $x < 200$

    Less than 200 tickets

    Profit: $7.50x > 1500$

          $x > 200$

    Greater than 200 tickets

77. b (commutative property of addition)

79. a (distributive property)

81. b and c (commutative and associative properties of addition)

2.9 Compound Inequalities

1-15. See the graph in the back of the textbook.

17. $3x - 1 < 5$ or $5x - 5 > 10$
  $3x < 6$ $5x > 15$
   $x < 2$ $x > 3$

   See the graph in the back of the textbook.

19. $x - 2 > -5$ and $x + 7 < 13$
     $x > -3$ $x < 6$

   See the graph in the back of the textbook.

21. $11x < 22$ or $12x > 36$
     $x < 2$ $x > 3$

   See the graph in the back of the textbook.

23. $3x - 5 < 10$ and $2x + 1 > -5$
     $3x < 15$ $2x > -6$
      $x < 5$ $x > -3$

   See the graph in the back of the textbook.

25. $2x - 3 < 8$ and $3x + 1 > -10$
     $2x < 11$ $3x > -11$
     $x < \dfrac{11}{2}$ $x > -\dfrac{11}{3}$

   See the graph in the back of the textbook.

27. $2x - 1 < 3$ and $3x - 2 > 1$
     $2x < 4$ $3x > 3$
      $x < 2$ $x > 1$

   See the graph in the back of the textbook.

29. $-1 \leq x - 5 \leq 2$
      $4 \leq x \leq 7$

   See the graph in the back of the textbook.

31. $-4 \leq 2x \leq 6$
     $-2 \leq x \leq 3$

   See the graph in the back of the textbook.

33. $-3 < 2x + 1 < 5$
     $-4 < 2x < 4$
     $-2 < x < 2$

   See the graph in the back of the textbook.

35. $0 \leq 3x + 2 \leq 7$
       $-2 \leq 3x \leq 5$
     $-\dfrac{2}{3} \leq x \leq \dfrac{5}{3}$

   See the graph in the back of the textbook.

37. $-7 < 2x + 3 < 11$

$\qquad -10 < 2x < 8$

$\qquad -5 < x < 4$

See the graph in the back of the textbook.

39. $-1 \le 4x + 5 \le 9$

$\qquad -6 \le 4x \le 4$

$\qquad -\dfrac{3}{2} \le x \le 1$

See the graph in the back of the textbook.

41. $-2 < x < 3$

43. $x \le -2$ or $x \ge 3$

45. (a) $2x + x > 10$; $x + 10 > 2$; $2x + 10 > x$

(b) $\begin{array}{ccc} 2x + x > 10 & x + 10 > 2x & 2x + 10 > x \\ 3x > 10 & 10 > x & x + 10 > 0 \\ x > \dfrac{10}{3} & x < 10 & x > -10 \end{array}$

$\qquad$ Therefore $\dfrac{10}{3} < x < 10$

47. See the graph in the back of the textbook.

49. Let $x =$ the number

$\qquad 10 < x + 5 < 20$

$\qquad 5 < x < 15$

The number is between 5 and 15.

51. Let $x =$ the number

$\qquad 5 < 2x - 3 < 7$

$\qquad 8 < 2x < 10$

$\qquad 4 < x < 5$

The number is between 4 and 5.

53. Let $w =$ the width, $w + 4 =$ the length. Using the formula for perimeter:

$$20 < 2w + 2(w + 4) < 30$$
$$20 < 2w + 2w + 8 < 30$$
$$20 < 4w + 8 < 30$$
$$12 < 4w < 22$$
$$3 < w < \dfrac{11}{2}$$

The width is between 3 inches and $\dfrac{11}{2} = 5\dfrac{1}{2}$ inches.

55. Simplifying the expression: $-|-5| = -(5) = -5$

57. Simplifying the expression: $-3 - 4(-2) = -3 + 8 = 5$

59. Simplifying the expression: $5|3 - 8| - 6|2 - 5| = 5|-5| - 6|-3| = 5(5) - 6(3) = 25 - 18 = 7$

61. Simplifying the expression: $5 - 2[-3(5 - 7) - 8] = 5 - 2[-3(-2) - 8] = 5 - 2(6 - 8) = 5 - 2(-2) = 5 + 4 = 9$

63. $-3 - (-9) = -3 + 9 = 6$

65. Applying the distributive property: $\frac{1}{2}(4x - 6) = \frac{1}{2} \cdot 4x - \frac{1}{2} \cdot 6 = 2x - 3$

67. The integers are: $-3, 0, 2$

Chapter 2 Review

1. $5x - 8x = (5 - 8)x = -3x$

3. $-a + 2 + 5a - 9 = -a + 5a + 2 - 9 = (-1 + 5)a + (-7) = 4a - 7$

5. $6 - 2(3y + 1) - 4 = 6 - 6y - 2 - 4 = -6y$

7. $7x - 2$,   Letting $x = 3$ $7(3) - 2 = 19$

9. $-x - 2x - 3x = -6x$,   Letting $x = 3$ $-6(3) = -18$

11. $-3x + 2$,   Letting $x = -2$ $-3(-2) + 2 = 6 + 2 = 8$

13.
$$x + 2 = -6$$
$$x + 2 + (-2) = -6 + (-2)$$
$$x = -8$$

15.
$$10 - 3y + 4y = 12$$
$$10 + y = 12$$
$$-10 + 10 + y = -10 + 12$$
$$y = 2$$

17.
$$2x = -10$$
$$\frac{1}{2}(2x) = \frac{1}{2}(-10)$$
$$x = -5$$

19.
$$\frac{x}{3} = 4$$
$$3\left(\frac{x}{3}\right) = 3(4)$$
$$x = 12$$

21.
$$3a - 2 = 5a$$
$$-3a + 3a - 2 = -3a + 5a$$
$$-2 = 2a$$
$$\frac{1}{2}(-2) = \frac{1}{2}(2a)$$
$$a = -1$$

23.
$$3x + 2 = 5x - 8$$
$$3x + (-5x) + 2 = 5x + (-5x) - 8$$
$$-2x + 2 = -8$$
$$-2x + 2 + (-2) = -8 + (-2)$$
$$-2x = -10$$
$$-\frac{1}{2}(-2x) = -\frac{1}{2}(-10)$$
$$x = 5$$

25.
$$0.7x - 0.1 = 0.5x - 0.1$$
$$0.7x + (-0.5x) - 0.1 = 0.5x + (-0.5x) - 0.1$$
$$0.2x - 0.1 = -0.1$$
$$0.2x - 0.1 + 0.1 = -0.1 + 0.1$$
$$0.2x = 0$$
$$\frac{0.2x}{0.2} = \frac{0}{0.2}$$
$$x = 0$$

27.
$$2(x - 5) = 10$$
$$2x - 10 = 10$$
$$2x - 10 + 10 = 10 + 10$$
$$2x = 20$$
$$\frac{1}{2}(2x) = \frac{1}{2}(20)$$
$$x = 10$$

29. $\dfrac{1}{2}(3t-2)+\dfrac{1}{2}=\dfrac{5}{2}$

$\dfrac{3}{2}t-1+\dfrac{1}{2}=\dfrac{5}{2}$

$\dfrac{3}{2}t-\dfrac{1}{2}=\dfrac{5}{2}$

$\dfrac{3}{2}t-\dfrac{1}{2}+\dfrac{1}{2}=\dfrac{5}{2}+\dfrac{1}{2}$

$\dfrac{3}{2}t=3$

$\dfrac{2}{3}\left(\dfrac{3}{2}t\right)=\dfrac{2}{3}(3)$

$t=2$

31. $\quad 2(3x+7)=4(5x-1)+18$

$6x+14=20x-4+18$

$6x+14=20x+14$

$6x+(-20x)+14=20x+(-20x)+14$

$-14x+14=14$

$-14x+14+(-14)=14+(-14)$

$-14x=0$

$-\dfrac{1}{14}(-14x)=-\dfrac{1}{14}(0)$

$x=0$

33. $4(5)-5y=20$

$20-5y=20$

$-5y=0$

$y=0$

35. $4(-5)-5y=20$

$-20-5y=20$

$-5y=40$

$y=-8$

37. $2x-5y=10$

$-5y=-2x+10$

$y=\dfrac{2}{5}x-2$

39. $\pi r^2 h=V$

$\dfrac{\pi r^2 h}{\pi r^2}=\dfrac{V}{\pi r^2}$

$h=\dfrac{V}{\pi r^2}$

41. $0.86(240)=x$

$x=206.4$

86% of 240 is 206.4

43. $2x+6=28$

$2x=22$

$x=11$

The number is 11.

45.

	Dollars Invested at 9%	Dollars Invested at 10%
Number of	x	$x+300$
Interest on	$0.09(x)$	$0.10(x+300)$

$0.09(x)+0.10(x+300)=125$

$0.09x+0.10x+30=125$

$0.19x+30=125$

$0.19x=95$

$x=500$

$x+300=800$

The man invested $500 at 9% and $800 at 10%.

47. $-2x < 4$

$$-\frac{1}{2}(-2x) > -\frac{1}{2}(4)$$

$$x > -2$$

49. $-\dfrac{a}{2} \le -3$

$$-2\left(-\frac{a}{2}\right) \ge -2(-3)$$

$$a \ge 6$$

51. $-4x + 5 > 37$

$$-4x > 32$$

$$x < -8$$

See the graph in the back of the textbook.

53. $2(3t + 1) + 6 \ge 5(2t + 4)$

$$6t + 2 + 6 \ge 10t + 20$$

$$6t + 8 \ge 10t + 20$$

$$-4t + 8 \ge 20$$

$$-4t \ge 12$$

$$t \le -3$$

See the graph in the back of the textbook.

55. $-5x \ge 25$ or $2x - 3 \ge 9$

$\quad\;\; x \le -5$ $2x \ge 12$

$\quad\;\; x \le -5$ $x \ge 6$

See the graph in the back of the textbook.

Cumulative Review: Chapters 1-2

1. $6 + 3(6 + 2) = 6 + 3(8) = 6 + 24 = 30$

3. $7 - 9 - 12 = 7 + (-9) + (-12) = 7 + (-21) = -14$

5. $\dfrac{1}{5}(10x) = \left(\dfrac{1}{5} \cdot 10\right)x = 2x$

7. $\left(-\dfrac{2}{3}\right)^3 = \left(-\dfrac{2}{3}\right)\left(-\dfrac{2}{3}\right)\left(-\dfrac{2}{3}\right) = -\dfrac{8}{27}$

9. $-\dfrac{3}{4} \div \dfrac{15}{16} = -\dfrac{3}{4} \cdot \dfrac{16}{15} = -\dfrac{48}{60} = -\dfrac{4}{5}$

11. $\dfrac{-4(-6)}{-9} = \dfrac{24}{-9} = -\dfrac{8}{3}$

13. $\dfrac{(5-3)^2}{5^2 - 3^2} = \dfrac{2^2}{25 - 9} = \dfrac{4}{16} = \dfrac{1}{4}$

15. $21 = 3 \cdot 7$

$35 = 5 \cdot 7$

$\text{LCM} = 3 \cdot 5 \cdot 7 = 105$

$$\frac{4}{21} - \frac{9}{35} = \frac{4 \cdot 5}{21 \cdot 5} - \frac{9 \cdot 3}{35 \cdot 3} = \frac{20}{105} - \frac{27}{105} = -\frac{7}{105} = -\frac{7}{3 \cdot 5 \cdot 7} = -\frac{1}{3 \cdot 5} = -\frac{1}{15}$$

17.
$$7x = 6x + 4$$
$$7x + (-6x) = 6x + (-6x) + 4$$
$$x = 4$$

19.
$$-\frac{3}{5}x = 30$$
$$-\frac{5}{3}\left(\frac{3}{5}x\right) = -\frac{5}{3}(30)$$
$$x = -50$$

21.
$$5x - 7 = x - 1$$
$$5x + (-x) - 7 = x + (-x) - 1$$
$$4x - 7 = -1$$
$$4x - 7 + 7 = -1 + 7$$
$$4x = 6$$
$$\frac{1}{4}(4x) = \frac{1}{4}(6)$$
$$x = \frac{3}{2}$$

23.
$$15 - 3(2t + 4) = 1$$
$$15 - 6t - 12 = 1$$
$$-6t + 3 = 1$$
$$-6t + 3 + (-3) = 1 + (-3)$$
$$-6t = -2$$
$$-\frac{1}{6}(-6t) = -\frac{1}{6}(-2)$$
$$t = \frac{1}{3}$$

25.
$$\frac{1}{3}(x - 6) = \frac{1}{4}(x + 8)$$
$$\frac{1}{3}x - 2 = \frac{1}{4}x + 2$$
$$12\left(\frac{1}{3}x - 2\right) = 12\left(\frac{1}{4}x + 2\right)$$
$$4x - 24 = 3x + 24$$
$$4x + (-3x) - 24 = 3x + (-3x) + 24$$
$$x - 24 = 24$$
$$x - 24 + 24 = 24 + 24$$
$$x = 48$$

27.
$$3x + 4y = 12$$
$$3x + (-3x) + 4y = 12 + (-3x)$$
$$4y = -3x + 12$$
$$\frac{1}{4}(4y) = \frac{1}{4}(-3x + 12)$$
$$y = -\frac{3}{4}x + 3$$

29.
$$-5x + 9 < -6$$
$$-5x + 9 + (-9) < 6 + (-9)$$
$$-5x < -15$$
$$-\frac{1}{5}(-5x) > -\frac{1}{5}(-15)$$
$$x > 3$$

See the graph in the back of the textbook

31.
$$-2 < x + 1 < 5$$
$$-3 < x < 4$$

See the graph in the back of the textbook.

33. The opposite of $-\dfrac{2}{3}$ is $\dfrac{2}{3}$, the reciprocal is $-\dfrac{3}{2}$, and the absolute value is $\left|-\dfrac{2}{3}\right| = \dfrac{2}{3}$.

35. The pattern is to add -3, so the next term is: $-5 + (-3) = -8$

37. $\dfrac{1}{4}(8x - 4) = \dfrac{1}{4} \cdot 8x - \dfrac{1}{4} \cdot 4 = 2x - 1$

39. $\dfrac{234}{312} = \dfrac{2 \cdot 3 \cdot 3 \cdot 13}{2 \cdot 2 \cdot 2 \cdot 3 \cdot 13} = \dfrac{3}{2 \cdot 2} = \dfrac{3}{4}$

41. Evaluating when $a = 3$ and $b = -2$: $a^2 - 2ab + b^2 = (3)^2 - 2(3)(-2) + (-2)^2 = 9 + 12 + 4 = 25$

43. Let $x =$ the number.
$$2x + 7 = 31$$
$$2x = 24$$
$$x = 12$$

45. Let $x =$ the acute angle, and $90°$ is the right angle. The sum of the
three angles is $180°$.
$$x + 42° + 90° = 180°$$
$$x + 132° = 180°$$
$$x = 48°$$
The other two angles are $48°$ and $90°$.

47. The other angle must be $90° - 25° = 65°$.

49.

	Dollars Invested at 5%	Dollars Invested at 6%
Number of	x	$x + 200$
Interest on	$0.05(x)$	$0.06(x + 200)$

$$0.05(x) + 0.06(x + 200) = 56$$
$$0.05x + 0.06x + 12 = 56$$
$$0.11x + 12 = 56$$
$$0.11x = 44$$
$$x = 400$$
$$x + 200 = 600$$
You have \$400 invested at 5% and \$600 invested at 6%.

Chapter 2 Test

1. $3x + 2 - 7x + 3 = 3x - 7x + 2 + 3 = -4x + 5$

2. $4a - 5 - a + 1 = 4a - a - 5 + 1 = 3a - 4$

3. $7 - 3(y + 5) - 4 = 7 - 3y - 15 - 4 = 7 - 3y - 15 - 4 = -3y - 12$

4. $8(2x + 1) - 5(x - 4) = 16x + 8 - 5x + 20 = 16x - 5x + 8 + 20 = 11x + 28$

5. Evaluating when $x = -5$: $2x - 3 - 7x = -5x - 3 = -5(-5) - 3 = 25 - 3 = 22$

6. Evaluating when $x = 2$ and $y = 3$: $x^2 + 2xy + y^2 = (2)^2 + 2(2)(3) + (3)^2 = 4 + 12 + 9 = 25$

7.
$$2x - 5 = 7$$
$$2x - 5 + 5 = 7 + 5$$
$$2x = 12$$
$$\frac{1}{2}(2x) = \frac{1}{2}(12)$$
$$x = 6$$

8.
$$2y + 4 = 5y$$
$$-2y + 2y + 4 = -2y + 5y$$
$$4 = 3y$$
$$\frac{1}{3}(4) = \frac{1}{3}(3y)$$
$$y = \frac{4}{3}$$

9.
$$\frac{1}{2}x - \frac{1}{10} = \frac{1}{5}x + \frac{1}{2}$$
$$10\left(\frac{1}{2}x - \frac{1}{10}\right) = 10\left(\frac{1}{5}x + \frac{1}{2}\right)$$
$$5x - 1 = 2x + 5$$
$$5x + (-2x) - 1 = 2x + (-2x) + 5$$
$$3x - 1 = 5$$
$$3x - 1 + 1 = 5 + 1$$
$$3x = 6$$
$$\frac{1}{3}(3x) = \frac{1}{3}(6)$$
$$x = 2$$

10.
$$\frac{2}{5}(5x - 10) = -5$$
$$2x - 4 = -5$$
$$2x - 4 + 4 = -5 + 4$$
$$2x = -1$$
$$\frac{1}{2}(2x) = \frac{1}{2}(-1)$$
$$x = -\frac{1}{2}$$

11.
$$-5(2x + 1) - 6 = 19$$
$$-10x - 5 - 6 = 19$$
$$-10x - 11 = 19$$
$$-10x - 11 + 11 = 19 + 11$$
$$-10x = 30$$
$$-\frac{1}{10}(-10x) = -\frac{1}{10}(30)$$
$$x = -3$$

12.
$$0.04x + 0.06(100 - x) = 4.6$$
$$0.04x + 6 - 0.06x = 4.6$$
$$-0.02x + 6 = 4.6$$
$$-0.02x + 6 + (-6) = 4.6 + (-6)$$
$$-0.02x = -1.4$$
$$\frac{-0.02x}{-0.02} = \frac{-1.4}{-0.02}$$
$$x = 70$$

13. $2(t-4)+3(t+5)=2t-2$

$\quad\quad 2t-8+3t+15=2t-2$

$\quad\quad\quad\quad\quad\quad 5t+7=2t-2$

$\quad 5t+(-2t)+7=2t+(-2t)-2$

$\quad\quad\quad\quad\quad\quad 3t+7=-2$

$\quad\quad 3t+7+(-7)=-2+(-7)$

$\quad\quad\quad\quad\quad\quad\quad 3t=-9$

$\quad\quad\quad\quad \dfrac{1}{3}(3t)=\dfrac{1}{3}(-9)$

$\quad\quad\quad\quad\quad\quad\quad\quad t=-3$

14. $\quad\quad 2x-4(5x+1)=3x+17$

$\quad\quad\quad 2x-20x-4=3x+17$

$\quad\quad\quad\quad\quad -18x-4=3x+17$

$\quad -18x+(-3x)-4=3x+(-3x)+17$

$\quad\quad\quad\quad\quad\quad -21x-4=17$

$\quad\quad\quad\quad -21x-4+4=17+4$

$\quad\quad\quad\quad\quad\quad\quad -21x=21$

$\quad\quad -\dfrac{1}{21}(21x)=-\dfrac{1}{21}(21)$

$\quad\quad\quad\quad\quad\quad\quad\quad x=-1$

15. $0.15(38)=x$

$\quad\quad 5.7=x$

15% of 38 is 5.7.

16. $0.12x=240$

$\quad \dfrac{0.12x}{0.12}=\dfrac{240}{0.12}$

$\quad\quad\quad x=2000$

12% of 2,000 is 240.

17. Substituting $y=-2$:

$\quad 2x-3(-2)=12$

$\quad\quad\quad 2x+6=12$

$\quad\quad\quad\quad 2x=6$

$\quad\quad\quad\quad\quad x=3$

18. Substituting $V=88$, $\pi=\dfrac{22}{7}$, and $r=3$:

$\quad\quad \dfrac{1}{3}\cdot\dfrac{22}{7}\cdot(3)^2 h=88$

$\quad\quad\quad\quad \dfrac{66}{7}h=88$

$\quad\quad \dfrac{7}{66}\left(\dfrac{66}{7}h\right)=\dfrac{7}{66}(88)$

$\quad\quad\quad\quad\quad h=\dfrac{28}{3}$ inches

19. Solving for y:

$\quad 2x+5y=20$

$\quad\quad 5y=-2x+20$

$\quad\quad\quad y=-\dfrac{2}{5}x+4$

20. Solving for v:

$\quad x+vt+16t^2=h$

$\quad\quad vt=h-x-16t^2$

$\quad\quad\quad v=\dfrac{h-x-16t^2}{t}$

21.

	Ten Years Ago	Now
Dave	$2x-10$	$2x$
Rick	$x-10$	x

$$2x-10+x-10=40$$
$$3x-20=40$$
$$3x=60$$
$$x=20$$
$$2x=40$$

Rick is 20 and Fred is 40.

22. Let $w=$ the width and $2w=$ the length.
$$2(w)+2(2w)=60$$
$$2w+4w=60$$
$$6w=60$$
$$w=10$$
$$2w=20$$

The width is 10 inches and the
length is 20 inches.

23.

	Dimes	Quarters
Number	$x+7$	x
Value (cents)	$10(x+7)$	$25(x)$

$$10(x+7)+25(x)=350$$
$$10x+70+25x=350$$
$$35x+70=350$$
$$35x=280$$
$$x=8$$
$$x+7=15$$

He has 8 quarters and 15 dimes
in his collection.

24.

	Dollars Invested at 7%	Dollars Invested at 9%
Number of	x	$x+600$
Interest on	$0.07(x)$	$0.09(x+600)$

$$0.07(x)+0.09(x+600)=182$$
$$0.07x+0.09x+54=182$$
$$0.16x+54=182$$
$$0.16x=128$$
$$x=800$$
$$x+600=1400$$

She has \$800 invested at 7% and \$1,400 invested at 9%.

25.
$$2x+3<5$$
$$2x+3+(-3)<5+(-3)$$
$$2x<2$$
$$\frac{1}{2}(2x)<\frac{1}{2}(2)$$
$$x<1$$

See the graph in the back of the textbook.

26.
$$-5a>20$$
$$-\frac{1}{5}(-5a)<-\frac{1}{5}(20)$$
$$a<-4$$

See the graph in the back of the textbook.

27.
$$0.4-0.2x\geq1$$
$$-0.4+0.4-0.2x\geq-0.4+1$$
$$-0.2x\geq0.6$$
$$\frac{-0.2x}{-0.2}\leq\frac{0.6}{-0.2}$$
$$x\leq-3$$

See the graph in the back of the textbook.

28.
$$4-5(m+1)\leq9$$
$$4-5m-5\leq9$$
$$-5m-1\leq9$$
$$-5m-1+1\leq9+1$$
$$-5m\leq10$$
$$-\frac{1}{5}(-5m)\geq-\frac{1}{5}(10)$$
$$m\geq-2$$

See the graph in the back of the textbook.

29. $3 - 4x \geq -5$ or $2x \geq 10$

    $-4x \geq -8$ $x \geq 5$

    $\quad x \leq 2$ $x \geq 5$

    See the graph in the back of the textbook.

30. $-7 < 2x - 1 < 9$

    $-6 < 2x < 10$

    $-3 < x < 5$

    See the graph in the back of the textbook.

CHAPTER 3

Graphing and Linear Systems

3.1 Paired Data and Graphing Ordered Pairs

1, 3,5,7,9,11,13,15,17 - See the graph of these ordered pairs in the back of the textbook.

19. (–4, 4)

21. (–4, 2)

23. (–3, 0)

25. (2, –2)

27. (–5, –5)

29. Yes, see the graph in the back of the textbook.

31. No, see the graph in the back of the textbook.

33. Yes, see the graph in the back of the textbook.

35. No, see the graph in the back of the textbook.

37. Yes, see the graph in the back of the textbook.

39. No, see the graph in the back of the textbook.

41. No, see the graph in the back of the textbook.

43. No, see the graph in the back of the textbook.

45. Every point on this line has a y-coordinate of -3. See the graph in the back of the textbook.

47. They are on the y-axis.

49. Any three: (0, 0), (5, 40), (10, 80), (15, 120), (20, 160), (25, 200), (30, 240), (35, 280), (40, 320)

51. See the graph in the back of the textbook.

53. See the graph in the back of the textbook.

55. See the graph in the back of the textbook.

57. 3, 10, 17, 24, 31   Add seven to the previous number to produce the new number.

59. 3, 1, $\frac{1}{3}$, $\frac{1}{9}$, $\frac{1}{27}$ Multiply the previous number by $\frac{1}{3}$ to produce the new number.

61. 7, 4, 1, –2, –5   Add –3 to the previous number to produce the new number.

63. 7, 21, 63, 189, 567 Multiply the previous number by 3 to produce the new number.

65. 5, 6, 8, 11, 15   Add the next consecutive counting number to the previous number to produce the new number.

3.2 Solutions to Linear Equations in Two Variables

1. Substituting $x = 0, x = 3,$ and $y = -6$

$$2(0) + y = 6 \qquad 2(3) + y = 6 \qquad 2x + (-6) = 6$$
$$0 + y = 6 \qquad\quad 6 + y = 6 \qquad\quad 2x = 12$$
$$y = 6 \qquad\qquad y = 0 \qquad\qquad x = 6$$

The ordered pairs are $(0, 6)$, $(3, 0)$, and $(6, -6)$.

3. Substituting $x = 0, y = 0,$ and $x = -4$:

$$3(0) + 4y = 12 \qquad 3x + 4(0) = 12 \qquad 3(-4) + 4y = 12$$
$$0 + 4y = 12 \qquad\quad 3x + 0 = 12 \qquad\quad -12 + 4y = 12$$
$$4y = 12 \qquad\qquad 3x = 12 \qquad\qquad 4y = 24$$
$$y = 3 \qquad\qquad\; x = 4 \qquad\qquad\;\; y = 6$$

The ordered pairs are $(0, 3)$, $(4, 0)$, and $(-4, 6)$.

5. Substituting $x = 1, y = 0, x = 5$:

$$y = 4(1) - 3 \qquad 0 = 4x - 3 \qquad y = 4(5) - 3$$
$$y = 4 - 3 \qquad\quad 3 = 4x \qquad\quad\; y = 20 - 3$$
$$y = 1 \qquad\qquad x = \frac{3}{4} \qquad\qquad y = 17$$

The ordered pairs are $(1,1)$, $\left(\dfrac{3}{4}, 0\right)$, and $(5, 17)$.

7. Substituting $x = 2, y = 6, x = 0$:

$$y = 7(2) - 1 \qquad 6 = 7x - 1 \qquad y = 7(0) - 1$$
$$y = 14 - 1 \qquad\quad 7 = 7x \qquad\quad\; y = 0 - 1$$
$$y = 13 \qquad\qquad x = 1 \qquad\qquad y = -1$$

The ordered pairs are $(2, 13)$, $(1, 6)$, and $(0, -1)$.

9. Substituting $y = 4, y = -3,$ and $y = 0$ results (in each case) in $x = -5$. The ordered pairs are $(-5, 4)$, $(-5, -3)$, and $(-5, 0)$.

11. See the table in the back of the textbook.

13. See the table in the back of the textbook.

15. See the table in the back of the textbook.

17. See the table in the back of the textbook.

19. See the table in the back of the textbook.

21. Substituting each ordered pair into the equation:

$(2,3):$ $2(2)-5(3)=4-15=-11\neq10$

$(0,-2):$ $2(0)-5(-2)=0+10=10$

$\left(\dfrac{5}{2},1\right):$ $2\left(\dfrac{5}{2}\right)-5(1)=5-5=0\neq10$

Only the ordered pair $(0,-2)$ is a solution.

23. Substituting each ordered pair into the equation:

$(1,5):$ $7(1)-2=7-2=5$

$(0,-2):$ $7(0)-2=0-2=-2$

$(-2,-16):$ $7(-2)-2=-14-2=-16$

All the ordered pairs $(1,5)$, $(0,-2)$ and $(-2,-16)$ are solutions.

25. Substituting each ordered pair into the equation:

$(1,6):$ $6(1)=6$

$(-2,-12):$ $6(-2)=-12$

$(0,0):$ $6(0)=0$

All the ordered pairs $(1,6)$, $(-2,-12)$ and $(0,0)$ are solutions.

27. Substituting each ordered pair into the equation:

$(1,1):$ $1+1=2\neq0$

$(2,-2):$ $2+(-2)=0$

$(3,3):$ $3+3=6\neq0$

Only the ordered pair $(2,-2)$ is a solution.

29. Since $x=3$, the ordered pair $(5,3)$ cannot be a solution. The ordered pairs $(3,0)$ and $(3,-3)$ are solutions.

31. Substituting $w=3$:

$$2l+2(3)=30$$
$$2l+6=30$$
$$2l=24$$
$$l-12$$

The length is 12 inches.

33. See the table and the graph in the back of the textbook.

35. See the table and the graph in the back of the textbook.

37. $y = 13 + 1.5x$ See the table and bar chart in the back of the textbook.

39. $y = 7 + 1.1x$ See the table and bar chart in the back of the textbook.

41. Substituting $x = 4$:

$$3(4) + 2y = 6$$
$$12 + 2y = 6$$
$$2y = -6$$
$$y = -3$$

43. Substituting $x = 0$: $y = -\dfrac{1}{3}(0) + 2 = 0 + 2 = 2$

45. Substituting $x = 2$: $y = \dfrac{3}{2}(2) - 3 = 3 - 3 = 0$

47. Solving for y:

$$5x + y = 4$$
$$y = -5x + 4$$

49. Solving for y:

$$3x - 2y = 6$$
$$-2y = -3x + 6$$
$$y = \dfrac{3}{2}x - 3$$

3.3 Graphing Linear Equations in Two Variables

1. The ordered pairs are (0, 4), (2, 2) and (4, 0). See the graph in the back of the textbook.

3. The ordered pairs are (0, 3), (2, 1), and $(4, -1)$. See the graph in the back of the textbook.

5. The ordered pairs are (0, 0), $(-2, -4)$ and (2, 4). See the graph in the back of the textbook.

7. The ordered pairs are $(-3, -1)$, (0, 0), and (3, 1). See the graph in the back of the textbook.

9. The ordered pairs are (0, 1), $(-1, -1)$ and (1, 3). See the graph in the back of the textbook.

11. The ordered pairs are (0, 4), $(-1, 4)$ and (2, 4). See the graph in the back of the textbook.

13. The ordered pairs are $(-2, 2)$, (0, 3) and (2, 4). See the graph in the back of the textbook.

15. The ordered pairs are $(-3, 3)$, (0, 1) and $(3, -1)$. See the graph in the back of the textbook.

17. Solving for y:

$$2x + y = 3$$
$$y = -2x + 3$$

The ordered pairs are $(-1, 5)$, $(0, 3)$, and $(1, 1)$.
See the graph in the back of the textbook.

19. Solving for y:

$$3x + 2y = 6$$
$$2y = -3x + 6$$
$$y = -\dfrac{3}{2}x + 3$$

The ordered pairs are $(0, 3)$, $(2, 0)$, and $(4, -3)$.
See the graph in the back of the textbook.

21. Solving for y:

$$-x + 2y = 6$$
$$2y = x + 6$$
$$y = \frac{1}{2}x + 3$$

The ordered pairs are $(-2, 2)$, $(0, 3)$, and $(2, 4)$.

See the graph in the back of the textbook.

23. Three solutions are $(-4, 2)$, $(0, 0)$, and $(4, -2)$. See the graph in the back of the textbook.

25. Three solutions are $(-1, -4)$, $(0, -1)$ and $(1, 2)$. See the graph in the back of the textbook.

27. Solving for y:

$$-2x + y = 1$$
$$y = 2x + 1$$

Solutions are $(-2, -3)$, $(0, 1)$, and $(2, 5)$.

See the graph in the back of the textbook.

29. Solving for y:

$$3x + 4y = 8$$
$$4y = -3x + 8$$
$$y = -\frac{3}{4}x + 2$$

Solutions are $(-4, 5)$, $(0, 2)$, and $(4, -1)$.

See the graph in the back of the textbook.

31. Three solutions are $(-2, -4)$, $(-2, 0)$, and $(-2, 4)$. See the graph in the back of the textbook.

33. Three solutions are $(-4, 2)$, $(0, 2)$, and $(4, 2)$. See the graph in the back of the textbook.

35. The ordered pairs are $(1, 4)$, $(2, 3)$, and $(3, 2)$. See the graph in the back of the textbook.

37. $y = x$: See the graph in the back of the textbook.

39. See the table in the back of the textbook.

41. The ordered pairs are $(-3, 3)$, $(-2, 2)$, $(-1, 1)$, $(0, 0)$, $(1, 1)$, $(2, 2)$ and $(3, 3)$. See the graph in the back of the textbook.

43. See the graph in the back of the textbook.

45. See the table and the graph in the back of the textbook.

47. See the table and the graph in the back of the textbook.

49.
$$3(x - 2) = 9$$
$$3x - 6 = 9$$
$$3x = 15$$
$$x = 5$$

51.
$$2(3x - 1) + 4 = -10$$
$$6x - 2 + 4 = -10$$
$$6x + 2 = -10$$
$$6x = -12$$
$$x = -2$$

53. $6 - 2(4x - 7) = -4$
$$6 - 8x + 14 = -4$$
$$-8x + 20 = -4$$
$$-8x = -24$$
$$x = 3$$

55. $6\left(\dfrac{1}{2}x + 4\right) = 6\left(\dfrac{2}{3}x + 5\right)$
$$3x + 24 = 4x + 30$$
$$-x + 24 = 30$$
$$-x = 6$$
$$x = -6$$

3.4 More on Graphing: Intercepts

1. To find the x-intercept, let $y = 0$:
$$2x + 0 = 4$$
$$2x = 4$$
$$x = 2$$
To find the y-intercept, let $x = 0$:
$$2(0) + y = 4$$
$$0 + y = 4$$
$$y = 4$$
See the graph in the back of the textbook.

3. To find the x-intercept, let $y = 0$:
$$-x + 0 = 3$$
$$-x = 3$$
$$x = -3$$
To find the y-intercept, let $x = 0$:
$$0 + y = 3$$
$$y = 3$$
See the graph in the back of the textbook.

5. To find the x-intercept, let $y = 0$:
$$-x + 2(0) = 2$$
$$-x = 2$$
$$x = -2$$
To find the y-intercept, let $x = 0$:
$$-0 + 2y = 2$$
$$2y = 2$$
$$y = 1$$
See the graph in the back of the textbook.

7. To find the x-intercept, let $y = 0$:
$$5x + 2(0) = 10$$
$$5x = 10$$
$$x = 2$$
To find the y-intercept, let $x = 0$:
$$5(0) + 2y = 10$$
$$2y = 10$$
$$y = 5$$
See the graph in the back of the textbook.

9. To find the x-intercept, let $y = 0$:
$$4x - 2(0) = 8$$
$$4x = 8$$
$$x = 2$$
To find the y-intercept, let $x = 0$:
$$4(0) - 2y = 8$$
$$-2y = 8$$
$$y = -4$$
See the graph in the back of the textbook.

11. To find the x-intercept, let $y = 0$:
$$-4x + 5(0) = 20$$
$$-4x = 20$$
$$x = -5$$
To find the y-intercept, let $x = 0$:
$$-4(0) + 5y = 20$$
$$5y = 20$$
$$y = 4$$
See the graph in the back of the textbook.

13. To find the x-intercept, let $y = 0$:

$$2x - 6 = 0$$
$$2x = 6$$
$$x = 3$$

To find the y-intercept, let $x = 0$:

$$y = 2(0) - 6$$
$$y = -6$$

See the graph in the back of the textbook.

15. To find the x-intercept, let $y = 0$:

$$2x + 2 = 0$$
$$2x = -2$$
$$x = -1$$

To find the y-intercept, let $x = 0$:

$$y = 2(0) + 2$$
$$y = 2$$

See the graph in the back of the textbook.

17. To find the x-intercept, let $y = 0$:

$$2x - 1 = 0$$
$$2x = 1$$
$$x = \frac{1}{2}$$

To find the y-intercept, let $x = 0$:

$$y = 2(0) - 1$$
$$y = -1$$

See the graph in the back of the textbook.

19. To find the x-intercept, let $y = 0$:

$$\frac{1}{2}x + 3 = 0$$
$$\frac{1}{2}x = -3$$
$$x = -6$$

To find the y-intercept, let $x = 0$:

$$y = \frac{1}{2}(0) + 3$$
$$y = 3$$

See the graph in the back of the textbook.

21. To find the x-intercept, let $y = 0$:

$$-\frac{1}{3}x - 2 = 0$$
$$-\frac{1}{3}x = 2$$
$$x = -6$$

To find the y-intercept, let $x = 0$:

$$y = -\frac{1}{3}(0) - 2$$
$$y = -2$$

See the graph in the back of the textbook.

23. Another point on the line is $(2, -4)$. See the graph in the back of the textbook.

25. Another point on the line is $(2, 4)$. See the graph in the back of the textbook.

27. Another point on the line is $(3, 1)$. See the graph in the back of the textbook.

29. Another point on the line is $(3, -1)$. See the graph in the back of the textbook.

31. Another point on the line is $(3, 2)$. See the graph in the back of the textbook.

33. The y-intercept is -4. See the graph in the back of the textbook.

35. The x-intercept is -3. See the graph in the back of the textbook.

37. The x and y-intercepts are both 3. See the graph in the back of the textbook.

39. See the table in the back of the textbook.

41. The x-intercept is 3. See the graph in the back of the textbook.

43. The y-intercept is 4. See the graph in the back of the textbook.

45. To find the α-intercept, let $\theta = 0$:
$$\alpha + 0 = 90$$
$$\alpha = 90$$
To find the θ-intercept, let $\alpha = 0$:
$$0 + \theta = 90$$
$$\theta = 90$$
See the graph in the back of the textbook.

47.
$$-3x \geq 12$$
$$-\frac{1}{3}(-3x) \leq -\frac{1}{3}(12)$$
$$x \leq -4$$

49.
$$-\frac{x}{3} \leq -1$$
$$-3\left(-\frac{x}{3}\right) \geq -3(-1)$$
$$x \geq 3$$

51.
$$-4x + 1 < 17$$
$$-4x + 1 - 1 < 17 - 1$$
$$-4x < 16$$
$$-\frac{1}{4}(-4x) > -\frac{1}{4}(16)$$
$$x > -4$$

3.5 Solving Linear Equations by Graphing

1. The intersection point is $(2, 1)$.
   See the graph in the back of the textbook.

3. The intersection point is $(-1, 2)$.
   See the graph in the back of the textbook.

5. The intersection point is $(3, 5)$.
   See the graph in the back of the textbook.

7. The intersection point is $(4, 3)$.
   See the graph in the back of the textbook.

9. The intersection point is $(0, -6)$.
   See the graph in the back of the textbook.

11. The intersection point is $(1, 0)$.
    See the graph in the back of the textbook.

13. The intersection point is $(0, 0)$.
    See the graph in the back of the textbook.

15. The intersection point is $(-5, -6)$.
    See the graph in the back of the textbook.

17. The intersection point is $(-1, -1)$.
    See the graph in the back of the textbook.

19. The intersection point is $(-3, 2)$.
    See the graph in the back of the textbook.

21. The intersection point is $(-3,5)$.
    See the graph in the back of the textbook.

23. The intersection point is $(-4,6)$.
    See the graph in the back of the textbook.

25. There is no intersection (the lines are parallel).
    See the graph in the back of the textbook.
    There is no solution.

27. The system is dependent (both lines are the same, they coincide). There are infinitely many solutions.
    See the graph in the back of the textbook.

29. The intersection point is $\left(\dfrac{1}{2},1\right)$.

    See the graph in the back of the textbook.

31. The intersection point is $(2,1)$.
    See the graph in the back of the textbook.

33. (a) The lines intersect when the number of hours worked is 25.
    (b) For less than 20 hours the line for Gigi's is higher, so she should choose Gigi's.
    (c) For more than 30 hours, the line for Marcy's is higher, so she should choose Marcy's.

35. (a) The lines intersect when the number of hours used is 4.
    (b) For less than 3 hours, the line for Computer Services is lower, so she should choose Computer Services.
    (c) For more than 6 hours, the line for ICM is lower, so she should choose ICM.

37. When $x = -3$: $2x - 9 = 2(-3) - 9 = -6 - 9 = -15$

39. When $x = -3$: $9 - 6x = 9 - 6(-3) = 9 + 18 = 27$

41. When $x = -3$: $4(3x + 2) + 1 = 12x + 8 + 1 = 12x + 9 = 12(-3) + 9 = -36 + 9 = -27$

43. When $x = -3$: $2x^2 + 3x + 4 = 2(-3)^2 + 3(-3) + 4 = 18 - 9 + 4 = 13$

3.6 The Addition Method

1. $x + y = 3$
   $\underline{x - y = 1}$
   $2x = 4$
   $x = 2$
   Substitute $x = 2$ into $x + y = 3$
   $2 + y = 3$
   $y = 1$
   The solution is $(2, 1)$.

3. $x + y = 10$
   $\underline{-x + y = 4}$
   $2y = 14$
   $y = 7$
   Substitute $y = 7$ into $x + y = 10$
   $x + 7 = 10$
   $x = 3$
   The solution is $(3, 7)$.

5. $x - y = 7$
    $\underline{-x - y = 3}$
    $\quad -2y = 10$
    $\qquad y = -5$
    Substitute $y = -5$ into $x - y = 7$
    $x - (-5) = 7$
    $\quad x + 5 = 7$
    $\qquad x = 2$
    The solution is $(2, -5)$.

7. $x + y = -1$
    $\underline{3x - y = -3}$
    $4x \quad = -4$
    $\quad x = -1$
    Substitute $x = -1$ into $x + y = -1$
    $-1 + y = -1$
    $\quad y = 0$
    The solution is $(-1, 0)$.

9. $3x + 2y = 1$
    $\underline{-3x + -2y = -1}$
    $\qquad 0 = 0$
    The lines coincide and there is an infinite number of solutions.

11. $3x - y = 4$ $\qquad$ (1)
     $2x + 2y = 24$ $\qquad$ (2)
     Multiply (1) by 2
     $6x - 2y = 8$
     $\underline{2x + 2y = 24}$
     $8x \qquad = 32$
     $\qquad x = 4$
     Substitute $x = 4$ into (2)
     $2(4) + 2y = 24$
     $\quad 8 + 2y = 24$
     $\qquad 2y = 16$
     $\qquad y = 8$
     The solution is $(4, 8)$.

13. $5x - 3y = 2$ $\qquad$ (1)
     $10x - y = 1$ $\qquad$ (2)
     Multiply (2) by -3
     $5x - 3y = -2$
     $\underline{-30x + 3y = -3}$
     $-25x \qquad = -5$
     $\qquad x = \dfrac{1}{5}$
     Substitute $x = \dfrac{1}{5}$ into (2)
     $10\left(\dfrac{1}{5}\right) - y = 1$
     $\quad 2 - y = 1$
     $\quad -y = -1$
     $\qquad y = 1$
     The solution is $\left(\dfrac{1}{5}, 1\right)$.

15. $11x - 4y = 11$ $\qquad$ (1)
     $5x + y = 5$ $\qquad$ (2)
     Multiply (2) by 4
     $11x - 4y = 11$
     $\underline{20x + 4y = 20}$
     $31x \qquad = 31$
     $\qquad x = 1$
     Substitute $x = 1$ into (2)
     $5(1) + y = 5$
     $\qquad y = 0$
     The solution is $(1, 0)$.

17. $3x - 5y = 7$ (1)

   $-x + y = -1$ (2)

Multiply (2) by 3

   $3x - 5y = 7$

   $\underline{-3x + 3y = -3}$

      $-2y = 4$

        $y = -2$

Substitute $y = -2$ into (2)

   $-x - 2 = -1$

     $-x = 1$

      $x = -1$

The solution is $(-1, -2)$.

19. $-x - 8y = -1$ (1)

   $-2x + 4y = 13$ (2)

Multiply (1) by -2

   $2x + 16y = 2$

   $\underline{-2x + 4y = 13}$

      $20y = 15$

       $y = 3/4$

Substitute $y = 3/4$ into (1)

   $-x - 8(3/4) = -1$

     $-x - 6 = -1$

       $-x = 5$

        $x = -5$

The solution is $(-5, \ 3/4)$.

21. $-3x - y = 7$ (1)

   $6x + 7y = 11$ (2)

Multiply (1) by 2

   $-6x - 2y = 14$

   $\underline{6x + 7y = 11}$

      $5y = 25$

      $y = 5$

Substitute $y = 5$ into (1)

   $-3x - 5 = 7$

     $-3x = 12$

       $x = -4$

The solution is $(-4, 5)$.

23. $6x - y = -8$

   $\underline{2x + y = -16}$

   $8x \quad\ = -24$

   $x \quad\ = -3$

Substitute $x = -3$ into $2x + y = -16$

   $2(-3) + y = -16$

     $y = -10$

The solution is $(-3, -10)$.

25. $x + 3y = 9$ (1)

   $2x - y = 4$ (2)

Multiply (2) by 3

   $x + 3y = 9$

   $\underline{6x - 3y = 12}$

   $7x \quad\ = 21$

     $x = 3$

Substitute $x = 3$ into (1)

   $3 + 3y = 9$

     $3y = 6$

      $y = 2$

The solution is $(3, 2)$.

27. $x - 6y = 3$ (1)

   $4x + 3y = 21$ (2)

Multiply (2) by 2

   $x - 6y = 3$

   $\underline{8x + 6y = 42}$

   $9x \quad\ = 45$

     $x = 5$

Substitute $x = 5$ into (2)

   $4(5) + 3y = 21$

     $y = 1/3$

The solution is $(5, \ 1/3)$.

29. $2x + 9y = 2$ (1)

       $5x + 3y = -8$ (2)

       Multiply (2) by -3

       $2x + 9y = 2$

       $\underline{-15x - 9y = 24}$

       $-13x \quad\quad = 26$

            $x = -2$

       Substitute $x = -2$ into (1)

       $2(-2) + 9y = 2$

         $-4 + 9y = 2$

            $9y = 6$

             $y = \dfrac{2}{3}$

       The solution is $\left(-2, \dfrac{2}{3}\right)$.

31. $(1/3)x + (1/4)y = 7/6$ (1)

       $(3/2)x - (1/3)y = 7/3$ (2)

       Multiply (1) by 24 and (2) by 18

       $8x + 6y = 28$

       $\underline{27x - 6y = 42}$

       $35x \quad\quad = 70$

            $x = 2$

       Substitute $x = 2$ into (2)

       $(3/2)2 - (1/3)y = 7/3$

            $9 - y = 7$

              $-y = -2$

               $y = 2$

       The solution is $(2, 2)$.

33. $3x + 2y = -1$ (1)

       $6x + 4y = 0$ (2)

       Multiply (1) by -2

       $-6x - 4y = 2$

       $\underline{6x + 4y = 0}$

            $0 = 2$

       The lines are parallel, there is no solution.

35. $11x + 6y = 17$ (1)

       $5x - 4y = 1$ (2)

       Multiply (1) by 2 and (2) by 3

       $22x + 12y = 34$

       $\underline{15x - 12y = 3}$

       $37x \quad\quad = 37$

            $x = 1$

       Substitute $x = 1$ into (1)

       $11(1) + 6y = 17$

         $11 + 6y = 17$

            $6y = 6$

             $y = 1$

       The solution is $(1, 1)$.

37. $\dfrac{1}{2}x + \dfrac{1}{6}y = \dfrac{1}{3}$ (1)

       $-x - \dfrac{1}{3}y = -\dfrac{1}{6}$ (2)

       Multiply (1) by 12

       Multiply (2) by 6

       $6x + 2y = 4$

       $\underline{-6x - 2y = -1}$

            $0 = 3$

       The lines are parallel, there is no solution.

39. $6x - 5y = 17$

       $\underline{-3x + 5y = 4}$

       $3x \quad\quad = 21$

       $x \quad\quad = 7$

       Substitute $x = 7$ into $-3x + 5y = 4$

       $5y = 3(7) + 4 = 25$

       $y = 5$

       The solution is $(7, 5)$.

41. $x + y = 22$ (1)

$0.05x + 0.10y = 1.70$ (2)

Multiply (1) by -5

Multiply (2) by 100

$-5x - 5y = -110$

$\underline{5x + 10y = 170}$

$5y = 60$

$y = 12$

Substitute $y = 12$ into (1)

$x + 12 = 22$

$x = 10$

The solution is $(10, 12)$.

45. $0.15x = 60$

$x = \dfrac{60}{0.15}$

$x = 400$

60 is 15% of 400.

43. $(0.25)(300) = x$

$x = 75$

75 is 25% of 300.

3.7 The Substitution Method

1. $x + y = 11$ (1)

 $y = 2x - 1$ (2)

 Substitute (2) into (1)

 $x + (2x - 1) = 11$

 $3x - 1 = 11$

 $3x = 12$

 $x = 4$

 Substitute $x = 4$ into (2)

 $y = 2(4) - 1 = 8 - 1 = 7$

 The solution is $(4, 7)$.

3. $x + y = 20$ (1)

 $y = 5x + 2$ (2)

 Substitute (2) into (1)

 $x + (5x + 2) = 20$

 $6x + 2 = 20$

 $6x = 18$

 $x = 3$

 Substitute $x = 3$ into (2)

 $y = 5(3) + 2 = 17$

 The solution is $(3, 17)$.

5. $-2x + y = -1$ (1)

 $y = -4x + 8$ (2)

 Substitute (2) into (1)

 $-2x + (-4x + 8) = -1$

 $-6x + 8 = -1$

 $x = \dfrac{-9}{-6}$

 $x = \dfrac{3}{2}$

 Substitute $x = \dfrac{3}{2}$ into (2)

 $y = -4\left(\dfrac{3}{2}\right) + 8 = -6 + 8 = 2$

 The solution is $\left(\dfrac{3}{2}, 2\right)$.

7. $3x - 2y = -2$ (1)

 $x = -y + 6$ (2)

 Substitute (2) into (1)

 $3(-y + 6) - 2y = -2$

 $-5y + 18 = -2$

 $-5y = -20$

 $y = 4$

 Substitute $y = 4$ into (2)

 $x = -4 + 6 = 2$

 The solution is $(2, 4)$.

9. $5x - 4y = -16$ (1)

 $y = 4$ (2)

 Substitute 4 for y in (1)

 $5x - 4(4) = -16$

 $5x - 16 = -16$

 $x = 0$

 $x = 0$

 The solution is $(0, 4)$.

11. $5x + 4y = 7$ (1)

 $y = -3x$ (2)

 Substitute (2) into (1)

 $5x + 4(-3x) = 7$

 $-7x = 7$

 $x = -1$

 Substitute $x = -1$ into (2)

 $y = -3(-1) = 3$

 The solution is $(-1, 3)$.

13. $x + 3y = 4 \qquad (1)$

$x - 2y = -1 \qquad (2)$

Solve (2) for x and substitute into (1)

$x = 2y - 1 \qquad (3)$

$(2y - 1) + 3y = 4$

$5y - 1 = 4$

$5y = 5$

$y = 1$

Substitute $y = 1$ into (3)

$x = 2(1) - 1 = 2 - 1 = 1$

The solution is $(1, 1)$.

15. $2x + y = 1 \qquad (1)$

$x - 5y = 17 \qquad (2)$

Solve (2) for x and substitute into (1)

$x = 5y + 17 \qquad (3)$

$2(5y + 17) + y = 1$

$11y + 34 = 1$

$11y = -33$

$y = -3$

Substitute $y = -3$ into (3)

$x = 5(-3) + 17 = 2$

The solution is $(2, -3)$.

17. $3x + 5y = -3 \qquad (1)$

$x - 5y = -5 \qquad (2)$

Solve (2) for x and substitute into (1)

$x = 5y - 5 \qquad (3)$

$3(5y - 5) + 5y = -3$

$15y - 15 + 5y = -3$

$20y - 15 = -3$

$20y = 12$

$y = \dfrac{12}{20}$

$y = \dfrac{3}{5}$

Substitute $y = \dfrac{3}{5}$ into (3)

$x - 5\left(\dfrac{3}{5}\right) = -5$

$x - 3 = -5$

$x = -2$

The solution is $\left(-2, \dfrac{3}{5}\right)$.

19. $5x + 3y = 0 \qquad (1)$

$x - 3y = -18 \qquad (2)$

Solve (2) for x and substitute into (1)

$x = 3y - 18 \qquad (3)$

$5(3y - 18) + 3y = 0$

$18y - 90 = 0$

$18y = 90$

$y = 5$

Substitute $y = 5$ into (3)

$x = 3(5) - 18 = -3$

The solution is $(-3, 5)$.

21. $-3x - 9y = 7$ (1)

$x + 3y = 12$ (2)

Solve (2) for x and substitute into (1)

$-3(-3y + 12) - 9y = 7$

$9y - 36 - 9y = 7$

$-36 = 7$ False

The lines are parallel and there is no solution.

23. $5x - 8y = 7$ (1)

$y = 2x - 5$ (2)

Substitute (2) into (1)

$5x - 8(2x - 5) = 7$

$5x - 16x + 40 = 7$

$-11x + 40 = 7$

$-11x = -33$

$x = 3$

Substitute $x = 3$ into (2)

$y = 2(3) - 5 = 1$

The solution is $(3, 1)$.

25. $7x - 6y = -1$ (1)

$x = 2y - 1$ (2)

Substitute (2) into (1)

$7(2y - 1) - 6y = -1$

$14y - 7 - 6y = -1$

$8y - 7 = -1$

$8y = 6$

$y = \dfrac{6}{8} = \dfrac{3}{4}$

Substitute $y = \dfrac{3}{4}$ into (2)

$x = 2\left(\dfrac{3}{4}\right) - 1 = \dfrac{3}{2} - 1 = \dfrac{1}{2}$

The solution is $\left(\dfrac{1}{2}, \dfrac{3}{4}\right)$.

27. $-3x + 2y = 6$ (1)

$y = 3x$ (2)

Substitute (2) into (1)

$-3x + 2(3x) = 6$

$3x = 6$

$x = 2$

Substitute $x = 2$ into (2)

$y = 3(2) = 6$

The solution is $(2, 6)$.

29. $5x - 6y = -4$ (1)

$x = y$ (2)

Substitute (2) into (1)

$5y - 6y = -4$

$-y = -4$

$y = 4$

Substitute $y = 4$ into (2)

$x = 4$

The solution is $(4, 4)$.

31. $3x + 3y = 9$ (1)

$y = 2x - 12$ (2)

Substitute (2) into (1)

$3x + 3(2x - 12) = 9$

$3x + 6x - 36 = 9$

$9x - 36 = 9$

$9x = 45$

$x = 5$

Substitute $x = 5$ into (2)

$y = 2(5) - 12 = -2$

The solution is $(5, -2)$.

33. $7x - 11y = 16$ **(1)**

          $y = 10$ **(2)**

Substitute (2) into (1)

$7x - 11(10) = 16$

   $7x - 110 = 16$

       $7x = 126$

         $x = 18$

The solution is $(18, 10)$.

35. $-4x + 4y = -8$ (1)

       $y = x - 2$ (2)

Substitute (2) into (1)

$-4x + 4(x - 2) = -8$

           $-8 = -8$

A true statement. The graphs coincide
and there is an infinite number of solutions.

37. $0.05x + 0.10y = 1.70$ (1)

          $y = 22 - x$ (2)

Substitute (2) into (1)

$0.05x + 0.10(22 - x) = 1.70$

   $0.05x + 2.2 - 0.10x = 1.70$

        $2.2 - 0.05x = 1.70$

           $-0.05x = -0.5$

               $x = 10$

Substitute $x = 10$ into (2)

$y = 22 - 10 = 12$

The solution is $(10, 12)$.

39. (a) The lines intersect at 1,000 miles.

    (b) For more than 1,200 miles, the line for the car
        is lower, so he should buy the car.

    (c) For fewer than 800 miles, the line for the truck
        is lower, so he should buy the truck.

    (d) Because the miles are ≥ 0.

41. Let x = number of 5-gallon bottles

Let y = amount of the charge

$y = 7 + (5x)(1.10) \rightarrow y = 7 + 5.5x$ (1)

$y = 5 + (5x)(1.15) \rightarrow y = 5 + 5.75x$ (2)

Substitute (2) into (1)

$5 + 5.75x = 7 + 5.5x$

      $0.25x = 2$

          $x = 8$

Substitute $x = 8$ into (1)

$y = 7 + 5.50(8) = 51$

Eight 5-gallon (total of 40 gallons) bottles
must be used in a month for the two
companies to charge the same amount of $51.

43. Let x = width, and $3x$ = length.

      $P = 2l + 2w$

    $24 = 2(3x) + 2x$

    $24 = 6x + 2x$

    $24 = 8x$

     $x = 3$ meters

The width is 3 meters and length is 9 meters.

45. Let x = number of nickels

Let $3 + x$ = number of dimes

$$0.05x + 0.10(3 + x) = 2.10$$
$$0.05x + 0.30 + 0.10x = 2.10$$
$$0.15x = 1.80$$
$$x = 12$$

There are 12 nickels and 15 dimes.

47. What number is 8% of 6000?

$$N = (0.08) \cdot (6000)$$
$$N = 480$$

49. Let x = amount invested at 8%

Let $2x$ = amount invested at 10%

$$0.08x + 0.10(2x) = 224$$
$$0.08x + 0.20x = 224$$
$$0.28x = 224$$
$$x = 800$$

He invested \$800 at 8% and \$1600 at 10%.

3.8 Applications

1. Let x = first number

Let y = second number

$$x + y = 25 \qquad (1)$$
$$y = 5 + x \qquad (2)$$

Substitute (2) into (1)

$$x + (5 + x) = 25$$
$$2x + 5 = 25$$
$$2x = 20$$
$$x = 10$$
$$y = 5 + 10 = 15$$

The two numbers are 10 and 15.

3. Let x = first number

Let y = second number

$$x + y = 15 \qquad (1)$$
$$x = 4y \qquad (2)$$

Substitute (2) into (1)

$$4y + y = 15$$
$$5y = 15$$
$$y = 3$$
$$x = 4(3) = 12$$

The two numbers are 12 and 3.

5. Let $x =$ larger number

Let $y =$ smaller number

$$x - y = 5 \qquad (1)$$
$$x = 2y + 1 \qquad (2)$$

Substitute (2) into (1)

$$2y + 1 - y = 5$$
$$y + 1 = 5$$
$$y = 4$$
$$x = 2(4) + 1 = 9$$

The two numbers are 4 and 9.

7. Let $x =$ first number

Let $y =$ second number

$$x = 4y + 5 \qquad (1)$$
$$x + y = 35 \qquad (2)$$

Substitute (1) into (2)

$$4y + 5 + y = 35$$
$$5y = 30$$
$$y = 6$$
$$x = 4(6) + 5 = 29$$

The two numbers are 29 and 6.

9. Let $x =$ amount invested at 6%

Let $y =$ amount invested at 8%

$$x + y = 20,000 \qquad (1)$$
$$0.06x + 0.08y = 1380 \qquad (2)$$

Multiply (1) by -0.06

$$-0.06x - 0.06y = -1200$$
$$\underline{0.06x + 0.08y = 1380}$$
$$0.02y = 180$$
$$y = 9000$$

Substitute $y = 9000$ into (1)

$$x + 9000 = 20,000$$
$$x = 11,000$$

Mr. Wilson invested $9000 at 8% and 11,000 at 6%.

11. Let $x =$ amount invested at 5%

Let $y =$ amount invested at 6%

$$x = 4y \qquad (1)$$
$$0.05x + 0.06y = 520 \qquad (2)$$

Substitute (1) into (2)

$$0.05(4y) + 0.06y = 520$$
$$0.26y = 520$$
$$y = 2000$$
$$x = 4(2000) = 8000$$

She invested $8000 at 5% and $2000 at 6%.

13. Let $x =$ number of nickels

Let $y =$ number of quarters

$$x + y = 14$$
$$0.05x + 0.25y = 2.30$$

Multiply (1) by -0.05

$$-0.05x - 0.05y = -0.7$$
$$\underline{0.05x + 0.25y = 2.30}$$
$$0.20y = 1.6$$
$$y = 8$$

Substitute $y - 8$ into (1)

$$x + 8 = 14$$
$$x = 6$$

Ron has 6 nickels and 8 quarters.

15. Let $x =$ number of dimes

Let $y =$ number of quarters

$$x + y = 21 \qquad (1)$$
$$0.10x + 0.25y = 3.45 \qquad (2)$$

Solve (1) for x

$$x = 21 - y \qquad (3)$$

Substitute into (2)

$$0.10(21 - y) + 0.25y = 3.45$$
$$2.1 - 0.10y + 0.25y = 3.45$$
$$0.15y = 1.35$$
$$y = 9$$

Substitute $y = 9$ into (3)

$$x = 21 - 9 = 12$$

Tom has 12 dimes and 9 quarters.

17. Let x = number of liters of 50% alcohol solution

Let y = number of liters of 20% alcohol solution

$x + y = 18$ (1)

$0.50x + 0.20y = 0.30(18)$ (2)

Multiply (1) by -0.20

$-0.20x - 0.20y = -3.6$ (3)

$\underline{0.50x + 0.20y = 5.4}$

$0.30x = 1.8$

$x = 6$

Substitute $x = 6$ into (1)

$6 + y = 18$

$y = 12$

The mixture contains 6 liters of 50% alcohol solution and 12 liters of 20% alcohol solution.

21. Let x = number of adult tickets

Let y = number of kids tickets

$x + y = 70$ (1)

$5.50x + 4.00y = 310$ (2)

Mujltiply (1) by -4

$-4.00x - 4.00y = -280$

$\underline{5.50x + 4.00y = 310}$

$1.5x = 30$

$x = 20$

Substitute $x = 20$ into (1)

$20 + y = 70$

$y = 50$

The matinee had 20 adult tickets sold and 50 kids tickets sold.

19. Let x = amount of 10% disinfectant solution

Let y = amount of 7% disinfectant solution

$x + y = 30$ (1)

$0.10x + 0.07y = 30(0.08)$ (2)

Solve (1) for x

$x = 30 - y$ (3)

Substitute into (2)

$0.10(30 - y) + 0.07y = 2.4$

$3.00 - 0.10y + 0.07y = 2.4$

$-0.03y = -0.60$

$y = 20$

$x = 30 - 20 = 10$

10 gallons of 10% solution and 20 gallons of 7% solution should be used.

23. Let x = width

Let y = length

$2x + 2y = 96$ (1)

$y = 2x$ (2)

Substitute (2) into (1)

$2x + 2(2x) = 96$

$6x = 96$

$x = 16$

$y = 2(16) = 32$

The width is 16 feet and the length is 32 feet.

25. Let x = number of $5 chips

   Let y = number of $25 chips

   $$x + y = 45 \qquad (1)$$

   $$5x + 25y = 465 \qquad (2)$$

   Multiply (1) by -5

   $$-5x - 5y = -225$$

   $$\underline{-5x + 25y = 465}$$

   $$20y = 240$$

   $$y = 12$$

   Substitute $y = 12$ into (1)

   $$x + 12 = 45$$

   $$x = 33$$

   The gambler has 33 $5 chips and 12 $25 chips.

27. Let x = number of shares at $11 a share

   Let y = number of shares at $20 a share

   $$x + y = 150 \qquad (1)$$

   $$11x + 20y = 2550 \qquad (2)$$

   Solve (1) for x

   $$x = 150 - y \qquad (3)$$

   Substitute into (2)

   $$11(150 - y) + 20y = 2550$$

   $$1650 - 11y + 20y = 2550$$

   $$9y = 900$$

   $$y = 100$$

   $$x = 150 - 100 = 50$$

   Mary Jo bought 50 shares at $11 a share and 100 shares at $20 a share.

29. $7 - 3(2x - 4) - 8 = 7 - 6x + 12 - 8 = -6x + 11$

31.
$$-\frac{3}{2}x = 12$$

$$-\frac{2}{3}\left(-\frac{3}{2}x\right) = -\frac{2}{3}(12)$$

$$x = -8$$

33. $8 - 2(x + 7) = 2$

   $$8 - 2x - 14 = 2$$

   $$-2x - 6 = 2$$

   $$-2x = 8$$

   $$x = -4$$

35. $2x + 2w = P$

   $$2w = P - 2l$$

   $$w = \frac{P - 2l}{2}$$

37.
$$3 - 2x > 5$$

$$-3 + 3 - 2x > 5 - 3$$

$$-2x > 2$$

$$-\frac{1}{2}(-2x) < -\frac{1}{2}(2)$$

$$x < -1$$

See the graph in the back of the textbook.

39.
$$3x - 2y \le 12$$

$$3x - 3x - 2y \le -3x + 12$$

$$-2y \le -3x + 12$$

$$-\frac{1}{2}(-2y) \ge -\frac{1}{2}(-3x + 12)$$

$$y \ge \frac{3}{2}x - 6$$

41. Let $w =$ the width and $3w + 5 =$ the length.
$$2(w) + 2(3w + 5) = 26$$
$$2w + 6w + 10 = 26$$
$$8w + 10 = 26$$
$$w = 2$$
$$3w + 5 = 3(2) + 5 = 11$$

The width is 2 inches and the
length is 11 inches.

Chapter 3 Review

1. Substituting $x = 4$, $x = 0$, $y = 3$, and $y = 0$:

$3(4) + y = 6$	$3(0) + y = 6$	$3x + 3 = 6$	$3x + 0 = 6$
$12 + y = 6$	$0 + y = 6$	$3x = 3$	$3x = 6$
$y = -6$	$y = 6$	$x = 1$	$x = 2$

The ordered pairs are $(4, -6)$, $(0, 6)$, $(1, 3)$, and $(2, 0)$.

3. Substituting $x = 4$, $y = -2$, and $y = 3$:

$y = 2(4) - 6$	$-2 = 2x - 6$	$3 = 2x - 6$
$y = 8 - 6$	$4 = 2x$	$9 = 2x$
$y = 2$	$x = 2$	$x = \dfrac{9}{2}$

The ordered pairs are $(4, 2)$, $(2, -2)$, and $(\frac{9}{2}, 3)$.

5. Substituting $x = 2$, $x = -1$, and $x = -3$ results (in each case) in $y = -3$. The ordered pairs are $(2, -3)$, $(-1, -3)$, and $(-3, -3)$.

7. Substituting each ordered pair into the equation:

$\left(-2, \dfrac{9}{2}\right):$ $3(-2) - 4\left(\dfrac{9}{2}\right) = -6 - 18 = -24 \neq 12$

$(0, 3):$ $3(0) - 4(3) = 0 - 12 = -12 \neq 12$

$\left(2, -\dfrac{3}{2}\right):$ $3(2) - 4\left(-\dfrac{3}{2}\right) = 6 + 6 = 12$

Only the ordered pair $\left(2, -\dfrac{3}{2}\right)$ is a solution.

9. See the graph in the back of the textbook.

11. See the graph in the back of the textbook.

13. See the graph in the back of the textbook.

15. The ordered pairs are $(-2, 0)$, $(0, -2)$, and $(1, -3)$.

    See the graph in the back of the textbook.

17. The ordered pairs are $(1, 1)$, $(0, -1)$, and $(-1, -3)$.

    See the graph in the back of the textbook.

19. See the graph in the back of the textbook.

21. See the graph in the back of the textbook.

23. To find the x-intercept, let $y = 0$:

    $$3x - 0 = 6$$
    $$3x = 6$$
    $$x = 2$$

    To find the y-intercept, let $x = 0$.

    $$3(0) - y = 6$$
    $$-y = 6$$
    $$y = -6$$

25. To find the x-intercept, let $y = 0$:

    $$0 = x - 3$$
    $$x = 3$$

    To find the y-intercept, let $x = 0$:

    $$y = 0 - 3$$
    $$y = -3$$

27. The intersection point is $(4, -2)$.

    See the graph in the back of the textbook.

29. The intersection point is $(3, -2)$.

    See the graph in the back of the textbook

31. The intersection point is $(2, 1)$.

    See the graph in the back of the textbook

33. Adding the two equations:

    $$2x = 2$$
    $$x = 1$$

    Substituting into the second equation:

    $$1 + y = -2$$
    $$y = -3$$

    The solution is $(1, -3)$.

35. Multiplying the first equation by 2:
$$10x - 6y = 4$$
$$-10x + 6y = -4$$
Adding the two equations:
$$0 = 0$$
Since this statement is true, the system is dependent. The two lines coincide.

37. Multiplying the second equation by -4:
$$-3x + 4y = 1$$
$$16x - 4y = 12$$
Adding the two equations:
$$13x = 13$$
$$x = 1$$
Substituting into the second equation:
$$-4(1) + y = -3$$
$$-4 + y = -3$$
$$y = 1$$
The solution is $(1,1)$.

39. Multiply the first equation by 3 and the second equation by 5:
$$-6x + 15y = -33$$
$$35x - 15y = -25$$
Adding the two equations:
$$29x = -58$$
$$x = -2$$
Substituting into the first equation:
$$-2(-2) + 5y = -11$$
$$4 + 5y = -11$$
$$5y = -15$$
$$y = -3$$
The solution is $(-2, -3)$.

41. Substituting into the first equation:
$$x + (-3x + 1) = 5$$
$$-2x + 1 = 5$$
$$-2x = 4$$
$$x = -2$$
Substituting into the second equation:
$$y = -3(-2) + 1 = 6 + 1 = 7$$

43. Substituting into the first equation:
$$4x - 3(3x + 7) = -16$$
$$4x - 9x - 21 = -16$$
$$-5x - 21 = -16$$
$$-5x = 5$$
$$x = -1$$
Substituting into the second equation:
$$y = 3(-1) + 7 = -3 + 7 = 4$$
The solution is $(-1, 4)$.

45. Solving the first equation for x:
$$x - 4y = 2$$
$$x = 4y + 2$$
Substituting into the second equation:
$$-3(4y + 2) + 12y = -8$$
$$-12y - 6 + 12y = -8$$
$$-6 = -8$$
Since this statement is false, there is no solution to the system. The two lines are parallel.

47. Solving the second equation for x:

$$x + 6y = -11$$
$$x = -6y - 11$$

Substituting into the second equation:

$$10(-6y - 11) - 5y = 20$$
$$-60y - 110 - 5y = 20$$
$$-65y - 110 = 20$$
$$-65y = 130$$
$$y = -2$$

Substituting into $x = -6y - 11$:

$$x = -6(-2) - 11 = 12 - 11 = 1$$

The solution is $(1, -2)$.

49. Let $x =$ smaller number
Let $y =$ larger number

$$x + y = 18$$
$$2x = 6 + y$$

Solving the first equation for y:

$$x + y = 18$$
$$y = -x + 18$$

Substituting into the second equation

$$2x = 6 + (-x + 18)$$
$$2x = -x + 24$$
$$3x = 24$$
$$x = 8$$

Substituting into the first equation:

$$8 + y = 18$$
$$y = 10$$

The two numbers are 8 and 10.

51. Let $x =$ amount invested at 4%
Let $y =$ amount invested at 5%

$$x + y = 12000$$
$$0.04x + 0.05y = 560$$

Multiplying the first equation by -0.04:

$$-0.04x - 0.04y = -480$$
$$0.04x + 0.05y = 560$$

Adding the two equations:

$$0.01y = 80$$
$$y = 8000$$

Substituting into the first equation:

$$x + 8000 = 12000$$
$$x = 4000$$

$4,000 was invested at 4% and $8,000 was invested at 5%.

53. Let $x =$ number of dimes
Let $y =$ number of nickels

$$x + y = 17$$
$$0.10x + 0.05y = 1.35$$

Multiplying the first equation by -0.05:

$$-0.05x - 0.05y = -0.85$$
$$0.10x + 0.05y = 1.35$$

Adding the two equations:

$$0.05x = 0.50$$
$$x = 10$$

Substituting into the first equation:

$$10 + y = 17$$
$$y = 7$$

Barbara has 10 dimes and 7 nickels.

55. Let x = liters of 20% alcohol solution

Let y = liters of 10% alcohol solution

$$x + y = 50$$

$$0.20x + 0.10y = 0.12(50)$$

Multiplying the first equation by -0.10:

$$-0.10x - 0.10y = -5$$

$$0.20x + 0.10y = 6$$

Adding the two equations:

$$0.10x = 1$$

$$x = 10$$

Substituting into the first equation:

$$10 + y = 50$$

$$y = 40$$

The solution contains 40 liters of 10% alcohol solution and 10 liters of 20% alcohol solution.

Cumulative Review: Chapters 1-3

1. $3 \cdot 4 + 5 = 12 + 5 = 17$

3. $7[8 + (-5)] + 3(-7 + 12) = 7(3) + 3(5) = 21 + 15 = 36$

5. $8 - 6(5 - 9) = 8 - 6(-4) = 8 + 24 = 32$

7. $\dfrac{2}{3} + \dfrac{3}{4} - \dfrac{1}{6} = \dfrac{2 \cdot 4}{3 \cdot 4} + \dfrac{3 \cdot 3}{4 \cdot 3} - \dfrac{1 \cdot 2}{6 \cdot 2} = \dfrac{8}{12} + \dfrac{9}{12} - \dfrac{2}{12} = \dfrac{15}{12} = \dfrac{5}{4}$

9. $-5 - 6 = -y - 3 + 2y$

$-11 = y - 3$

$-11 + 3 = y - 3 + 3$

$y = -8$

11. $3(x - 4) = 9$

$3x - 12 = 9$

$3x - 12 + 12 = 9 + 12$

$3x = 21$

$\dfrac{1}{3}(3x) = \dfrac{1}{3}(21)$

$x = 7$

13. $0.3x + 0.7 \le -2$

$0.3x + 0.7 - 0.7 \le -2 - 0.7$

$0.3x \le -2.7$

$\dfrac{0.3x}{0.3} \le \dfrac{-2.7}{0.3}$

$x \le -9$

See the graph in the back of the textbook.

15. See the graph in the back of the textbook.

17. See the graph in the back of the textbook.

19. The two lines are parallel.
    See the graph in the back of the textbook.

21. Multiplying the first equation by -2:
$$-2x - 2y = -14$$
$$2x + 2y = 14$$
Adding the two equations:
$$0 = 0$$
Since this statement is true, the system is dependent. The two lines coincide.

23. Multiplying the second equation by 3:
$$2x + 3y = 13$$
$$3x - 3y = -3$$
Adding the two equations:
$$5x = 10$$
$$x = 2$$
Substituting into the first equation:
$$2(2) + 3y = 13$$
$$4 + 3y = 13$$
$$y = 3$$
The solution is $(2, 3)$.

25. Multiplying the second equation by -2:
$$2x + 5y = 33$$
$$-2x + 6y = 0$$
Adding the two equations:
$$11y = 33$$
$$y = 3$$
Substituting into the second equation:
$$x - 3(3) = 0$$
$$x - 9 = 0$$
$$x = 9$$
The solution is $(9, 3)$.

27. Multiplying the second equation by 7:
$$3x - 7y = 12$$
$$14x + 7y = 56$$
Adding the two equations:
$$17x = 68$$
$$x = 4$$
Substituting into the second equation:
$$2(4) + y = 8$$
$$8 + y = 8$$
$$y = 0$$
The solution is $(4, 0)$.

29. Substituting the value of y from the second equation into the first equation:
$$2x - 3(5x + 2) = 7$$
$$2x - 15x - 6 = 7$$
$$-13x - 6 = 7$$
$$-13x = 13$$
$$x = -1$$
Substituting into the second equation:
$$y = 5(-1) + 2 = -5 + 2 = -3$$
The solution is $(-1, -3)$.

31. Commutative property of addition

33. The quotient is: $\dfrac{-30}{6} = -5$

35. $p \cdot 82 = 20.5$

$$p = \frac{20.5}{82}$$

$$p = 0.25 = 25\%$$

25% of 82 is 20.5.

37. Simplifying, then evaluating when $x = 3$: $-3x + 7 + 5x = 2x + 7 = 2(3) + 7 = 6 + 7 = 13$

39. When $x = -2$: $4x - 5 = 4(-2) - 5 = -8 - 5 = -13$

41. $5 - (-8) = 5 + 8 = 13$

43. To find the x-intercept, let $y = 0$.

$3x - 4(0) = 12$

$3x = 12$

$x = 4$

To find the y-intercept, let $x = 0$:

$3(0) - 4y = 12$

$-4y = 12$

$y = -3$

45. $2x - 3y = 7$

When $x = -1$

$2(-1) - 3y = 7$

$-2 - 3y = 7$

$-3y = 9$

$y = -3$

47. Let $w =$ the width and $2w + 5 =$ the length.

$$2(w) + 2(2w + 5) = 44$$

$$2w + 4w + 10 = 44$$

$$6w + 10 = 44$$

$$6w = 34$$

$$w = \frac{17}{3}$$

$$2w + 5 = 2\left(\frac{17}{3}\right) + 5 = \frac{34}{3} + 5 = \frac{49}{3}$$

The width is $\dfrac{17}{3}$ cm and the

length is $\dfrac{49}{3}$ cm.

49.

	Dollars Invested at 8%	Dollars Invested at 6%
Number of	$x + 900$	x
Interest on	$0.08(x + 900)$	$0.06(x)$

$$0.08(x + 900) + 0.06(x) = 8240$$
$$0.08x + 72 + 0.06x = 240$$
$$0.14x + 72 = 240$$
$$0.14x = 168$$
$$x = 1200$$
$$x + 900 = 2100$$

Barbara invested \$1,200 at 6% and \$2,100 at 8%.

Chapter 3 Test

1. Substituting $x = 0$, $y = 0$, $x = 10$, and $y = -3$:

$$2(0) - 5y = 10 \qquad 2x - 5(0) = 10 \qquad 2(10) - 5y = 10 \qquad 2x - 5(-3) = 10$$
$$0 - 5y = 10 \qquad 2x - 0 = 10 \qquad 20 - 5y = 10 \qquad 2x + 15 = 10$$
$$-5y = 10 \qquad 2x = 10 \qquad -5y = -10 \qquad 2x = -5$$
$$y = -2 \qquad\qquad x = 5 \qquad\qquad y = 2 \qquad\qquad x = -\frac{5}{2}$$

The ordered pairs are $(0, -2)$, $(5, 0)$, $(10, 2)$ and $\left(-\frac{5}{2}, -3\right)$.

2. Substituting each ordered pair into the equation:

$(2, 5)$: $4(2) - 3 = 8 - 3 = 5$

$(0, -3)$: $4(0) - 3 = 0 - 3 = -3$

$(3, 0)$: $4(3) - 3 = 12 - 3 = 9 \neq 0$

$(-2, 11)$: $4(-2) - 3 = -8 - 3 = -11 \neq 11$

The ordered pairs $(2, 5)$ and $(0, -3)$ are solutions.

3. See the graph in the back of the textbook.   4. See the graph in the back of the textbook.

5. To find the x-intercept, let $y = 0$:

$$3x - 5(0) = 15$$
$$3x = 15$$
$$x = 5$$

To find the y-intercept, let $x = 0$:

$$3(0) - 5y = 15$$
$$-5y = 15$$
$$y = -3$$

6. To find the x-intercept, let $y = 0$:

$$0 = \frac{3}{2}x + 1$$
$$-1 = \frac{3}{2}x$$
$$x = -\frac{2}{3}$$

To find the y-intercept, let $x = 0$:

$$y = \frac{3}{2}(0) + 1 = 1$$

7. The intersection point is $(1, 2)$.

   See the graph in the back of the textbook.

8. The intersection point is $(3, -2)$.

   See the graph in the back of the textbook

9. Adding the two equations:

$$3x = -9$$
$$x = -3$$

Substituting into the first equation:

$$-3 - y = 1$$
$$-y = 4$$
$$y = -4$$

The solution is $(-3, -4)$.

10. Multiplying the first equation by -1:

$$-2x - y = -7$$
$$3x + y = 12$$

Adding the two equations:

$$x = 5$$

Substituting into the first equation:

$$2(5) + y = 7$$
$$10 + y = 7$$
$$y = -3$$

The solution is $(5, -3)$.

11. Multiplying the second equation by 4:

$$7x + 8y = -2$$
$$12x - 8y = 40$$

Adding the two equations:

$$19x = 38$$
$$x = 2$$

Substituting into the first equation:

$$7(2) + 8y = -2$$
$$14 + 8y = -2$$
$$8y = -16$$
$$y = -2$$

The solution is $(2, -2)$.

12. Multiply the first equation by -3 and the second equation by 2:

$$-18x + 30y = -18$$
$$18x - 30y = 18$$

Adding the two equations:

$$0 = 0$$

Since this equation is true, the system is dependent. The two lines coincide.

13. Substituting the value of y from the second equation into the first equation:

$$3x + 2(2x + 3) = 20$$
$$3x + 4x + 6 = 20$$
$$7x + 6 = 20$$
$$7x = 14$$
$$x = 2$$

Substituting into the second equation:

$$y = 2(2) + 3 = 4 + 3 = 7$$

The solution is $(2, 7)$.

14. Substituting x from the second equation into the first equation:

$$3(y + 1) - 6y = -6$$
$$3y + 3 - 6y = -6$$
$$-3y + 3 = -6$$
$$-3y = -9$$
$$y = 3$$

Substituting into the second equation:

$$x = 3 + 1 = 4$$

The solution is $(4, 3)$.

15. Solving the second equation for y:

$$-3x + y = 3$$
$$y = 3x + 3$$

Substituting into the first equation:

$$7x - 2(3x + 3) = -4$$
$$7x - 6x - 6 = -4$$
$$x - 6 = -4$$
$$x = 2$$

Substituting into $y = 3x + 3$:

$$y = 3(2) + 3 = 6 + 3 = 9$$

The solution is $(2, 9)$.

16. Solving the second equation for x:

$$x + 3y = -8$$
$$x = -3y - 8$$

Substituting into the first equation:

$$2(-3y - 8) - 3y = -7$$
$$-6y - 16 - 3y = -7$$
$$-9y - 16 = -7$$
$$-9y = 9$$
$$y = -1$$

Substituting into $x = -3y - 8$:

$$x = -3(-1) - 8 = 3 - 8 = -5$$

The solution is $(-5, -1)$.

17. Let x and y represent the two numbers. The system of equation is:

$$x + y = 12$$
$$x - y = 2$$

Adding the two equations:

$$2x = 14$$
$$x = 7$$

Substituting into the first equation:

$$7 + y = 12$$
$$y = 5$$

The two numbers are 7 and 5.

18. Let x and y represent the two numbers. The system of equation is:

$$x + y = 15$$
$$y = 6 + 2x$$

Substituting into the first equation:

$$x + 6 + 2x = 15$$
$$3x + 6 = 15$$
$$3x = 9$$
$$x = 3$$

Substituting into the second equation:

$$y = 6 + 2(3) = 6 + 6 = 12$$

The two numbers are 3 and 12.

19. Let x = amount invested at 9%

 Let y = amount invested at 11%

$$x + y = 10000$$

$$0.09x + 0.11y = 980$$

 Multiplying the first equation by -0.09:

$$-0.09x - 0.09y = -900$$

$$0.09x + 0.11y = 980$$

 Adding the two equations:

$$0.02y = 80$$

$$y = 4000$$

 Substituting into the first equation:

$$x + 4000 = 10000$$

$$x = 6000$$

 Dr. Stork should invest $6,000 at 9%.

20. Let x = number of nickels

 Let y = number of quarters

$$x + y = 12$$

$$0.05x + 0.25y = 1.60$$

 Multiplying the first equation by -0.05:

$$-0.05x - 0.05y = -0.60$$

$$0.05x + 0.25y = 1.60$$

 Adding the two equations:

$$0.20y = 1.00$$

$$y = 5$$

 Substituting into the first equation:

$$x + 5 = 12$$

$$x = 7$$

 Diane has 7 nickels and 5 quarters.

CHAPTER 4

Exponents and Polynomials

4.1 Multiplication with Exponents

1. The base is 4 and the exponent is 2. $4^2 = 4 \cdot 4 = 16$

3. The base is 0.3 and the exponent is 2. $(0.3)^2 = (0.3) \cdot (0.3) = 0.09$

5. The base is 4 and the exponent is 3. $4^3 = 4 \cdot 4 \cdot 4 = 64$

7. The base is -5 and the exponent is 2. $(-5)^2 = (-5) \cdot (-5) = 25$

9. The base is 2 and the exponent is 3. $-2^3 = -2 \cdot 2 \cdot 2 = -8$

11. The base is 3 and the exponent is 4. $3^4 = 3 \cdot 3 \cdot 3 \cdot 3 = 81$

13. The base is $\frac{2}{3}$ and the exponent is 2. $\left(\frac{2}{3}\right)^2 = \left(\frac{2}{3}\right) \cdot \left(\frac{2}{3}\right) = \frac{4}{9}$

15. The base is $\frac{1}{2}$ and the exponent is 4. $\left(\frac{1}{2}\right)^4 = \left(\frac{1}{2}\right) \cdot \left(\frac{1}{2}\right) \cdot \left(\frac{1}{2}\right) \cdot \left(\frac{1}{2}\right) = \frac{1}{16}$

17. (a) See the table in the back of the textbook. (b) For numbers larger than 1, the square of the number is greater than the number.

19. $x^4 \cdot x^5 = x^{4+5} = x^9$

21. $y^{10} \cdot y^{20} = y^{10+20} = y^{30}$

23. $2^5 \cdot 2^4 \cdot 2^3 = 2^{5+4+3} = 2^{12}$

25. $x^4 \cdot x^6 \cdot x^8 \cdot x^{10} = x^{4+6+8+10} = x^{28}$

27. $(x^2)^5 = x^{2 \cdot 5} = x^{10}$

29. $(5^4)^3 = 5^{4 \cdot 3} = 5^{12}$

31. $(y^3)^3 = y^{3 \cdot 3} = y^9$

33. $(2^5)^{10} = 2^{5 \cdot 10} = 2^{50}$

35. $(a^3)^x = a^{3x}$

37. $(b^x)^y = b^{xy}$

39. $(4x)^2 = 4^2 \cdot x^2 = 16x^2$

41. $(2y)^5 = 2^5 \cdot y^5 = 32y^5$

43. $(-3x)^4 = (-3)^4 \cdot x^4 = 81x^4$

45. $(0.5ab)^2 = (0.5)^2 \cdot a^2 b^2 = 0.25a^2 b^2$

47. $(4xyz)^3 = 4^3 \cdot x^3 y^3 z^3 = 64x^3 y^3 z^3$

49. $(2x^4)^3 = 2^3(x^4)^3 = 8x^{12}$

51. $(4a^3)^2 = 4^2(a^3)^2 = 16a^6$

53. $(x^2)^3(x^4)^2 = x^6 \cdot x^8 = x^{14}$

55. $(a^3)^1(a^2)^4 = a^3 \cdot a^8 = a^{11}$

57. $(2x)^3(2x)^4 = (2x)^7 = 2^7 x^7 = 128x^7$

59. $(3x^2)^3(2x)^4 = 3^3 x^6 \cdot 2^4 x^4 = 27x^6 \cdot 16x^4 = 432x^{10}$

61. $(4x^2 y^3)^2 = 4^2 x^4 y^6 = 16x^4 y^6$

63. $\left(\frac{2}{3}a^4 b^5\right)^3 = \left(\frac{2}{3}\right)^3 a^{12} b^{15} = \frac{8}{27}a^{12} b^{15}$

65. See the table and the graph in the back of the textbook.

67. See the table in the back of the textbook. They will create a smoother curve.

69. $43,200 = 4.32 \times 10^4$

71. $570 = 5.7 \times 10^2$

73. $238,000 = 2.38 \times 10^5$

75. $2.49 \times 10^3 = 2,490$

77. $3.52 \times 10^2 = 352$

79. $2.8 \times 10^4 = 28,000$

81. $V = (3 \text{ in.})^3 = 27 \text{ inches}^3$

83. $V = (2.5 \text{ in.})^3 = 15.6 \text{ inches}^3$

85. $V = (8 \text{ in.})(4.5 \text{ in.})(1 \text{ in.}) = 36 \text{ inches}^3$

87. Yes. If the box was $7 \text{ ft} \times 3 \text{ ft} \times 2 \text{ ft}$ you could fit inside of it.

89. $650,000,000 \text{ seconds} = 6.5 \times 10^8 \text{ seconds}$

91. $7.4 \times 10^5 \text{ dollars} = \$740,000$

93. $1.8 \times 10^5 \text{ dollars} = \$180,000$

Use $d = \pi \cdot s \cdot c \cdot \left(\frac{1}{2} \cdot b\right)^2$ in Exercises 95, and 97.

95. $d = (3.14)(3.11)(8)\left(\frac{3.35}{2}\right)^2 = 219 \text{ inches}^3$

97. $d = (3.14)(2.99)(6)\left(\frac{3.59}{2}\right)^2 = 182 \text{ inches}^3$

99. (a) Stage 1: $\frac{4}{3}L_0 = \left(\frac{4}{3}\right)(1 \text{ foot}) = \frac{4}{3}$ feet (b) Stage 2: $\left(\frac{4}{3}\right)^2 L_0 = \frac{16}{9}$ feet

   (c) See the drawing in the back of the textbook. Stage 3: $\left(\frac{4}{3}\right)^3 L_0 = \left(\frac{4}{3}\right)^3 = \frac{64}{27}$ feet

   (d) Stage 10: $\left(\frac{4}{3}\right)^{10} L_0 = \left(\frac{4}{3}\right)^{10}$ feet

101. $4 - 7 = 4 + (-7) = -3$

103. $4 - (-7) = 4 + 7 = 11$

105. $15 - 20 = 15 + (-20) = -5$

107. $-15 - (-20) = -15 + 20 = 5$

4.2 Division with Exponents

1. $3^{-2} = \frac{1}{3^2} = \frac{1}{9}$

3. $6^{-2} = \frac{1}{6^2} = \frac{1}{36}$

5. $8^{-2} = \frac{1}{8^2} = \frac{1}{64}$

7. $5^{-3} = \frac{1}{5^3} = \frac{1}{125}$

9. $2x^{-3} = 2 \cdot \frac{1}{x^3} = \frac{2}{x^3}$

11. $(2x)^{-3} = \frac{1}{(2x)^3} = \frac{1}{8x^3}$

13. $(5y)^{-2} = \frac{1}{(5y)^2} = \frac{1}{25y^2}$

15. $10^{-2} = \frac{1}{10^2} = \frac{1}{100}$

17. See the table in the back of the textbook.

19. $\frac{5^1}{5^3} = 5^{1-3} = 5^{-2} = \frac{1}{5^2} = \frac{1}{25}$

21. $\frac{x^{10}}{x^4} = x^{10-4} = x^6$

23. $\frac{4^3}{4^0} = 4^{3-0} = 4^3 = 64$

25. $\frac{(2x)^7}{(2x)^4} = (2x)^{7-4} = (2x)^3 = 2^3 x^3 = 8x^3$

27. $\frac{6^{11}}{6} = \frac{6^{11}}{6^1} = 6^{11-1} = 6^{10} = 60,\,466,\,176$

29. $\frac{6}{6^{11}} = \frac{6^1}{6^{11}} = 6^{1-11} = 6^{-10} = \frac{1}{6^{10}} = \frac{1}{60,466,176}$

31. $\frac{2^{-5}}{2^3} = 2^{-5-3} = 2^{-8} = \frac{1}{2^8} = \frac{1}{256}$

33. $\frac{2^5}{2^{-3}} = 2^{5-(-3)} = 2^{5+3} = 2^8 = 256$

35. $\frac{(3x)^{-5}}{(3x)^{-8}} = (3x)^{-5-(-8)} = (3x)^{-5+8} = (3x)^3 = 3^3 x^3 = 27x^3$

37. $(3xy)^4 = 3^4 x^4 y^4 = 81x^4 y^4$

39. $10^\circ = 1$

41. $(2a^2 b)^1 = 2a^2 b$

43. $(7y^3)^{-2} = \frac{1}{(7y^3)^2} = \frac{1}{49y^6}$

45. $x^{-3} \cdot x^{-5} = x^{-3-5} = x^{-8} = \frac{1}{x^8}$

47. $y^7 \cdot y^{-10} = y^{7-10} = y^{-3} = \frac{1}{y^3}$

49. $\frac{(x^2)^3}{x^4} = \frac{x^6}{x^4} = x^{6-4} = x^2$

51. $\frac{(a^4)^3}{(a^3)^2} = \frac{a^{12}}{a^6} = a^{12-6} = a^6$

53. $\frac{y^7}{(y^2)^8} = \frac{y^7}{y^{16}} = y^{7-16} = y^{-9} = \frac{1}{y^9}$

55. $\left(\frac{y^7}{y^2}\right)^8 = (y^{7-2})^8 = (y^5)^8 = y^{40}$

57. $\dfrac{\left(x^{-2}\right)^3}{x^{-5}} = \dfrac{x^{-6}}{x^{-5}} = x^{-6-(-5)} = x^{-6+5} = x^{-1} = \dfrac{1}{x}$

59. $\left(\dfrac{x^{-2}}{x^{-5}}\right)^3 = \left(x^{-2+5}\right)^3 = \left(x^3\right)^3 = x^9$

61. $\dfrac{\left(a^3\right)^2\left(a^4\right)^5}{\left(a^5\right)^2} = \dfrac{a^6 \cdot a^{20}}{a^{10}} = \dfrac{a^{26}}{a^{10}} = a^{26-10} = a^{16}$

63. $\dfrac{\left(a^{-2}\right)^3\left(a^4\right)^2}{\left(a^{-3}\right)^{-2}} = \dfrac{a^{-6} \cdot a^8}{a^6} = \dfrac{a^2}{a^6} = a^{2-6} = a^{-4} = \dfrac{1}{a^4}$

65. See the table and the graph in the back of the textbook.

67. $0.0048 = 4.8 \times 10^{-3}$

69. $25 = 2.5 \times 10^1$

71. $0.000009 = 9 \times 10^{-6}$

73. See the table in the back of the textbook.

75. $4.23 \cdot 10^{-3} = 0.00423$

77. $8 \times 10^{-5} = 0.00008$

79. $4.2 \times 10^0 = 4.2$

81. 2×10^{-3} seconds $= 0.002$ seconds

83. 0.006 inches $= 6 \times 10^{-3}$ inches

85. $25 \times 10^3 = 2.5 \times 10^4$

87. $23.5 \times 10^4 = 2.35 \times 10^5$

89. $0.82 \times 10^{-3} = 8.2 \times 10^{-4}$

91. The area of the smaller square is $(10 \text{ in.})^2 = 100$ inches^2, while the area of the larger square is $(20 \text{ in.})^2 = 400$ inches^2. It would take 4 smaller squares to cover the larger square.

93. The area of the smaller square is x^2, while the area of the larger square is $(2x)^2 = 4x^2$. It would take 4 smaller squares to cover the larger square.

95. The volume of the smaller box is $(6 \text{ in.})^3 = 216$ inches^3, while the volume of the larger box is $(12 \text{ in.})^3 = 1,728$ inches^3. Thus 8 smaller boxes will fit inside the larger box $(8 \cdot 216 = 1,728)$.

97. The volume of the smaller box is x^3, while the volume of the larger box is $(2x)^3 = 8x^3$. Thus 8 smaller boxes will fit inside the larger box.

99. $4x + 3x = (4 + 3)x = 7x$

101. $5a - 3a = (5 - 3)a = 2a$

103. $4y + 5y + y = (4 + 5 + 1)y = 10y$

4.3 Operations with Monomials

1. $\left(3x^4\right)\left(4x^3\right)=12x^{4+3}=12x^7$

3. $\left(-2y^4\right)\left(8y^7\right)=-16y^{4+7}=-16y^{11}$

5. $(8x)(4x)=32x^{1+1}=32x^2$

7. $\left(10a^3\right)(10a)\left(2a^2\right)=200a^{3+1+2}=200a^6$

9. $\left(6ab^2\right)\left(-4a^2b\right)=-24a^{1+2}b^{2+1}=-24a^3b^3$

11. $\left(4x^2y\right)\left(3x^3y^3\right)\left(2xy^4\right)=24x^{2+3+1}y^{1+3+4}=24x^6y^8$

13. $\dfrac{15x^3}{5x^2}=\dfrac{15}{5}\cdot\dfrac{x^3}{x^2}=3x$

15. $\dfrac{18y^9}{3y^{12}}=\dfrac{18}{3}\cdot\dfrac{y^9}{y^{12}}=6\cdot\dfrac{1}{y^3}=\dfrac{6}{y^3}$

17. $\dfrac{32a^3}{64a^4}=\dfrac{32}{64}\cdot\dfrac{a^3}{a^4}=\dfrac{1}{2}\cdot\dfrac{1}{a}=\dfrac{1}{2a}$

19. $\dfrac{21a^2b^3}{-7ab^5}=\dfrac{21}{-7}\cdot\dfrac{a^2}{a}\cdot\dfrac{b^3}{b^5}=-3\cdot a\cdot\dfrac{1}{b^2}=-\dfrac{3a}{b^2}$

21. $\dfrac{3x^3y^2z}{27xy^2z^3}=\dfrac{3}{27}\cdot\dfrac{x^3}{x}\cdot\dfrac{y^2}{y^2}\cdot\dfrac{z}{z^3}=\dfrac{1}{9}\cdot x^2\cdot\dfrac{1}{z^2}=\dfrac{x^2}{9z^2}$

23. See the table in the back of the textbook.

25. $\left(3\times10^3\right)\left(2\times10^5\right)=6\times10^8$

27. $\left(3.5\times10^4\right)\left(5\times10^{-6}\right)=17.5\times10^{-2}=1.75\times10^{-1}$

29. $\left(5.5\times10^{-3}\right)\left(2.2\times10^{-4}\right)=12.1\times10^{-7}=1.21\times10^{-6}$

31. $\dfrac{8.4\times10^5}{2\times10^2}=4.2\times10^3$

33. $\dfrac{6\times10^8}{2\times10^{-2}}=3\times10^{10}$

35. $\dfrac{2.5\times10^{-6}}{5\times10^{-4}}=0.5\times10^{-2}=5.0\times10^{-3}$

37. $3x^2+5x^2=(3+5)x^2=8x^2$

39. $8x^5-19x^5=(8-19)x^5=-11x^5$

41. $2a+a-3a=(2+1-3)a=0a=0$

43. $10x^3-8x^3+2x^3=(10-8+2)x^3=4x^3$

45. $20ab^2-19ab^2+30ab^2=(20-19+30)ab^2=31ab^2$

47. See the table in the back of the textbook.

49. $\dfrac{\left(3x^2\right)\left(8x^5\right)}{6x^4}=\dfrac{24x^7}{6x^4}=\dfrac{24}{6}\cdot\dfrac{x^7}{x^4}=4x^3$

51. $\dfrac{\left(9a^2b\right)\left(2a^3b^4\right)}{18a^5b^7}=\dfrac{18a^5b^5}{18a^5b^7}=\dfrac{18}{18}\cdot\dfrac{a^5}{a^5}\cdot\dfrac{b^5}{b^7}=1\cdot\dfrac{1}{b^2}=\dfrac{1}{b^2}$

53. $\dfrac{\left(4x^3y^2\right)\left(9x^4y^{10}\right)}{\left(3x^5y\right)\left(2x^6y\right)}=\dfrac{36x^7y^{12}}{6x^{11}y^2}=\dfrac{36}{6}\cdot\dfrac{x^7}{x^{11}}\cdot\dfrac{y^{12}}{y^2}=6\cdot\dfrac{1}{x^4}\cdot y^{10}=\dfrac{6y^{10}}{x^4}$

55. $\dfrac{\left(6\times10^8\right)\left(3\times10^5\right)}{9\times10^7}=\dfrac{18\times10^{13}}{9\times10^7}=2\times10^6$

57. $\dfrac{\left(5\times10^3\right)\left(4\times10^{-5}\right)}{2\times10^{-2}}=\dfrac{20\times10^{-2}}{2\times10^{-2}}=10=1\times10^1$

59. $\dfrac{\left(2.8\times10^{-7}\right)\left(3.6\times10^{4}\right)}{2.4\times10^{3}} = \dfrac{10.08\times10^{-3}}{2.4\times10^{3}} = 4.2\times10^{-6}$

61. $\dfrac{18x^{4}}{3x} + \dfrac{21x^{7}}{7x^{4}} = 6x^{3} + 3x^{3} = 9x^{3}$

63. $\dfrac{45a^{6}}{9a^{4}} - \dfrac{50a^{8}}{2a^{6}} = 5a^{2} - 25a^{2} = -20a^{2}$

65. $\dfrac{6x^{7}y^{4}}{3x^{2}y^{2}} + \dfrac{8x^{5}y^{8}}{2y^{6}} = 2x^{5}y^{2} + 4x^{5}y^{2} = 6x^{5}y^{2}$

67. $4^{x}\cdot4^{5} = 4^{7}$

$4^{x+5} = 4^{7}$

$x + 5 = 7$

$x = 2$

69. $\left(7^{3}\right)^{x} = 7^{12}$

$7^{3x} = 7^{12}$

$3x = 12$

$x = 4$

71. $\left(a+b\right)^{2} = \left(4+5\right)^{2} = 9^{2} = 81$

$a^{2} + b^{2} = 4^{2} + 5^{2} = 16 + 25 = 41$

Note that the values are not equal.

73. $\left(a+b\right)^{2} = \left(3+4\right)^{2} = 7^{2} = 49$

$a^{2} + 2ab + b^{2} = 3^{2} + 2\left(3\right)\left(4\right) + 4^{2} = 9 + 24 + 16 = 49$

Note that the values are equal.

75. Let $x = $ width

Let $2x = $ length

$P = 2\left(x\right) + 2\left(2x\right) = 2x + 4x = 6x$

$A = x\cdot2x = 2x^{2}$

77. Let $x = $ width

Let $2x = $ length

$V = \left(x\right)\left(2x\right)\left(4\right) = 8x^{2}$ inches^{3}

79. (a) $V = \left(8.5\right)\left(55\right)\left(10.1\right) = 4,700$ feet$^{3} = 4.7\times10^{3}$ feet^{3}

(b) $n = \dfrac{4700}{0.15} = 31,000$ boxes $= 3.1\times10^{4}$ boxes

81. $y = x^{2}$ $\left(-4,16\right), \left(-2,4\right), \left(-1,1\right), \left(0,0\right), \left(1,1\right), \left(2,4\right), \left(4,16\right)$

See the graph in the back of the textbook.

83. $y = 2x^{2}$ $\left(-3,18\right), \left(-2,8\right), \left(-1,2\right), \left(0,0\right), \left(1,2\right), \left(2,8\right), \left(3,18\right)$

See the graph in the back of the textbook.

85. $y = \dfrac{1}{2}x^{2}$ $\left(-4,8\right), \left(-2,2\right), \left(-1,1/2\right), \left(0,0\right), \left(1,1/2\right), \left(2,2\right), \left(4,8\right)$

See the graph in the back of the textbook.

87. $y = \dfrac{1}{4}x^{2}$ $\left(-4,4\right), \left(-2,1\right), \left(-1,1/4\right), \left(0,0\right), \left(1,1/4\right), \left(2,1\right), \left(4,4\right)$

See the graph in the back of the textbook.

89. Evaluating when $x = -2$: $-2x + 5 = -2(-2) + 5 = 4 + 5 = 9$

91. Evaluating when $x = -2$: $x^2 + 5x + 6 = (-2)^2 + 5(-2) + 6 = 4 - 10 + 6 = 0$

93. The ordered pairs are $(-2, -2)$, $(0, 2)$, and $(2, 6)$. See the graph in the back of the textbook.

95. The ordered pairs are $(-3, 0)$, $(0, 1)$, and $(3, 2)$. See the graph in the back of the textbook.

4.4 Addition and Subtraction of Polynomials

1. This is a trinomial of degree 3.
3. This is a trinomial of degree 3.
5. This is a trinomial of degree 1.
7. This is a binomial of degree 2.
9. This is a monomial of degree 2.
11. This is a monomial of degree 0.

13. $\left(2x^2 + 3x + 4\right) + \left(3x^2 + 2x + 5\right) = \left(2x^2 + 3x^2\right) + \left(3x + 2x\right) + \left(4 + 5\right) = 5x^2 + 5x + 9$

15. $\left(3a^2 - 4a + 1\right) + \left(2a^2 - 5a + 6\right) = \left(3a^2 + 2a^2\right) + \left(-4a - 5a\right) + \left(1 + 6\right) = 5a^2 - 9a + 7$

17. $x^2 + 4x + 2x + 8 = x^2 + \left(4x + 2x\right) + 8 = x^2 + 6x + 8$

19. $6x^2 - 3x - 10x + 5 = 6x^2 + \left(-3x - 10x\right) + 5 = 6x^2 - 13x + 5$

21. $x^2 - 3x + 3x - 9 = x^2 + \left(-3x + 3x\right) - 9 = x^2 - 9$

23. $3y^2 - 5y - 6y + 10 = 3y^2 + \left(-5y - 6y\right) + 10 = 3y^2 - 11y + 10$

25. $\left(6x^3 - 4x^2 + 2x\right) + \left(9x^2 - 6x + 3\right) = 6x^3 + \left(-4x^2 + 9x^2\right) + \left(2x - 6x\right) + 3 = 6x^3 + 5x^2 - 4x + 3$

27. $\left(\frac{2}{3}x^2 - \frac{1}{5}x - \frac{3}{4}\right) + \left(\frac{4}{3}x^2 - \frac{4}{5}x + \frac{7}{4}\right) = \left(\frac{2}{3}x^2 + \frac{4}{3}x^2\right) + \left(-\frac{1}{5}x - \frac{4}{5}x\right) + \left(-\frac{3}{4} + \frac{7}{4}\right) = 2x^2 - x + 1$

29. $\left(a^2 - a - 1\right) - \left(-a^2 + a + 1\right) = a^2 - a - 1 + a^2 - a - 1 = 2a^2 - 2a - 2$

31. $\left(\frac{5}{9}x^3 + \frac{1}{3}x^2 - 2x + 1\right) - \left(\frac{2}{3}x^3 + x^2 + \frac{1}{2}x - \frac{3}{4}\right) = \frac{5}{9}x^3 + \frac{1}{3}x^2 - 2x + 1 - \frac{2}{3}x^3 - x^2 - \frac{1}{2}x + \frac{3}{4}$

$$= \left(\frac{5}{9} - \frac{2}{3}\right)x^3 + \left(\frac{1}{3} - 1\right)x^2 - \left(2 + \frac{1}{2}\right)x + 1 + \frac{3}{4}$$

$$= -\frac{1}{9}x^3 - \frac{2}{3}x^2 - \frac{5}{2}x + \frac{7}{4}$$

33. $\left(4y^2 - 3y + 2\right) + \left(5y^2 + 12y - 4\right) - \left(13y^2 - 6y + 20\right)$

$\quad = 4y^2 - 3y + 2 + 5y^2 + 12y - 4 - 13y^2 + 6y - 20$

$\quad = \left(4y^2 + 5y^2 - 13y^2\right) + \left(-3y + 12y + 6y\right) + \left(2 - 4 - 20\right)$

$\quad = -4y^2 + 15y - 22$

35. $\left(11x^2 - 10x + 13\right) - \left(10x^2 + 23x - 50\right) = 11x^2 - 10x + 13 - 10x^2 - 23x + 50$

$$= \left(11x^2 - 10x^2\right) + \left(-10x - 23x\right) + \left(13 + 50\right)$$
$$= x^2 - 33x + 63$$

37. $\left(11y^2 + 11y + 11\right) - \left(3y^2 + 7y - 15\right) = 11y^2 + 11y + 11 - 3y^2 - 7y + 15$

$$= \left(11y^2 - 3y^2\right) + \left(11y - 7y\right) + \left(11 + 15\right)$$
$$= 8y^2 + 4y + 26$$

39. $\left(25x^2 - 50x + 75\right) + \left(50x^2 - 100x - 150\right) = \left(25x^2 + 50x^2\right) + \left(-50x - 100x\right) + \left(75 - 150\right) = 75x^2 - 150x - 75$

41. $\left(3x - 2\right) + \left(11x + 5\right) - \left(2x + 1\right) = 3x - 2 + 11x + 5 - 2x - 1 = \left(3x + 11x - 2x\right) + \left(-2 + 5 - 1\right) = 12x + 2$

43. When $x = 3$: $x^2 - 2x + 1 = (3)^2 - 2(3) + 1 = 9 - 6 + 1 = 4$

45. When $y = 10$: $(y - 5)^2 = (10 - 5)^2 = (5)^2 = 25$

47. When $a = 2$: $a^2 + 4a + 4 = (2)^2 + 4(2) + 4 = 4 + 8 + 4 = 16$

49. $V_x = \dfrac{4}{3}\pi r^3 = \dfrac{4}{3}\pi(3)^3 = 36\pi$

$V_c = \pi r^2 h = \pi(3)^2(6) = 54\pi$

Volume filled with padding $= V_c - V_s$

$V_c - V_s = 54\pi - 36\pi = 18\pi$

51. $3x(-5x) = -15x^2$

53. $2x(3x^2) = 6x^{1+2} = 6x^3$

55. $3x^2(2x^2) = 6x^{2+2} = 6x^4$

4.5 Multiplication with Polynomials

1. $2x(3x+1) = 2x(3x)+2x(1) = 6x^2+2x$

3. $2x^2(3x^2-2x+1) = 2x^2(3x^2)+2x^2(-2x)+2x^2(1) = 6x^4-4x^3+2x^2$

5. $2ab(a^2-ab+1) = 2ab(a^2)-2ab(ab)+2ab(1) = 2a^3b-2a^2b^2+2ab$

7. $y^2(3y^2+9y+12) = 3y^4+9y^3+12y^2$

9. $4x^2y(2x^3y+3x^2y^2+8y^3) = 4x^2y(2x^3y)+4x^2y(3x^2y^2)+4x^2y(8y^3) = 8x^5y^2+12x^4y^3+32x^2y^4$

11. Foil Method: $(x+3)(x+4) = x^2+3x+4x+12 = x^2+7x+12$

13. Foil Method: $(x+6)(x+1) = x^2+6x+1x+6 = x^2+7x+6$

15. Foil Method: $\left(x+\dfrac{1}{2}\right)\left(x+\dfrac{3}{2}\right) = x^2+\dfrac{1}{2}x+\dfrac{3}{2}x+\dfrac{3}{4} = x^2+2x+\dfrac{3}{4}$

17. Foil Method: $(a+5)(a-3) = a^2-3a+5a-15 = a^2+2a-15$

19. Foil Method: $(x-a)(y+b) = xy-ay+bx-ab$

21. Foil Method: $(x+6)(x-6) = x^2+6x-6x-36 = x^2-36$

23. Foil Method: $\left(y+\dfrac{5}{6}\right)\left(y-\dfrac{5}{6}\right) = y^2+\dfrac{5}{6}y-\dfrac{5}{6}y-\dfrac{25}{36} = y^2-\dfrac{25}{36}$

25. Foil Method: $(2x-3)(x-4) = 2x^2-3x-8x+12 = 2x^2-11x+12$

27. Foil Method: $(a+2)(2a-1) = 2a^2+4a-a-2 = 2a^2+3a-2$

29. Foil Method: $(2x-5)(3x-2) = 6x^2-15x-4x+10 = 6x^2-19x+10$

31. Foil Method: $(2x+3)(a+4) = 2ax+3a+8x+12$

33. Foil Method: $(5x-4)(5x+4) = 25x^2-20x+20x-16 = 25x^2-16$

35. Foil Method: $\left(2x-\dfrac{1}{2}\right)\left(x+\dfrac{3}{2}\right) = 2x^2-\dfrac{1}{2}x+3x-\dfrac{3}{4} = 2x^2+\dfrac{5}{2}x-\dfrac{3}{4}$

37. Foil Method: $(1-2a)(3-4a) = 3-6a-4a+8a^2 = 3-10a+8a^2$

39. See the rectangle in the back of the textbook.
41. See the rectangle in the back of the textbook.

43.
$$
\begin{array}{r}
a^2-3a+2 \\
a-3 \\
\hline
a^3-3a^2+2a \\
-3a^2+9a-6 \\
\hline
a^3-6a^2+11a-6
\end{array}
$$

45.
$$
\begin{array}{r}
x^2-2x+4 \\
x+2 \\
\hline
x^3-2x^2+4x \\
2x^2-4x+8 \\
\hline
x^3 \qquad\qquad +8
\end{array}
$$

47.
$$
\begin{array}{r}
x^2+8x+9 \\
2x+1 \\
\hline
2x^3+16x^2+18x \\
x^2+8x+9 \\
\hline
2x^3+17x^2+26x+9
\end{array}
$$

49.
$$
\begin{array}{r}
5x^2+2x+1 \\
x^2-3x+5 \\
\hline
5x^4+2x^3+x^2 \\
-15x^3-6x^2-3x \\
25x^2+10x+5 \\
\hline
5x^4-13x^3+20x^2+7x+5
\end{array}
$$

51. $\left(x^2+3\right)\left(2x^2-5\right)=2x^4-5x^2+6x^2-15=2x^4+x^2-15$

53. $\left(3a^4+2\right)\left(2a^2+5\right)=6a^6+15a^4+4a^2+10$

55. $(x+3)(x+4)=x^2+3x+4x+12=x^2+7x+12$

$x^2+7x+12$

$\underline{\qquad x+5\qquad}$

x^3+7x^2+12x

$\underline{\qquad 5x^2+35x+60}$

$x^3+12x^2+47x+60$

57. Let $x=$ width

Let $2x+5=$ length

$A=x\left(2x+5\right)=2x^2+5x$

59. Let x and $x+1=$ width and length, respectively.

$A=x\left(x+1\right)=x^2+x$

61. $x=1200-100p$

$R=xp=\left(1200-100p\right)p$

$R=1200p-100p^2$

63. $x=1700-100p$

$R=xp=\left(1700-100p\right)p$

$R=1700p-100p^2$

65. The intersection point is $(3,1)$.

See the graph in the back of the textbook.

67. Multiply the first equation by 3 and the

second equation by -2:

$6x+9y=-3$

$-6x-10y=4$

Adding the two equations:

$-y=1$

$y=-1$

Substituting into the first equation:

$2x+3(-1)=-1$

$2x-3=-1$

$2x=2$

$x=1$

The solution is $(1,-1)$.

69. Substituting $y=3x+1$ from the second equation

into the first equation:

$2x-6\left(3x+1\right)=2$

$2x-18x-6=2$

$-16x-6=2$

$-16x=8$

$x=-\dfrac{1}{2}$

Substituting into the second equation:

$y=3\left(-\dfrac{1}{2}\right)+1=-\dfrac{3}{2}+1=-\dfrac{1}{2}$

The solution is $\left(-\dfrac{1}{2},-\dfrac{1}{2}\right)$.

71. Let $x =$ the number of dimes

   Let $y =$ the number of quarters

   $$x + y = 11$$

   $$0.10x + 0.25y = 1.85$$

   Multiplying the first equation by -0.10:

71. Continued

   $$-0.10x - 0.10y = -1.10$$

   $$0.10x + 0.25y = 1.85$$

   Adding the two equations:

   $$0.15y = 0.75$$

   $$y = 5$$

   Substituting into the first equation:

   $$x + 5 = 11$$

   $$x = 6$$

   Amy has 6 dimes and 5 quarters.

4.6 Binomial Squares and Other Special Products

1. Foil Method: $(x-2)^2 = (x-2)(x-2) = x^2 - 2x - 2x + 4 = x^2 - 4x + 4$

3. Foil Method: $(a+3)^2 = (a+3)(a+3) = a^2 + 3a + 3a + 9 = a^2 + 6a + 9$

5. Foil Method: $(x-5)^2 = (x-5)(x-5) = x^2 - 5x - 5x + 25 = x^2 - 10x + 25$

7. Foil Method: $\left(a - \dfrac{1}{2}\right)^2 = \left(a - \dfrac{1}{2}\right)\left(a - \dfrac{1}{2}\right) = a^2 - \dfrac{1}{2}a - \dfrac{1}{2}a + \dfrac{1}{4} = a^2 - a + \dfrac{1}{4}$

9. Foil Method: $(x+10)^2 = (x+10)(x+10) = x^2 + 10x + 10x + 100 = x^2 + 20x + 100$

11. $(a+0.8)^2 = a^2 + 2(a)(0.8) + (0.8)^2 = a^2 + 1.6a + 0.64$

13. $(2x-1)^2 = (2x)^2 - 2(2x)(1) + (1)^2 = 4x^2 - 4x + 1$

15. $(4a+5)^2 = (4a)^2 + 2(4a)(5) + (5)^2 = 16a^2 + 40a + 25$

17. $(3x-2)^2 = 9x^2 + 2(3x)(-2) + 4 = 9x^2 - 12x + 4$

19. $(3a+5b)^2 = (3a)^2 + 2(3a)(5b) + (5b)^2 = 9a^2 + 30ab + 25b^2$

21. $(4x-5y)^2 = 16x^2 + 2(4x)(-5y) + 25y^2 = 16x^2 - 40xy + 25y^2$

23. $(7m+2n)^2 = (7m)^2 + 2(7m)(2n) + (2n)^2 = 49m^2 + 28mn + 4n^2$

25. $(6x-10y)^2 = 36x^2 + 2(6x)(-10y) + 100y^2 = 36x^2 - 120xy + 100y^2$

27. $(x^2+5)^2 = (x^2)^2 + 2(x^2)(5) + (5)^2 = x^4 + 10x^2 + 25$

29. $(a^2+1)^2 = a^4 + 2(a^2)(1) + 1^2 = a^4 + 2a^2 + 1$

31. See the table in the back of the textbook.

33. See the table in the back of the textbook.

35. $(a+5)(a-5) = a^2 + 5a - 5a - 25 = a^2 - 25$

37. $(y-1)(y+1) = y^2 - y + y - 1 = y^2 - 1$

39. $(9+x)(9-x) = (9)^2 - (x)^2 = 81 - x^2$

41. $(2x+5)(2x-5) = (2x)^2 - (5)^2 = 4x^2 - 25$

43. $\left(4x + \dfrac{1}{3}\right)\left(4x - \dfrac{1}{3}\right) = (4x)^2 - \left(\dfrac{1}{3}\right)^2 = 16x^2 - \dfrac{1}{9}$

45. $(2a+7)(2a-7) = (2a)^2 - (7)^2 = 4a^2 - 49$

47. $(6-7x)(6+7x) = (6)^2 - (7x)^2 = 36 - 49x^2$

49. $(x^2+3)(x^2-3) = (x^2)^2 - (3)^2 = x^4 - 9$

51. $(a^2+4)(a^2-4) = (a^2)^2 - (4)^2 = a^4 - 16$

53. $(5y^4-8)(5y^4+8) = (5y^4)^2 - (8)^2 = 25y^8 - 64$

55. $(x+3)(x-3)+(x+5)(x-5)=(x^2-9)+(x^2-25)=2x^2-34$

57. $(2x+3)^2-(4x-1)^2=(4x^2+12x+9)-(16x^2-8x+1)=4x^2+12x+9-16x^2+8x-1=-12x^2+20x+8$

59. $(a+1)^2-(a+2)^2+(a+3)^2=(a^2+2a+1)-(a^2+4a+4)+(a^2+6a+9)$

$$=a^2+2a+1-a^2-4a-4+a^2+6a+9$$

$$=a^2+4a+6$$

61. $(2x+3)^3=(2x+3)(2x+3)^2$

$$=(2x+3)(4x^2+12x+9)$$

$$=8x^3+24x^2+18x+12x^2+36x+27$$

$$=8x^3+36x^2+54x+27$$

63. $49(51)=(50-1)(50+1)=(50)^2-(1)^2=2,500-1=2,499$

65. $(x+3)^2=(2+3)^2=(5)^2=25$

$x^2+6x+9=(2)^2+6(2)+9=4+12+9=25$

67. Let x and $x+1=$ the two consecutive integers.

$(x)^2+(x+1)^2=x^2+(x^2+2x+1)=2x^2+2x+1$

69. Let x, $x+1$ and $x+2=$ the three consecutive integers.

$(x)^2+(x+1)^2+(x+2)^2=x^2+(x^2+2x+1)+(x^2+4x+4)=3x^2+6x+5$

71. Verifying the areas: $(a+b)^2=a^2+ab+ab+b^2=a^2+2ab+b^2$

73. See the rectangle in the back of the textbook.

75. $\dfrac{15x^2y}{3xy}=\dfrac{15}{3}\cdot\dfrac{x^2}{x}\cdot\dfrac{y}{y}=5x$

77. $\dfrac{35a^6b^8}{70a^2b^{10}}=\dfrac{35}{70}\cdot\dfrac{a^6}{a^2}\cdot\dfrac{b^8}{b^{10}}=\dfrac{1}{2}\cdot a^4\cdot\dfrac{1}{b^2}=\dfrac{a^4}{2b^2}$

79. The intersection point is $(3,-1)$.

See the graph in the back of the textbook.

81. The intersection point is $(-1,1)$.

See the graph in the back of the textbook.

4.7 Dividing a Polynomial by a Monomial

1. $\dfrac{5x^2 - 10x}{5x} = \dfrac{5x^2}{5x} - \dfrac{10x}{5x} = x - 2$

3. $\dfrac{15x - 10x^3}{5x} = \dfrac{15x}{5x} - \dfrac{10x^3}{5x} = 3 - 2x^2$

5. $\dfrac{25x^2 y - 10xy}{5x} = \dfrac{25x^2 y}{5x} - \dfrac{10xy}{5x} = 5xy - 2y$

7. $\dfrac{35x^5 - 30x^4 + 25x^3}{5x} = \dfrac{35x^5}{5x} - \dfrac{30x^4}{5x} + \dfrac{25x^3}{5x} = 7x^4 - 6x^3 + 5x^2$

9. $\dfrac{50x^5 - 25x^3 + 5x}{5x} = \dfrac{50x^5}{5x} - \dfrac{25x^3}{5x} + \dfrac{5x}{5x} = 10x^4 - 5x^2 + 1$

11. $\dfrac{8a^2 - 4a}{-2a} = \dfrac{8a^2}{-2a} + \dfrac{-4a}{-2a} = -4a + 2$

13. $\dfrac{16a^5 + 24a^4}{-2a} = \dfrac{16a^5}{-2a} + \dfrac{24a^4}{-2a} = -8a^4 - 12a^3$

15. $\dfrac{8ab + 10a^2}{-2a} = \dfrac{8ab}{-2a} + \dfrac{10a^2}{-2a} = -4b - 5a$

17. $\dfrac{12a^3 b - 6a^2 b^2 + 14ab^3}{-2a} = \dfrac{12a^3 b}{-2a} - \dfrac{6a^2 b^2}{-2a} + \dfrac{14ab^3}{-2a} = -6a^2 b + 3ab^2 - 7b^3$

19. $\dfrac{a^2 + 2ab + b^2}{-2a} = \dfrac{a^2}{-2a} + \dfrac{2ab}{-2a} + \dfrac{b^2}{-2a} = -\dfrac{a}{2} - b - \dfrac{b^2}{2a}$

21. $\dfrac{6x + 8y}{2} = \dfrac{6x}{2} + \dfrac{8y}{2} = 3x + 4y$

23. $\dfrac{7y - 21}{-7} = \dfrac{7y}{-7} + \dfrac{-21}{-7} = -y + 3$

25. $\dfrac{10xy - 8x}{2x} = \dfrac{10xy}{2x} - \dfrac{8x}{2x} = 5y - 4$

27. $\dfrac{x^2 y - x^3 y^2}{x} = \dfrac{x^2 y}{x} - \dfrac{x^3 y^2}{x} = xy - x^2 y^2$

29. $\dfrac{x^2 y - x^3 y^2}{-x^2 y} = \dfrac{x^2 y}{-x^2 y} - \dfrac{x^3 y^2}{-x^2 y} = -1 + xy$

31. $\dfrac{a^2 b^2 - ab^2}{-ab^2} = \dfrac{a^2 b^2}{-ab^2} + \dfrac{-ab^2}{-ab^2} = -a + 1$

33. $\dfrac{x^3 - 3x^2 y + xy^2}{x} = \dfrac{x^3}{x} - \dfrac{3x^2 y}{x} + \dfrac{xy^2}{x} = x^2 - 3xy + y^2$

35. $\dfrac{10a^2 - 15a^2 b + 25a^2 b^2}{5a^2} = \dfrac{10a^2}{5a^2} - \dfrac{15a^2 b}{5a^2} + \dfrac{25a^2 b}{5a^2} = 2 - 3b + 5b^2$

37. $\dfrac{26x^2 y^2 - 13xy}{-13xy} = \dfrac{26x^2 y^2}{-13xy} - \dfrac{13xy}{-13xy} = -2xy + 1$

39. $\dfrac{4x^2 y^2 - 2xy}{4xy} = \dfrac{4x^2 y^2}{4xy} - \dfrac{2xy}{4xy} = xy - \dfrac{1}{2}$

41. $\dfrac{5a^2 x - 10ax^2 + 15a^2 x^2}{20a^2 x^2} = \dfrac{5a^2 x}{20a^2 x^2} - \dfrac{10ax^2}{20a^2 x^2} + \dfrac{15a^2 x^2}{20a^2 x^2} = \dfrac{1}{4x} - \dfrac{1}{2a} + \dfrac{3}{4}$

43. $\dfrac{16x^5 + 8x^2 + 12x}{12x^3} = \dfrac{16x^5}{12x^3} + \dfrac{8x^2}{12x^3} + \dfrac{12x}{12x^3} = \dfrac{4x^2}{3} + \dfrac{2}{3x} + \dfrac{1}{x^2}$

45. $\dfrac{9a^{5m} - 27a^{3m}}{3a^{2m}} = \dfrac{9a^{5m}}{3a^{2m}} - \dfrac{27a^{3m}}{3a^{2m}} = 3a^{5m-2m} - 9a^{3m-2m} = 3a^{3m} - 9a^m$

47. $\dfrac{10x^{5m} - 25x^{3m} + 35x^m}{5x^m} = \dfrac{10x^{5m}}{5x^m} - \dfrac{25x^{3m}}{5x^m} + \dfrac{35x^m}{5x^m} = 2x^{5m-m} - 5x^{3m-m} + 7x^{m-m} = 2x^{4m} - 5x^{2m} + 7$

49. $\dfrac{2x^3(3x+2) - 3x^2(2x-4)}{2x^2} = \dfrac{6x^4 + 4x^3 - 6x^3 + 12x^2}{2x^2}$

$$= \dfrac{6x^4}{2x^2} + \dfrac{4x^3}{2x^2} - \dfrac{6x^3}{2x^2} + \dfrac{12x^2}{2x^2}$$

$$= 3x^2 + 2x - 3x + 6$$

$$= 3x^2 - x + 6$$

51. $\dfrac{(x+2)^2 - (x-2)^2}{2x} = \dfrac{(x^2 + 4x + 4) - (x^2 - 4x + 4)}{2x} = \dfrac{x^2 + 4x + 4 - x^2 + 4x - 4}{2x} = \dfrac{8x}{2x} = 4$

53. $\dfrac{(x+5)^2 + (x+5)(x-5)}{2x} = \dfrac{x^2 + 10x + 25 + x^2 - 25}{2x} = \dfrac{2x^2 + 10x}{2x} = \dfrac{2x^2}{2x} + \dfrac{10x}{2x} = x + 5$

55. When $x = 2$:

$\dfrac{10x + 15}{5} = \dfrac{10(2) + 15}{5} = \dfrac{20 + 15}{5} = \dfrac{35}{5} = 7$

$2x + 3 = 2(2) + 3 = 4 + 3 = 7$

57. When $x = 10$:

$\dfrac{3x + 8}{2} = \dfrac{3(10) + 8}{2} = \dfrac{30 + 8}{2} = \dfrac{38}{2} = 19$

$3x + 4 = 3(10) + 4 = 30 + 4 = 34$

Thus, $\dfrac{3x + 8}{2} \neq 3x + 4$.

59. Adding the two equations:

$$2x = 14$$
$$x = 7$$

Substituting into the first equation:

$$7 + y = 6$$
$$y = -1$$

The solution is $(7, -1)$.

61. Multiplying the second equation by 3:

$$2x - 3y = -5$$
$$3x + 3y = 15$$

Adding the two equations:

$$5x = 10$$
$$x = 2$$

Substituting into the second equation:

$$2 + y = 5$$
$$y = 3$$

The solution is $(2, 3)$.

63. Substituting y from the second equation into the first equation:

$$x + 2x - 1 = 2$$
$$3x - 1 = 2$$
$$3x = 3$$
$$x = 1$$

Substituting into the second equation:

$$y = 2(1) - 1 = 2 - 1 = 1$$

The solution is $(1, 1)$.

65. Substituting y from the second equation into the first equation:

$$4x + 2(-2x + 4) = 8$$
$$4x - 4x + 8 = 8$$
$$8 = 8$$

Since this statement is true, the system is dependent. The two lines coincide.

4.8 Dividing a Polynomial by a Polynomial

1.
$$\require{enclose}
\begin{array}{r}
x - 2 \\
x - 3 \enclose{longdiv}{x^2 - 5x + 6} \\
\underline{x^2 - 3x} \\
-2x + 6 \\
\underline{-2x + 6} \\
0
\end{array}$$

The answer is $x - 2$

3.
$$\begin{array}{r}
a + 4 \\
a + 5 \enclose{longdiv}{a^2 + 9a + 20} \\
\underline{a^2 + 5a} \\
4a + 20 \\
\underline{4a + 20} \\
0
\end{array}$$

The answer is $a + 4$

5.
$$\begin{array}{r}
x - 3 \\
x - 3 \enclose{longdiv}{x^2 - 6x + 9} \\
\underline{x^2 - 3x} \\
-3x + 9 \\
\underline{-3x + 9} \\
0
\end{array}$$

The answer is $x - 3$

7.
$$\begin{array}{r}
x + 3 \\
2x - 1 \enclose{longdiv}{2x^2 + 5x - 3} \\
\underline{2x^2 - x} \\
6x - 3 \\
\underline{6x - 3} \\
0
\end{array}$$

The answer is $x + 3$

9.
$$\begin{array}{r} a-5 \\ 2a+1{\overline{\smash{\big)}\,2a^2-9a-5}} \\ \underline{2a^2+a} \\ -10a-5 \\ \underline{-10a-5} \\ 0 \end{array}$$

The answer is $a-5$

11.
$$\begin{array}{r} x+2 \\ x+3{\overline{\smash{\big)}\,x^2+5x+8}} \\ \underline{x^2+3x} \\ 2x+8 \\ \underline{2x+6} \\ 2 \end{array}$$

The answer is $x+2+\dfrac{2}{x+3}$

13.
$$\begin{array}{r} a-2 \\ a+5{\overline{\smash{\big)}\,a^2+3a+2}} \\ \underline{a^2+5a} \\ -2a+2 \\ \underline{-2a-10} \\ 12 \end{array}$$

The answer is $a-2+\dfrac{12}{a+5}$

15.
$$\begin{array}{r} x+4 \\ x-2{\overline{\smash{\big)}\,x^2+2x+1}} \\ \underline{x^2-2x} \\ 4x+1 \\ \underline{4x-8} \\ 9 \end{array}$$

The answer is $x+4+\dfrac{9}{x-2}$

17.
$$\begin{array}{r} x+4 \\ x+1{\overline{\smash{\big)}\,x^2+5x-6}} \\ \underline{x^2+x} \\ 4x-6 \\ \underline{4x+4} \\ -10 \end{array}$$

The answer is $x+4-\dfrac{10}{x+1}$

19.
$$\begin{array}{r} a+1 \\ a+2{\overline{\smash{\big)}\,a^2+3a+1}} \\ \underline{a^2+2a} \\ a+1 \\ \underline{a+2} \\ -1 \end{array}$$

The answer is $a+1+\dfrac{-1}{a+2}$

21.
$$\begin{array}{r} x-3 \\ 2x+4{\overline{\smash{\big)}\,2x^2-2x+5}} \\ \underline{2x^2+4x} \\ -6x+5 \\ \underline{-6x-12} \\ 17 \end{array}$$

The answer is $x-3+\dfrac{17}{2x+4}$

23.
$$\begin{array}{r} 3a-2 \\ 2a+3{\overline{\smash{\big)}\,6a^2+5a+1}} \\ \underline{6a^2+9a} \\ -4a+1 \\ \underline{-4a-6} \\ 7 \end{array}$$

The answer is $3a-2+\dfrac{7}{2a+3}$

25.
$$3a-5 \overline{)6a^3 - 13a^2 - 4a + 15} \quad \frac{2a^2 - a - 3}{}$$
$$\underline{6a^3 - 10a^2}$$
$$-3a^2 - 4a$$
$$\underline{-3a^2 + 5a}$$
$$-9a + 15$$
$$\underline{-9a + 15}$$
$$0$$

The answer is $2a^2 - a - 3$

27.
$$x+1 \overline{)x^3 + 0x^2 + 4x + 5} \quad \frac{x^2 - x + 5}{}$$
$$\underline{x^3 + x^2}$$
$$-x^2 + 4x + 5$$
$$\underline{-x^2 - x}$$
$$5x + 5$$
$$\underline{5x + 5}$$
$$0$$

The answer is $x^2 - x + 5$

29.
$$x-1 \overline{)x^3 + 0x^2 + 0x - 1} \quad \frac{x^2 + x + 1}{}$$
$$\underline{x^3 - x^2}$$
$$x^2 + 0x$$
$$\underline{x^2 - x}$$
$$x - 1$$
$$\underline{x - 1}$$
$$0$$

The answer is $x^2 + x + 1$

31.
$$x-2 \overline{)x^3 + 0x^2 + 0x - 8} \quad \frac{x^2 + 2x + 4}{}$$
$$\underline{x^3 - 2x^2}$$
$$2x^2 + 0x$$
$$\underline{2x^2 - 4x}$$
$$4x - 8$$
$$\underline{4x - 8}$$
$$0$$

The answer is $x^2 + 2x + 4$

33. Let x and $y =$ the two numbers
$$x + y = 25$$
$$y = 4x$$
Substituting into the first equation:
$$x + 4x = 25$$
$$5x = 25$$
$$x = 5$$
$$y = 4(5) = 20$$
The two numbers are 5 and 20.

35. Let $x =$ amount invested at 8%
Let $y =$ amount invested at 9%
$$x + y = 1200 \qquad (1)$$
$$0.08x + 0.09y = 100 \qquad (2)$$
Solve (1) for x:
$$x = 1200 - y \qquad (3)$$
Substituting (3) into (2)
$$0.08(1200 - y) + 0.09y = 100$$
$$96 - 0.08y + 0.09y = 100$$
$$96 + 0.01y = 100$$
$$0.01y = 4$$
$$y = 400$$
Substitute $y = 400$ into (3)
$$x = 1200 - 400 = 800$$
You invested $800 at 8% and $400 at 9%.

37. Let x = number of $5 bills
 Let y = number of $10 bills
$$y = x + 4 \quad (1)$$
$$5x + 10y = 160 \quad (2)$$
 Substitute (1) into (2)
$$5x + 10(x + 4) = 160$$
$$5x + 10x + 40 = 160$$
$$15x + 40 = 160$$
$$15x = 120$$
$$x = 8$$
 Substitute $x = 8$ into (1)
$$y = 8 + 4 = 12$$
 You have 8 $5 bills and 12 $10 bills.

39. Let x = gallons of 20% solution
 Let y = gallons of 60% solution
$$x + y = 16 \quad (1)$$
$$0.20x + 0.60y = 16(0.35) \quad (2)$$
 Solve (1) for x :
$$x = 16 - y \quad (3)$$
 Substituting (3) into (2)
$$0.20(16 - y) + 0.60y = 5.6$$
$$3.2 - 0.20y + 0.60y = 5.6$$
$$3.2 + 0.4y = 5.6$$
$$0.4y = 2.4$$
$$y = 6$$
 Substitute $y = 6$ into (3)
$$x = 16 - 6 = 10$$
 10 gallons of 20% antifreeze solution must be mixed with 6 gallons of 60% antifreeze solution.

Chapter 4 Review

1. $(-1)^3 = (-1)(-1)(-1) = -1$

3. $\left(\frac{3}{7}\right)^2 = \left(\frac{3}{7}\right)\left(\frac{3}{7}\right) = \frac{9}{49}$

5. $x^{15} \cdot x^7 \cdot x^5 \cdot x^3 = x^{15+7+5+3} = x^{30}$

7. $(2^6)^4 = 2^{6 \cdot 4} = 2^{24}$

9. $(-2xyz)^3 = (-2)^3 x^3 y^3 z^3 = -8x^3 y^3 z^3$

11. $4x^{-5} = 4\left(\frac{1}{x^5}\right) = \frac{4}{x^5}$

13. $\dfrac{a^9}{a^3} = a^{9-3} = a^6$

15. $\dfrac{x^9}{x^{-6}} = x^{9-(-6)} = x^{15}$

17. $(-3xy)^0 = 1$

19. $\left(3x^3 y^2\right)^2 = 3^2 \left(x^3\right)^2 \left(y^2\right)^2 = 9x^{3(2)} y^{2(2)} = 9x^6 y^4$

21. $(-3xy^2)^{-3} = \dfrac{1}{\left(-3xy^2\right)^3}$
$$= \dfrac{1}{(-3)^3 x^3 \left(y^2\right)^3}$$
$$= \dfrac{1}{-27x^3 y^{2(3)}}$$
$$= \dfrac{-1}{27x^3 y^6}$$

23. $\dfrac{\left(x^{-3}\right)^3 \left(x^6\right)^{-1}}{\left(x^{-5}\right)^{-4}} = \dfrac{x^{-3(3)} x^{6(-1)}}{x^{-5(-4)}}$

$\qquad\qquad = \dfrac{x^{-9} x^{-6}}{x^{20}}$

$\qquad\qquad = \dfrac{x^{-15}}{x^{20}}$

$\qquad\qquad = \dfrac{1}{x^{20} x^{15}}$

$\qquad\qquad = \dfrac{1}{x^{35}}$

25. $\dfrac{\left(10x^3 y^5\right)\left(21x^2 y^6\right)}{\left(7xy^3\right)\left(5x^9 y\right)} = \dfrac{10 \cdot 21 x^3 x^2 y^5 y^6}{7 \cdot 5 xx^9 y^3 y}$

$\qquad\qquad = \dfrac{210 x^{3+2} y^{5+6}}{35 x^{1+9} y^{3+1}}$

$\qquad\qquad = \dfrac{6x^5 y^{11}}{x^{10} y^4}$

$\qquad\qquad = 6x^{5-10} y^{11-4}$

$\qquad\qquad = 6x^{-5} y^7$

$\qquad\qquad = \dfrac{6y^7}{x^5}$

27. $\dfrac{8x^8 y^3}{2x^3 y} - \dfrac{10x^6 y^9}{5xy^7} = 4x^{8-3} y^{3-1} - 2x^{6-1} y^{9-7}$

$\qquad\qquad = 4x^5 y^2 - 2x^5 y^2$

$\qquad\qquad = (4-2)x^5 y^2$

$\qquad\qquad = 2x^5 y^2$

29. $\dfrac{4.6 \times 10^5}{2 \times 10^{-3}} = 2.3 \times 10^{5-(-3)}$

$\qquad\qquad = 2.3 \times 10^8$

31. $\left(3a^2 - 5a + 5\right) + \left(5a^2 - 7a - 8\right) = 3a^2 + 5a^2 - 5a - 7a + 5 - 8$

$\qquad\qquad\qquad = 8a^2 - 12a - 3$

33. $\left(4x^2 - 3x - 2\right) - \left(8x^2 + 3x - 2\right) = 4x^2 - 3x - 2 - 8x^2 - 3x + 2$

$\qquad\qquad\qquad = -4x^2 - 6x$

35. $3x(4x - 7) = 3x(4x) + 3x(-7) = 12x^2 - 21x$

37. $\begin{array}{r} a^2 + 5a - 4 \\ a + 1 \\ \hline a^3 + 5a^2 - 4a \\ a^2 + 5a - 4 \\ \hline a^3 + 6a^2 + a - 4 \end{array}$

39. $(3x - 7)(2x - 5) = 3x(2x) + 3x(-5) + (-7)(2x) + (-7)(-5)$

$\qquad\qquad = 6x^2 - 15x - 14x + 35$

$\qquad\qquad = 6x^2 - 29x + 35$

41. $(a^2 - 3)(a^2 + 3) = (a^2)^2 - (3)^2 = a^4 - 9$

43. $(3x + 4)^2 = (3x)^2 + 2(3x)(4) + 4^2 = 9x^2 + 24x + 16$

45. $\dfrac{10ab}{-5a} + \dfrac{20a^2}{-5a} = -2b - 4a$

47.
$$\begin{array}{r} x+9 \\ x+6\overline{)x^2+15x+54} \\ -\ - \\ \underline{x^2+6x} \\ 9x+54 \\ -\ - \\ \underline{9x+54} \\ 0 \end{array}$$

49.
$$\begin{array}{r} x^2-4x+16 \\ x+4\overline{)x^3+0x^2+0x+64} \\ -\ - \\ \underline{x^3+4x^2} \\ -4x^2+0x \\ +\ \ \ + \\ \underline{-4x^2-16x} \\ 16x+64 \\ -\ \ \ - \\ \underline{16x+64} \\ 0 \end{array}$$

51.
$$\begin{array}{r} x^2-4x+5 \\ 2x+1\overline{)2x^3-7x^2+6x+10} \\ -\ \ - \\ \underline{2x^3+x^2} \\ -8x^2+6x \\ +\ \ \ + \\ \underline{-8x^2-4x} \\ 10x+10 \\ -\ \ \ - \\ \underline{10x+5} \\ 5 \end{array}$$

53. $V = 3x^3,\ x = 2$

$V = 3(2)^3 = 3(8) = 24$ cubic feet.

Yes, the volume is greater than
the volume of your refrigerator.

55. $A = s^2$

$225 = s^2$

$15 = s$

Maximum diameter = 15 feet

Maximum radius = 7 1/2 feet

The largest trampoline is the 6 foot radius

Cumulative Review: Chapters 1-4

1. $-\left(-\dfrac{3}{4}\right) = \dfrac{3}{4}$

3. $6 \cdot 7 + 7 \cdot 9 = 42 + 63 = 105$

5. $6(4a+2) - 3(5a-1) = 24a + 12 - 15a + 3 = 24a - 15a + 12 + 3 = 9a + 15$

7. $-15 - (-3) = -15 + 3 = -12$

9. $(-9)(-5) = 45$

11. $(2y)^4 = 2^4 y^4 = 16y^4$

13. $\dfrac{\left(12xy^5\right)^3 \left(16x^2 y^2\right)}{\left(8x^3 y^3\right)\left(3x^5 y\right)} = \dfrac{1728x^3 y^{15} \cdot 16x^2 y^2}{24x^8 y^4} = \dfrac{27648x^5 y^{17}}{24x^8 y^4} = \dfrac{1152y^{13}}{x^3}$

15. $(5x-1)^2 = (5x)^2 - 2(5x)(1) + (1)^2 = 25x^2 - 10x + 1$

17. $\begin{array}{r} x^2 + x + 1 \\ x - 1 \\ \hline x^3 + x^2 + x \\ -x^2 - x - 1 \\ \hline x^3 - 1 \end{array}$

19. $6a - 5 = 4a$

$-5 = -2a$

$a = \dfrac{5}{2}$

21. $2(3x+5) + 8 = 2x + 10$

$6x + 10 + 8 = 2x + 10$

$6x + 18 = 2x + 10$

$4x + 18 = 10$

$4x = -8$

$x = -2$

23. $-4x > 28$

$\dfrac{-4x}{-4} < \dfrac{28}{-4}$

$x < -7$

See the graph in the back of the textbook.

25. $3(2t-5) - 7 \le 5(3t+1) + 5$

$6t - 15 - 7 \le 15t + 5 + 5$

$6t - 22 \le 15t + 10$

$-9t - 22 \le 10$

$-9t \le 32$

$t \ge -32/9$

See the graph in the back of the textbook.

27. $\dfrac{15x^5 - 10x^2 + 20x}{5x^5} = \dfrac{15x^5}{5x^5} - \dfrac{10x^2}{5x^5} + \dfrac{20x}{5x^5} = 3 - \dfrac{2}{x^3} + \dfrac{4}{x^4}$

29. See the graph in the back of the textbook.

31. The intersection point is $(0,1)$.

See the graph in the back of the textbook.

33. The system is dependent. The two lines coincide.

See the graph in the back of the textbook.

35. Multiplying the first equation by -3:
$$-3x - 6y = -15$$
$$3x + 6y = 14$$
Adding the two equations:
$$0 = -1$$
Since this statement is false, there is no solution to the system. The two lines are parallel.

37. Multiply the first equation by 12 and the second equation by 5:
$$12\left(\frac{1}{6}x + \frac{1}{4}y\right) = 12(1) \qquad 5\left(\frac{6}{5}x - y\right) = 5\left(\frac{8}{5}\right)$$
$$2x + 3y = 12 \qquad\qquad 6x - 5y = 8$$
The system of equation is:
$$2x + 3y = 12$$
$$6x - 5y = 8$$
Multiplying the first equation by -3:
$$-6x - 9y = -36$$
$$6x - 5y = 8$$
Adding the two equations:
$$-14y = -28$$
$$y = 2$$
Substituting into $2x + 3y = 12$:
$$2x + 3(2) = 12$$
$$2x + 6 = 12$$
$$2x = 6$$
$$x = 3$$
The solution is $(3, 2)$.

39. Solving the second equation for x:
$$2y = x - 3$$
$$x = 2y + 3$$
Substituting into the first equation:
$$4(2y + 3) + 5y = 25$$
$$8y + 12 + 5y = 25$$
$$13y + 12 = 25$$
$$13y = 13$$
$$y = 1$$
Substituting into the second equation
$$x = 2(1) + 3 = 5$$
The solution is $(5, 1)$.

41. When $x = 3$: $8x - 3 = 8(3) - 3 = 24 - 3 = 21$

43. The irrational numbers are $-\sqrt{2}$ and π.

45. Substituting each ordered pair:
$$(0, 3): \quad 3(0) - 4(3) = 0 - 12 = -12 \neq 12$$
$$(4, 0): \quad 3(4) - 4(0) = 12 - 0 = 12$$
$$\left(\frac{16}{3}, 1\right): \quad 3\left(\frac{16}{3}\right) - 4(1) = 16 - 4 = 12$$
The ordered pairs $(4, 0)$ and $\left(\frac{16}{3}, 1\right)$ are solutions to the equation.

47. $186,000 = 1.86 \times 10^5$

49. $(2 \times 10^5)(3 \times 10^{-8}) = 6 \times 10^{5-8} = 6 \times 10^{-3}$

Chapter 4 Test

1. $(-3)^4 = (-3)(-3)(-3)(-3) = 81$

2. $\left(\dfrac{3}{4}\right)^2 = \left(\dfrac{3}{4}\right)\left(\dfrac{3}{4}\right) = \dfrac{9}{16}$

3. $\begin{aligned}\left(3x^3\right)^2\left(2x^4\right)^3 &= 3^2 \cdot x^{3 \cdot 2} \cdot 2^3 \cdot x^{4 \cdot 3}\\ &= 9x^6 \cdot 8x^{12}\\ &= 9 \cdot 8x^{6+12}\\ &= 72x^{18}\end{aligned}$

4. $3^{-2} = \dfrac{1}{3^2} = \dfrac{1}{9}$

5. $\left(3a^4b^2\right)^0 = 1$

6. $\begin{aligned}\dfrac{a^{-3}}{a^{-5}} &= a^{-3-(-5)}\\ &= a^{-3+5}\\ &= a^2\end{aligned}$

7. $\begin{aligned}\dfrac{\left(x^{-2}\right)^3\left(x^{-3}\right)^{-5}}{\left(x^{-4}\right)^{-2}} &= \dfrac{x^{-2(3)}x^{-3(-5)}}{x^{-4(-2)}}\\ &= \dfrac{x^{-6}x^{15}}{x^8}\\ &= \dfrac{x^9}{x^8}\\ &= x\end{aligned}$

8. $0.0278 = 2.78 \times 10^{-2}$

9. $2.43 \times 10^5 = 2.43 \times 100,000 = 243,000$

10. $\begin{aligned}\dfrac{35x^2y^4z}{70x^6y^2z} &= \dfrac{35}{70} \cdot \dfrac{x^2}{x^6} \cdot \dfrac{y^4}{y^2} \cdot \dfrac{z}{z}\\ &= \dfrac{1}{2} \cdot \dfrac{1}{x^4} \cdot \dfrac{y^2}{1} \cdot 1\\ &= \dfrac{y^2}{2x^4}\end{aligned}$

11. $\dfrac{(6a^2b)(9a^3b^2)}{18a^4b^3} = \dfrac{54a^5b^3}{18a^4b^3}$

$\qquad = \dfrac{54}{18} \cdot \dfrac{a^5}{a^4} \cdot \dfrac{b^3}{b^3}$

$\qquad = \dfrac{3}{1} \cdot \dfrac{a}{1} \cdot 1$

$\qquad = 3a$

12. $\dfrac{24x^7}{3x^2} + \dfrac{14x^9}{7x^4} = 8x^5 + 2x^5 = 10x^5$

13. $\dfrac{(2.4 \times 10^5)(4.5 \times 10^{-2})}{1.2 \times 10^{-6}} = \dfrac{(2.4)(4.5)}{1.2} \times \dfrac{10^5 \cdot 10^{-2}}{10^{-6}}$

$\qquad\qquad\qquad = 9.0 \times 10^9$

14. $8x^2 - 4x + 6x + 2 = 8x^2 + 2x + 2$

15. $(5x^2 - 3x + 4) - (2x^2 - 7x - 2)$

$\quad = 5x^2 - 3x + 4 - 2x^2 + 7x + 2$

$\quad = (5x^2 - 2x^2) + (-3x + 7x) + (4 + 2)$

$\quad = 3x^2 + 4x + 6$

16. $(6x - 8) - (3x - 4) = 6x - 8 - 3x + 4$

$\qquad\qquad\qquad = (6x - 3x) + (-8 + 4)$

$\qquad\qquad\qquad = 3x - 4$

17. $2y^2 - 3y - 4$ when $y = -2$

$\quad 2(-2)^2 - 3(-2) - 4 = 2 \cdot 4 + 6 - 4$

$\qquad\qquad\qquad\quad = 8 + 6 - 4$

$\qquad\qquad\qquad\quad = 10$

18. $2a^2(3a^2 - 5a + 4) = 2a^2(3a^2) + 2a^2(-5a) + 2a^2(4)$

$\qquad\qquad\qquad = 6a^4 - 10a^3 + 8a^2$

19. $\left(x + \dfrac{1}{2}\right)\left(x + \dfrac{1}{3}\right) = x(x) + x\left(\dfrac{1}{3}\right) + x\left(\dfrac{1}{2}\right) + \left(\dfrac{1}{2}\right)\left(\dfrac{1}{3}\right)$

$\qquad\qquad\qquad = x^2 + \dfrac{1}{3}x + \dfrac{1}{2}x + \dfrac{1}{6}$

$\qquad\qquad\qquad = x^2 + \dfrac{5}{6}x + \dfrac{1}{6}$

20. $(4x - 5)(2x + 3) = 4x(2x) + 4x(3) + (-5)(2x) + (-5)(3)$

$\qquad\qquad\qquad = 8x^2 + 12x - 10x - 15$

$\qquad\qquad\qquad = 8x^2 + 2x - 15$

21. $\begin{array}{r} x^2 + 3x + 9 \\ x - 3 \\ \hline x^3 + 3x^2 + 9x \\ -3x^2 - 9x - 27 \\ \hline x^3 \qquad\quad - 27 \end{array}$

22. $(x + 5)^2 = (x + 5)(x + 5)$

$\qquad\quad = x^2 + 2 \cdot 5x + 5^2$

$\qquad\quad = x^2 + 10x + 25$

23. $(3a - 2b)^2 = (3a)^2 + 2(3a)(-2b) + (-2b)^2$

$\qquad\qquad = 9a^2 - 12ab + 4b^2$

24. $(3x - 4y)(3x + 4y) = (3x)^2 - (4y)^2$

$\qquad\qquad\qquad = 9x^2 - 16y^2$

25. $\left(a^2 - 3\right)\left(a^2 + 3\right) = \left(a^2\right)^2 - 3^2$

$\qquad\qquad\qquad = a^4 - 9$

26. $\dfrac{10x^3 + 15x^2 - 5x}{5x} = \dfrac{10x^3}{5x} + \dfrac{15x^2}{5x} - \dfrac{5x}{5x}$

$\qquad\qquad\qquad = 2x^2 + 3x - 1$

27. $\dfrac{8x^2 - 6x - 5}{2x - 3}$

$$
\begin{array}{r}
4x + 3 \\
2x-3{\overline{\smash{\big)}\,8x^2 - 6x - 5}} \\
-\quad + \\
\underline{8x^2 - 12x} \\
6x - 5 \\
-\quad + \\
\underline{6x - 9} \\
4
\end{array}
$$

The answer is: $4x + 3 + \frac{4}{2x-3}$

28. $\dfrac{3x^3 - 2x + 1}{x - 3}$

$$
\begin{array}{r}
3x^2 + 9x + 25 \\
x-3{\overline{\smash{\big)}\,3x^3 + 0x^2 - 2x + 1}} \\
-\quad + \\
\underline{3x^3 - 9x^2} \\
9x^2 - 2x \\
-\quad + \\
\underline{9x^2 - 27x} \\
25x + 1 \\
-\quad + \\
\underline{25x - 75} \\
76
\end{array}
$$

The answer is: $3x^2 + 9x + 25 + \frac{76}{x-3}$

29. $V = s^3$

$V = (2.5)^3$

$V = (2.5)(2.5)(2.5)$

$V = 15.625 \text{ centimeters}^3$

30. x = width

$5x$ = length

$\dfrac{x}{5}$ = height

$V = W \cdot L \cdot H$

$V = x(5x)\left(\dfrac{x}{5}\right)$

$V = x^3$

CHAPTER 5

Factoring

5.1 The Greatest Common Factor and Factoring by Grouping

1. $15x + 25 = 5(3x + 5)$

3. $6a + 9 = 3(2a + 3)$

5. $4x - 8y = 4(x - 2y)$

7. $3x^2 - 6x - 9 = 3(x^2 - 2x - 3)$

9. $3a^2 - 3a - 60 = 3(a^2 - a - 20)$

11. $24y^2 - 52y + 24 = 4(6y^2 - 13y + 6)$

13. $9x^2 - 8x^3 = x^2(9 - 8x)$

15. $13a^2 - 26a^3 = 13a^2(1 - 2a)$

17. $21x^2y - 28xy^2 = 7xy(3x - 4y)$

19. $22a^2b^2 - 11ab^2 = 11ab^2(2a - 1)$

21. $7x^3 + 21x^2 - 28x = 7x(x^2 + 3x - 4)$

23. $121y^4 - 11x^4 = 11(11y^4 - x^4)$

25. $100x^4 - 50x^3 + 25x^2 = 25x^2(4x^2 - 2x + 1)$

27. $8a^2 + 16b^2 + 32c^2 = 8(a^2 + 2b^2 + 4c^2)$

29. $4a^2b - 16ab^2 + 32a^2b^2 = 4ab(a - 4b + 8ab)$

31. $121a^3b^2 - 22a^2b^3 + 33a^3b^3 = 11a^2b^2(11a - 2b + 3ab)$

33. $12x^2y^3 - 72x^5y^3 - 36x^4y^4 = 12x^2y^3(1 - 6x^3 - 3x^2y)$

35. $xy + 5x + 3y + 15 = x(y + 5) + 3(y + 5) = (y + 5)(x + 3)$

37. $xy + 6x + 2y + 12 = x(y + 6) + 2(y + 6) = (y + 6)(x + 2)$

39. $ab + 7a - 3b - 21 = a(b + 7) - 3(b + 7) = (b + 7)(a - 3)$

41. $ax - bx + ay - by = x(a - b) + y(a - b) = (a - b)(x + y)$

43. $2ax + 6x - 5a - 15 = 2x(a + 3) - 5(a + 3) = (a + 3)(2x - 5)$

45. $3xb - 4b - 6x + 8 = b(3x - 4) - 2(3x - 4) = (3x - 4)(b - 2)$

47. $x^2 + ax + 2x + 2a = x(x + a) + 2(x + a) = (x + a)(x + 2)$

49. $x^2 - ax - bx + ab = x(x - a) - b(x - a) = (x - a)(x - b)$

51. $ax + ay + bx + by + cx + cy = a(x + y) + b(x + y) + c(x + y) = (x + y)(a + b + c)$

53. $6x^2 + 9x + 4x + 6 = 3x(2x + 3) + 2(2x + 3) = (2x + 3)(3x + 2)$

55. $20x^2 - 2x + 50x - 5 = 2x(10x - 1) + 5(10x - 1) = (10x - 1)(2x + 5)$

57. $20x^2 + 4x + 25x + 5 = 4x(5x + 1) + 5(5x + 1) = (5x + 1)(4x + 5)$

59. $x^3 + 2x^2 + 3x + 6 = x^2(x + 2) + 3(x + 2) = (x + 2)(x^2 + 3)$

61. $6x^3 - 4x^2 + 15x - 10 = 2x^2(3x - 2) + 5(3x - 2) = (3x - 2)(2x^2 + 5)$

63. Its greatest common factor is $3 \cdot 2 = 6$.

65. The correct factoring is: $12x^2 + 6x + 3 = 3(4x^2 + 2x + 1)$

67. $A = 1000 + 1000r = 1000(1 + r)$

69. $A = 1,000,000 + 1,000,000r$

   (a) $A = 1,000,000(1 + r)$

   (b) $A = 1,000,000(1 + 0.3) = 1,000,000(1.3)$

   $= 1,300,000$ bacteria

71. $A = 7,000 - 7,000r$

    (a) $A = 7,000(1-r)$

    (b) $A = 7,000(1-0.23)$

        $= 7,000(0.77)$

        $= 5390$ names

73. $(x+7)(x-2) = x^2 + 7x - 2x - 14 = x^2 + 5x - 14$

75. $(x-3)(x+2) = x^2 - 3x + 2x - 6 = x^2 - x - 6$

77.
$$\begin{array}{r} x^2 - 3x + 9 \\ x + 3 \\ \hline x^3 - 3x^2 + 9x \\ 3x^2 - 9x + 27 \\ \hline x^3 + 27 \end{array}$$

79.
$$\begin{array}{r} x^2 + 4x - 3 \\ 2x + 1 \\ \hline 2x^3 + 8x^2 - 6x \\ x^2 + 4x - 3 \\ \hline 2x^3 + 9x^2 - 2x - 3 \end{array}$$

5.2 Factoring Trinomials

1. $x^2 + 7x + 12 = (x+3)(x+4)$

3. $x^2 + 3x + 2 = (x+2)(x+1)$

5. $a^2 + 10a + 21 = (a+7)(a+3)$

7. $x^2 - 7x + 10 = (x-5)(x-2)$

9. $y^2 - 10y + 21 = (y-7)(y-3)$

11. $x^2 - x - 12 = (x-4)(x+3)$

13. $y^2 + y - 12 = (y+4)(y-3)$

15. $x^2 + 5x - 14 = (x+7)(x-2)$

17. $r^2 - 8r - 9 = (r-9)(r+1)$

19. $x^2 - x - 30 = (x-6)(x+5)$

21. $a^2 + 15a + 56 = (a+7)(a+8)$

23. $y^2 - y - 42 = (y-7)(y+6)$

25. $x^2 + 13x + 42 = (x+7)(x+6)$

27. $2x^2 + 6x + 4 = 2(x^2 + 3x + 2) = 2(x+2)(x+1)$

29. $3a^2 - 3a - 60 = 3(a^2 - a - 20) = 3(a-5)(a+4)$

31. $100x^2 - 500x + 600 = 100(x^2 - 5x + 6) = 100(x-3)(x-2)$

33. $100p^2 - 1300p + 4000 = 100(p^2 - 13p + 40) = 100(p-5)(p-8)$

35. $x^4 - x^3 - 12x^2 = x^2(x^2 - x - 12) = x^2(x-4)(x+3)$

37. $2r^3 + 4r^2 - 30r = 2r(r^2 + 2r - 15) = 2r(r+5)(r-3)$

39. $2y^4 - 6y^3 - 8y^2 = 2y^2(y^2 - 3y - 4) = 2y^2(y-4)(y+1)$

41. $x^5 + 4x^4 + 4x^3 = x^3(x^2 + 4x + 4) = x^3(x+2)(x+2) = x^3(x+2)^2$

43. $3y^4 - 12y^3 - 15y^2 = 3y^2(y^2 - 4y - 5) = 3y^2(y-5)(y+1)$

45. $4x^4 - 52x^3 + 144x^2 = 4x^2(x^2 - 13x + 36) = 4x^2(x-9)(x-4)$

47. $x^2 + 5xy + 6y^2 = (x + 2y)(x + 3y)$

49. $x^2 - 9xy + 20y^2 = (x - 4y)(x - 5y)$

51. $a^2 + 2ab - 8b^2 = (a + 4b)(a - 2b)$

53. $a^2 - 10ab + 25b^2 = (a - 5b)(a - 5b) = (a - 5b)^2$

55. $a^2 + 10ab + 25b^2 = (a + 5b)(a + 5b) = (a + 5b)^2$

57. $x^2 + 2xa - 48a^2 = (x + 8a)(x - 6a)$

59. $x^2 - 5xb - 36b^2 = (x - 9b)(x + 4b)$

61. $x^4 - 5x^2 + 6 = (x^2 - 2)(x^2 - 3)$

63. $x^2 - 80x - 2000 = (x - 100)(x + 20)$

65. $x^2 - x - \dfrac{1}{4} = \left(x - \dfrac{1}{2}\right)\left(x - \dfrac{1}{2}\right) = \left(x - \dfrac{1}{2}\right)^2$

67. $x^2 + 0.6x + 0.08 = (x + 0.4)(x + 0.2)$

69. Use long division to find the other factor:

$$
\begin{array}{r}
x + 16 \\
x + 8 \overline{\smash{)}\, x^2 + 24x + 128} \\
\underline{x^2 + 8x} \\
16x + 128 \\
\underline{16x + 128} \\
0
\end{array}
$$

The other factor is $x + 16$.

71. $(4x + 3)(x - 1) = 4x^2 + 3x - 4x - 3 = 4x^2 - x - 3$

73. $(6a + 1)(a + 2) = 6a^2 + a + 12a + 2 = 6a^2 + 13a + 2$

75. $(3a + 2)(2a + 1) = 3a(2a) + 3a(1) + 2(2a) + 2(1) = 6a^2 + 3a + 4a + 2 = 6a^2 + 7a + 2$

77. $(6a + 2)(a + 1) = 6a^2 + 6a + 2a + 2 = 6a^2 + 8a + 2$

79. $(5x^2 + 5x - 4) - (3x^2 - 2x + 7) = 5x^2 + 5x - 4 - 3x^2 + 2x - 7 = 2x^2 + 7x - 11$

81. $(7x + 3) - (4x - 5) = 7x + 3 - 4x + 5 = 3x + 8$

83. $(5x^2 - 5) - (2x^2 - 4x) = 5x^2 - 5 - 2x^2 + 4x = 3x^2 + 4x - 5$

5.3 More Trinomials to Factor

1. $2x^2 + 7x + 3 = (2x+1)(x+3)$

3. $2a^2 - a - 3 = (2a-3)(a+1)$

5. $3x^2 + 2x - 5 = (3x+5)(x-1)$

7. $3y^2 - 14y - 5 = (3y+1)(y-5)$

9. $6x^2 + 13x + 6 = (3x+2)(2x+3)$

11. $4x^2 - 12xy + 9y^2 = (2x-3y)(2x-3y) = (2x-3y)^2$

13. $4y^2 - 11y - 3 = (4y+1)(y-3)$

15. $20x^2 - 41x + 20 = (4x-5)(5x-4)$

17. $20a^2 + 48ab - 5b^2 = (10a-b)(2a+5b)$

19. $20x^2 - 21x - 5 = (4x-5)(5x+1)$

21. $12m^2 + 16m - 3 = (6m-1)(2m+3)$

23. $20x^2 + 37x + 15 = (4x+5)(5x+3)$

25. $12a^2 - 25ab + 12b^2 = (3a-4b)(4a-3b)$

27. $3x^2 - xy - 14y^2 = (3x-7y)(x+2y)$

29. $14x^2 + 29x - 15 = (2x+5)(7x-3)$

31. $6x^2 - 43x + 55 = (3x-5)(2x-11)$

33. $15t^2 - 67t + 38 = (5t-19)(3t-2)$

35. $4x^2 + 2x - 6 = 2(2x^2 + x - 3) = 2(2x+3)(x-1)$

37. $24a^2 - 50a + 24 = 2(12a^2 - 25a + 12) = 2(4a-3)(3a-4)$

39. $10x^3 - 23x^2 + 12x = x(10x^2 - 23x + 12) = x(5x-4)(2x-3)$

41. $6x^4 - 11x^3 - 10x^2 = x^2(6x^2 - 11x - 10) = x^2(3x+2)(2x-5)$

43. $10a^3 - 6a^2 - 4a = 2a(5a^2 - 3a - 2) = 2a(5a+2)(a-1)$

45. $15x^3 - 102x^2 - 21x = 3x(5x^2 - 34x - 7) = 3x(5x+1)(x-7)$

47. $35y^3 - 60y^2 - 20y = 5y(7y^2 - 12y - 4) = 5y(7y+2)(y-2)$

49. $15a^4 - 2a^3 - a^2 = a^2(15a^2 - 2a - 1) = a^2(5a+1)(3a-1)$

51. $24x^2y - 6xy - 45y = 3y(8x^2 - 2x - 15) = 3y(4x+5)(2x-3)$

53. $12x^2y - 34xy^2 + 14y^3 = 2y(6x^2 - 17xy + 7y^2) = 2y(2x-y)(3x-7y)$

55. When $x = 2$:

$2x^2 + 7x + 3 = 2(2)^2 + 7(2) + 3 = 8 + 14 + 3 = 25$

$(2x+1)(x+3) = (2\cdot2+1)(2+3) = (5)(5) = 25$

57. $(2x+3)(2x-3) = (2x)^2 - (3)^2 = 4x^2 - 9$

59. $(x+3)(x-3)(x^2+9) = (x^2-9)(x^2+9) = (x^2)^2 - (9)^2 = x^4 - 81$

61. $h = 8 + 62t - 16t^2$

$= 2(4 + 31t - 8t^2)$

$= 2(4-t)(1+8t)$

See the table in the back of the textbook.

63. $V = x(99 - 40x + 4x^2)$

   $= x(9 - 2x)(11 - 2x)$

   Dimensions of box = dimensions of cardboard $-2x$; $11 - 2x$ and $9 - 2x$

   Dimensions of cardboard: 11 inches by 9 inches.

65. $(x+3)(x-3) = x^2 - (3)^2 = x^2 - 9$ 67. $(6a+1)(6a-1) = (6a)^2 - (1)^2 = 36a^2 - 1$

69. $(x+4)^2 = x^2 + 2(x)(4) + (4)^2 = x^2 + 8x + 16$

71. $(2x+3)^2 = (2x)^2 + 2(2x)(3) + (3)^2 = 4x^2 + 12x + 9$

5.4 Special Factorings

1. $x^2 - 9 = (x+3)(x-3)$ 3. $a^2 - 36 = (a+6)(a-6)$

5. $x^2 - 49 = (x+7)(x-7)$ 7. $4a^2 - 16 = 4(a+2)(a-2)$

9. $9x^2 + 25$ Cannot be factored   11. $25x^2 - 169 = (5x+13)(5x-13)$

13. $9a^2 - 16b^2 = (3a+4b)(3a-4b)$ 15. $9 - m^2 = (3+m)(3-m)$

17. $25 - 4x^2 = (5+2x)(5-2x)$ 19. $2x^2 - 18 = 2(x^2 - 9) = 2(x+3)(x-3)$

21. $32a^2 - 128 = 32(a^2 - 4) = 32(a+2)(a-2)$ 23. $8x^2 y - 18y = 2y(4x^2 - 9) = 2y(2x+3)(2x-3)$

25. $a^4 - b^4 = (a^2 + b^2)(a^2 - b^2) = (a^2 + b^2)(a+b)(a-b)$

27. $16m^4 - 81 = (4m^2 + 9)(4m^2 - 9) = (4m^2 + 9)(2m+3)(2m-3)$

29. $3x^3 y - 75xy^3 = 3xy(x^2 - 25y^2) = 3xy(x+5y)(x-5y)$

31. $x^2 - 2x + 1 = (x-1)(x-1) = (x-1)^2$ 33. $x^2 + 2x + 1 = (x+1)(x+1) = (x+1)^2$

35. $a^2 - 10a + 25 = (a-5)(a-5) = (a-5)^2$ 37. $y^2 + 4y + 4 = (y+2)(y+2) = (y+2)^2$

39. $x^2 - 4x + 4 = (x-2)(x-2) = (x-2)^2$ 41. $m^2 - 12m + 36 = (m-6)(m-6) = (m-6)^2$

43. $4a^2 + 12a + 9 = (2a+3)(2a+3) = (2a+3)^2$ 45. $49x^2 - 14x + 1 = (7x-1)(7x-1) = (7x-1)^2$

47. $9y^2 - 30y + 25 = (3y-5)(3y-5) = (3y-5)^2$ 49. $x^2 + 10xy + 25y^2 = (x+5y)(x+5y) = (x+5y)^2$

51. $9a^2 + 6ab + b^2 = (3a+b)(3a+b) = (3a+b)^2$

53. $3a^2 + 18a + 27 = 3(a^2 + 6a + 9) = 3(a+3)(a+3) = 3(a+3)^2$

55. $2x^2 + 20xy + 50y^2 = 2(x^2 + 10xy + 25y^2) = 2(x+5y)(x+5y) = 2(x+5y)^2$

57. $5x^3 + 30x^2y + 45xy^2 = 5x(x^2 + 6xy + 9y^2) = 5x(x+3y)(x+3y) = 5x(x+3y)^2$

59. $x^2 + 6x + 9 - y^2 = (x+3)^2 - y^2 = (x+3+y)(x+3-y)$

61. $x^2 + 2xy + y^2 - 9 = (x+y)^2 - 9 = (x+y+3)(x+y-3)$

63. Since $(x+7)^2 = x^2 + 14x + 49$, the value is $b = 14$.

65. Since $(x+5)^2 = x^2 + 10x + 25$, the value is $c = 25$.

67. (a) $A = x^2 - 4^2 = x^2 - 16$

    (b) $A = (x+4)(x-4)$

    (c) See the figure in the back of the textbook.

69. $A = a^2 - b^2 = (a+b)(a-b)$

71.

$$\begin{array}{r} x-2 \\ x-3\overline{\smash{)}x^2-5x+8} \\ \underline{x^2-3x} \\ -2x+8 \\ \underline{-2x+6} \\ 2 \end{array}$$

$$\frac{x^2-5x+8}{x-3} = x-2+\frac{2}{x-3}$$

73.

$$\begin{array}{r} 3x-2 \\ 2x+3\overline{\smash{)}6x^2+5x+3} \\ \underline{6x^2+9x} \\ -4x+3 \\ \underline{-4x-6} \\ 9 \end{array}$$

$$3x-2+\frac{9}{2x+3}$$

5.5 Factoring: A General Review

1. $x^2 - 81 = (x+9)(x-9)$

3. $x^2 + 2x - 15 = (x+5)(x-3)$

5. $x^2 + 6x + 9 = (x+3)(x+3) = (x+3)^2$

7. $y^2 - 10y + 25 = (y-5)(y-5) = (y-5)^2$

9. $2a^3b + 6a^2b + 2ab = 2ab(a^2 + 3a + 1)$

11. $x^2 + x + 1$ Cannot be factored

13. $12a^2 - 75 = 3(4a^2 - 25) = 3(2a+5)(2a-5)$

15. $9x^2 - 12xy + 4y^2 = (3x-2y)(3x-2y) = (3x-2y)^2$

17. $4x^3 + 16xy^2 = 4x(x^2 + 4y^2)$

19. $2y^3 + 20y^2 + 50y = 2y(y^2 + 10y + 25) = 2y(y+5)(y+5) = 2y(y+5)^2$

21. $a^6 + 4a^4b^2 = a^4(a^2 + 4b^2)$

23. $xy + 3x + 4y + 12 = x(y+3) + 4(y+3) = (y+3)(x+4)$

25. $x^4 - 16 = (x^2 + 4)(x^2 - 4) = (x^2 + 4)(x+2)(x-2)$

27. $xy - 5x + 2y - 10 = x(y-5) + 2(y-5) = (y-5)(x+2)$

29. $5a^2 + 10ab + 5b^2 = 5(a^2 + 2ab + b^2) = 5(a+b)(a+b) = 5(a+b)^2$

31. $x^2 + 49$ Cannot be factored

33. $3x^2 + 15xy + 18y^2 = 3(x^2 + 5xy + 6y^2) = 3(x+2y)(x+3y)$

35. $2x^2 + 15x - 38 = (2x+19)(x-2)$

37. $100x^2 - 300x + 200 = 100(x^2 - 3x + 2) = 100(x-2)(x-1)$

39. $x^2 - 64 = (x+8)(x-8)$

41. $x^2 + 3x + ax + 3a = x(x+3) + a(x+3) = (x+3)(x+a)$

43. $49a^7 - 9a^5 = a^5(49a^2 - 9) = a^5(7a+3)(7a-3)$

45. $49x^2 + 9y^2$ Cannot be factored

47. $25a^3 + 20a^2 + 3a = a(25a^2 + 20a + 3) = a(5a+3)(5a+1)$

49. $xa - xb + ay - by = x(a-b) + y(a-b) = (a-b)(x+y)$

51. $48a^4b - 3a^2b = 3a^2b(16a^2 - 1) = 3a^2b(4a+1)(4a-1)$

53. $20x^4 - 45x^2 = 5x^2(4x^2 - 9) = 5x^2(2x+3)(2x-3)$

55. $3x^2 + 35xy - 82y^2 = (3x + 41y)(x - 2y)$

57. $16x^5 - 44x^4 + 30x^3 = 2x^3(8x^2 - 22x + 15) = 2x^3(2x-3)(4x-5)$

59. $2x^2 + 2ax + 3x + 3a = 2x(x+a) + 3(x+a) = (x+a)(2x+3)$

61. $y^4 - 1 = (y^2 + 1)(y^2 - 1) = (y^2 + 1)(y+1)(y-1)$

63. $12x^4y^2 + 36x^3y^3 + 27x^2y^4 = 3x^2y^2(4x^2 + 12xy + 9y^2) = 3x^2y^2(2x+3y)(2x+3y) = 3x^2y^2(2x+3y)^2$

65. $3x - 6 = 9$
$3x = 15$
$x = 5$

67. $2x + 3 = 0$
$2x = -3$
$x = -\dfrac{3}{2}$

69. $4x + 3 = 0$
$4x = -3$
$x = -\dfrac{3}{4}$

71. $x^8 \cdot x^7 = x^{8+7} = x^{15}$

73. $(3x^3)^2(2x^4)^3 = 9x^6 \cdot 8x^{12} = 72x^{18}$

75. $57,600 = 5.76 \times 10^4$

5.6 Solving Equations by Factoring

1. Set each factor equal to 0:
$$(x+2)(x-1)=0$$
$$x+2=0 \quad \text{or} \quad x-1=0$$
$$x=-2 \qquad\qquad x=1$$

3. Set each factor equal to 0:
$$a-4=0 \qquad a-5=0$$
$$a=4 \qquad\quad a=5$$

5. Set each factor equal to 0:
$$x(x+1)(x-3)=0$$
$$x=0 \quad \text{or} \quad x+1=0 \quad \text{or} \quad x-3=0$$
$$x=0 \qquad\qquad x=-1 \qquad\qquad x=3$$

7. Set each factor equal to 0:
$$3x+2=0 \qquad 2x+3=0$$
$$3x=-2 \qquad\quad x=-3$$
$$x=-\frac{2}{3} \qquad\quad x=-\frac{3}{2}$$

9. Set each factor equal to 0:
$$m(3m+4)(3m-4)=0$$
$$m=0 \quad \text{or} \quad 3m+4=0 \quad \text{or} \quad 3m-4=0$$
$$m=0 \qquad\qquad 3m=-4 \qquad\qquad 3m=4$$
$$m=0 \qquad\qquad m=-\frac{4}{3} \qquad\qquad m=\frac{4}{3}$$

11. Set each factor equal to 0:
$$2y=0 \qquad 3y+1=0 \qquad 5y+3=0$$
$$y=0 \qquad 3y=-1 \qquad 5y=-3$$
$$\qquad\qquad y=-\frac{1}{3} \qquad\quad y=-\frac{3}{5}$$

13. Solve by factoring:
$$x^2+3x+2=0$$
$$(x+2)(x+1)=0$$
$$x+2=0 \quad \text{or} \quad x+1=0$$
$$x=-2 \qquad\qquad x=-1$$

15. Solve by factoring:
$$x^2-9x+20=0$$
$$(x-4)(x-5)=0$$
$$x-4=0 \qquad\quad x-5=0$$
$$x=4 \qquad\qquad x=5$$

17. Solve by factoring:
$$a^2-2a-24=0$$
$$(a+4)(a-6)=0$$
$$a+4=0 \quad \text{or} \quad a-6=0$$
$$a=-4 \qquad\qquad a=6$$

19. Solve by factoring:
$$100x^2-500x+600=0$$
$$100(x^2-5x+6)=0$$
$$100(x-2)(x-3)=0$$
$$x-2=0 \qquad\quad x-3=0$$
$$x=2 \qquad\qquad x=3$$

21. Solve by factoring:
$$x^2=-6x-9$$
$$x^2+6x+9=0$$
$$(x+3)(x+3)=0$$
$$x+3=0 \quad \text{or} \quad x+3=0$$
$$x=-3 \qquad\qquad x=-3$$

23. Solve by factoring:
$$a^2-16=0$$
$$(a+4)(a-4)=0$$
$$a+4=0 \qquad\quad a-4=0$$
$$a=-4 \qquad\qquad a=4$$

25. Solve by factoring:

$$2x^2 + 5x - 12 = 0$$

$$(x+4)(2x-3) = 0$$

$$x + 4 = 0 \quad \text{or} \quad 2x - 3 = 0$$

$$x = -4 \qquad\qquad 2x = 3$$

$$x = -4 \qquad\qquad x = \frac{3}{2}$$

27. Solve by factoring:

$$9x^2 + 12x + 4 = 0$$

$$(3x+2)(3x+2) = 0$$

$$x = -\frac{2}{3}$$

29. Solve by factoring:

$$a^2 + 25 = 10a$$

$$a^2 - 10a + 25 = 0$$

$$(a-5)(a-5) = 0$$

$$a - 5 = 0 \quad \text{or} \quad a - 5 = 0$$

$$a = 5 \qquad\qquad a = 5$$

31. Solve by factoring:

$$2x^2 = 3x + 20$$

$$2x^2 - 3x - 20 = 0$$

$$(2x+5)(x-4) = 0$$

$$2x + 5 = 0 \qquad\qquad x - 4 = 0$$

$$2x = -5 \qquad\qquad x = 4$$

$$x = -\frac{5}{2} \qquad\qquad x = 4$$

33. Solve by factoring:

$$3m^2 = 20 - 7m$$

$$3m^2 + 7m - 20 = 0$$

$$(m+4)(3m-5) = 0$$

$$m + 4 = 0 \quad \text{or} \quad 3m - 5 = 0$$

$$m = -4 \qquad\qquad 3m = 5$$

$$m = -4 \qquad\qquad m = \frac{5}{3}$$

35. Solve by factoring:

$$4x^2 - 49 = 0$$

$$(2x+7)(2x-7) = 0$$

$$2x + 7 = 0 \qquad\qquad 2x - 7 = 0$$

$$2x = -7 \qquad\qquad 2x = 7$$

$$x = -\frac{7}{2} \qquad\qquad x = \frac{7}{2}$$

37. Solve by factoring:

$$x^2 + 6x = 0$$

$$x(x+6) = 0$$

$$x = 0 \quad \text{or} \quad x + 6 = 0$$

$$x = 0 \qquad\qquad x = -6$$

39. Solve by factoring:

$$x^2 - 3x = 0$$

$$x(x-3) = 0$$

$$x = 0 \qquad\qquad x - 3 = 0$$

$$x = 0 \qquad\qquad x = 3$$

41. Solve by factoring:

$$2x^2 = 8x$$

$$2x^2 - 8x = 0$$

$$2x(x-4) = 0$$

$$2x = 0 \quad \text{or} \quad x - 4 = 0$$

$$x = 0 \qquad\qquad x = 4$$

43. Solve by factoring:

$$3x^2 = 15x$$

$$3x^2 - 15x = 0$$

$$3x(x-5) = 0$$

$$3x = 0 \qquad\qquad x - 5 = 0$$

$$x = 0 \qquad\qquad x = 5$$

45. Solve by factoring:

$$1400 = 400 + 700x - 100x^2$$
$$100x^2 - 700x + 1000 = 0$$
$$100(x^2 - 7x + 10) = 0$$
$$100(x - 2)(x - 5) = 0$$
$$x - 2 = 0 \quad \text{or} \quad x - 5 = 0$$
$$x = 2 \qquad\qquad x = 5$$

47. Solve by factoring:

$$6x^2 = -5x + 4$$
$$6x^2 + 5x - 4 = 0$$
$$(3x + 4)(2x - 1) = 0$$
$$3x + 4 = 0 \qquad 2x - 1 = 0$$
$$3x = -4 \qquad\quad 2x = 1$$
$$x = -\frac{4}{3} \qquad\quad x = \frac{1}{2}$$

49. Solve by factoring:

$$x(2x - 3) = 20$$
$$2x^2 - 3x = 20$$
$$2x^2 - 3x - 20 = 0$$
$$(2x + 5)(x - 4) = 0$$
$$2x + 5 = 0 \quad \text{or} \quad x - 4 = 0$$
$$2x = -5 \qquad\quad x = 4$$
$$x = -\frac{5}{2} \qquad\quad x = 4$$

51. Solve by factoring:

$$t(t + 2) = 80$$
$$t^2 + 2t = 80$$
$$t^2 + 2t - 80 = 0$$
$$(t + 10)(t - 8) = 0$$
$$t + 10 = 0 \qquad t - 8 = 0$$
$$t = -10 \qquad\quad t = 8$$

53. Solve by factoring:

$$4000 = (1300 - 100p)p$$
$$4000 = 1300p - 100p^2$$
$$100p^2 - 1300p + 4000 = 0$$
$$100(p^2 - 13p + 40) = 0$$
$$100(p - 5)(p - 8) = 0$$
$$p - 5 = 0 \qquad p - 8 = 0$$
$$p = 5 \qquad\quad p = 8$$

55. Solve by factoring:

$$x(14 - x) = 48$$
$$14x - x^2 = 48$$
$$-x^2 + 14x - 48 = 0$$
$$x^2 - 14x + 48 = 0$$
$$(x - 6)(x - 8) = 0$$
$$x - 6 = 0 \qquad x - 8 = 0$$
$$x = 6 \qquad\quad x = 8$$

57. Solve by factoring:

$$(x + 5)^2 = 2x + 9$$
$$x^2 + 10x + 25 = 2x + 9$$
$$x^2 + 8x + 16 = 0$$
$$(x + 4)(x + 4) = 0$$
$$x + 4 = 0 \qquad x + 4 = 0$$
$$x = -4 \qquad\quad x = -4$$

59. Solve by factoring:

$$(y - 6)^2 = y - 4$$
$$y^2 - 12y + 36 = y - 4$$
$$y^2 - 13y + 40 = 0$$
$$(y - 5)(y - 8) = 0$$
$$y - 5 = 0 \qquad y - 8 = 0$$
$$y = 5 \qquad\quad y = 8$$

61. Solve by factoring:

$$10^2 = (x+2)^2 + x^2$$
$$100 = x^2 + 4x + 4 + x^2$$
$$2x^2 + 4x - 96 = 0$$
$$2(x^2 + 2x - 48) = 0$$
$$2(x+8)(x-6) = 0$$

$$x + 8 = 0 \qquad x - 6 = 0$$
$$x = -8 \qquad x = 6$$

63. Solve by factoring:

$$2x^3 + 11x^2 + 12x = 0$$
$$x(2x^2 + 11x + 12) = 0$$
$$x(2x+3)(x+4) = 0$$

$$x = 0 \qquad 2x + 3 = 0 \qquad x + 4 = 0$$
$$x = 0 \qquad 2x = -3 \qquad x = -4$$
$$x = 0 \qquad x = -\frac{3}{2} \qquad x = -4$$

65. Solve by factoring:

$$4y^3 - 2y^2 - 30y = 0$$
$$y(4y^2 - 2y - 30) = 0$$
$$y(4y+10)(y-3) = 0$$

$$y = 0 \quad \text{or} \quad 4y + 10 = 0 \quad \text{or} \quad y - 3 = 0$$
$$y = 0 \qquad 4y = -10 \qquad y = 3$$
$$y = 0 \qquad y = -\frac{5}{2} \qquad y = 3$$

67. Solve by factoring:

$$8x^3 + 16x^2 = 10x$$
$$8x^3 + 16x^2 - 10x = 0$$
$$2x(4x^2 + 8x - 5) = 0$$
$$2x(2x-1)(2x+5) = 0$$

$$2x = 0 \qquad 2x - 1 = 0 \qquad 2x + 5 = 0$$
$$x = 0 \qquad 2x = 1 \qquad 2x = -5$$
$$x = 0 \qquad x = \frac{1}{2} \qquad x = -\frac{5}{2}$$

69. Solve by factoring:

$$20a^3 = -18a^2 + 18a$$
$$20a^3 + 18a^2 - 18a = 0$$
$$2a(10a^2 + 9a - 9) = 0$$
$$2a(2a+3)(5a-3) = 0$$

$$2a = 0 \quad \text{or} \quad 2a + 3 = 0 \quad \text{or} \quad 5a - 3 = 0$$
$$a = 0 \qquad 2a = -3 \qquad 5a = 3$$
$$a = 0 \qquad a = -\frac{3}{2} \qquad a = \frac{3}{5}$$

71. Solve by factoring:

$$x^3 + 3x^2 - 4x - 12 = 0$$
$$x^2(x+3) - 4(x+3) = 0$$
$$(x+3)(x^2 - 4) = 0$$
$$(x+3)(x+2)(x-2) = 0$$

$$x + 3 = 0 \qquad x + 2 = 0 \qquad x - 2 = 0$$
$$x = -3 \qquad x = -2 \qquad x = 2$$

73. Solve by factoring:

$$x^3 + x^2 - 16x - 16 = 0$$
$$x^2(x+1) - 16(x+1) = 0$$
$$(x^2 - 16)(x+1) = 0$$
$$(x+4)(x-4)(x+1) = 0$$

$$x + 4 = 0 \quad \text{or} \quad x - 4 = 0 \quad \text{or} \quad x + 1 = 0$$
$$x = -4 \qquad x = 4 \qquad x = -1$$

75. Let $x =$ the cost of the suit

Let $5x =$ the cost of the bicycle

$$x + 5x = 90$$
$$6x = 90$$
$$x = 15$$
$$5x = 75$$

The suit costs $15 and the bicycle costs $75.

77. Let $x =$ the cost of the lot

Let $4x =$ the cost of the house

$$x + 4x = 3000$$
$$5x = 3000$$
$$x = 600$$
$$4x = 2400$$

The lot costs $600 and the house costs $2,400.

79. Simplifying using properties of exponents:

$$2^{-3} = \frac{1}{2^3} = \frac{1}{8}$$

81. Simplifying using properties of exponents:

$$\frac{x^5}{x^{-3}} = x^{5-(-3)} = x^8$$

83. Simplifying using properties of exponents:

$$\frac{\left(x^2\right)^3}{\left(x^{-3}\right)^4} = \frac{x^6}{x^{-12}} = x^{6-(-12)} = x^{6+12} = x^{18}$$

85. $0.0056 = 5.6 \times 10^{-3}$

5.7 Applications

1. Let x and $x + 2 =$ the two consecutive even integers.

$$x(x+2) = 80$$
$$x^2 + 2x = 80$$
$$x^2 + 2x - 80 = 0$$
$$(x+10)(x-8) = 0$$
$$x = -10, 8$$
$$x + 2 = -8, 10$$

The two numbers are either -10 and -8, or 8 and 10.

3. Let x and $x + 2 =$ the two consecutive even integers.

$$x(x+2) = 99$$
$$x^2 + 2x = 99$$
$$x^2 + 2x - 99 = 0$$
$$(x+11)(x-9) = 0$$
$$x = -11, 9$$
$$x + 2 = -9, 11$$

The two numbers are either -11 and -9, or 9 and 11.

5. Let x and $x + 2 =$ the two consecutive even integers.

$$x(x+2) = 5(x+x+2) - 10$$
$$x^2 + 2x = 5(2x+2) - 10$$
$$x^2 + 2x = 10x + 10 - 10$$
$$x^2 + 2x = 10x$$
$$x^2 - 8x = 0$$
$$x(x-8) = 0$$
$$x = 0, 8$$
$$x + 2 = 2, 10$$

The two numbers are either 0 and 2, or 8 and 10.

7. Let x and $14 - x =$ the two numbers.

$$x(14-x) = 48$$
$$14x - x^2 = 48$$
$$0 = x^2 - 14x + 48$$
$$0 = (x-8)(x-6)$$
$$x = 8, 6$$
$$14 - x = 6, 8$$

The two numbers are 6 and 8.

9. Let x and $5x + 2 =$ the two numbers.
$$x(5x + 2) = 24$$
$$5x^2 + 2x = 24$$
$$5x^2 + 2x - 24 = 0$$
$$(5x + 12)(x - 2) = 0$$
$$x = -\frac{12}{5}, \ 2$$
$$5x + 2 = -10, 12$$

The two numbers are either $-\frac{12}{5}$ and -10, or 2 and 12.

11. Let x and $4x =$ the two numbers.
$$x(4x) = 4(x + 4x)$$
$$4x^2 = 4(5x)$$
$$4x^2 = 20x$$
$$4x^2 - 20x = 0$$
$$4x(x - 5) = 0$$
$$x = 0, 5$$
$$4x = 0, 20$$

The two numbers are either 0 and 0, or 5 and 20.

13.
Let $w =$ the width
Let $w + 1 =$ the length
$$w(w + 1) = 12$$
$$w^2 + w = 12$$
$$w^2 + w - 12 = 0$$
$$(w + 4)(w - 3) = 0$$
$$w = 3, \ w = -4 \text{ is impossible}$$
$$w + 1 = 4$$

The width is 3 inches and the length is 4 inches.

15.
Let $b =$ the base
Let $2b =$ the height
$$\frac{1}{2}b(2b) = 9$$
$$b^2 = 9$$
$$b^2 - 9 = 0$$
$$(b + 3)(b - 3) = 0$$
$$b = 3$$
$$b = -3 \text{ is impossible}$$

The base is 3 inches.

17. Let x and $x + 2 =$ the two legs
$$x^2 + (x + 2)^2 = 10^2$$
$$x^2 + x^2 + 4x + 4 = 100$$
$$2x^2 + 4x + 4 = 100$$
$$2x^2 + 4x - 96 = 0$$
$$2(x^2 + 2x - 48) = 0$$
$$2(x + 8)(x - 6) = 0$$
$$x = 6$$
$$x = -8 \text{ is impossible}$$
$$x + 2 = 8$$

The legs are 6 inches and 8 inches.

19. Let $x =$ the longer leg
Let $x + 1 =$ the hypotenuse
$$5^2 + x^2 = (x + 1)^2$$
$$25 + x^2 = x^2 + 2x + 1$$
$$25 = 2x + 1$$
$$24 = 2x$$
$$x = 12$$

The longer leg is 12 meters.

21. Setting $C = \$1,400$:
$$1400 = 400 + 700x - 100x^2$$
$$100x^2 - 700x + 1000 = 0$$
$$100(x^2 - 7x + 10) = 0$$
$$100(x - 5)(x - 2) = 0$$
$$x = 2, 5$$

The company can manufacture either 200 items or 500 items.

23. Setting $C = \$2,200$:
$$2200 = 600 + 1000x - 100x^2$$
$$100x^2 - 1000x + 1600 = 0$$
$$100(x^2 - 10x + 16) = 0$$
$$100(x - 2)(x - 8) = 0$$
$$x = 2, 8$$

The company can manufacture either 200 videotapes or 800 videotapes.

25. The revenue is given by:
$$R = xp = (1200 - 100p)p$$
Setting $R = \$3,200$:
$$3200 = (1200 - 100p)p$$
$$3200 = 1200p - 100p^2$$
$$100p^2 - 1200p + 3200 = 0$$
$$100(p^2 - 12p + 32) = 0$$
$$100(p-4)(p-8) = 0$$
$$p = 4, 8$$

The company should sell the ribbons for either $4 or $8.

27. The revenue is given by:
$$R = xp = (1700 - 100p)p$$
Setting $R = \$7,000$:
$$7000 = (1700 - 100p)p$$
$$7000 = 1700p - 100p^2$$
$$100p^2 - 1700p + 7000 = 0$$
$$100(p^2 - 17p + 70) = 0$$
$$100(p-7)(p-10) = 0$$
$$p = 7, 10$$

The calculators should be sold for either $7 or $10.

29. (a) Let $x =$ distance of the base from the wall

Let $2x + 2 =$ point on the wall

Use the Pythagorean Theorem
$$x^2 + (2x+2)^2 = 13^2$$
$$x^2 + 4x^2 + 8x + 4 = 169$$
$$5x^2 + 8x - 165 = 0$$
$$(x-5)(5x+33) = 0$$
$$x = 5 \qquad x = -\frac{33}{5} \text{ is impossible}$$

The base of the ladder is 5 feet from

(b) $2x + 2 = 2(5) + 2 = 12$

The ladder reaches up 12 feet.

31. $h(t) = -16t^2 + 396t + 100$

(a) $0 = -16t^2 + 396t + 100$
$$0 = -4(4t^2 - 99t - 25)$$
$$0 = 4t^2 - 99t - 25$$
$$0 = (4t+1)(t-25)$$
$$t = 25 \quad \text{or} \quad t = -1/4 \text{ (not allowed)}$$

It will reach the ground in 25 seconds.

(b) See the table in the back of the textbook.

33. $(5x^3)^2(2x^6)^3 = 25x^6 \cdot 8x^{18} = 200x^{24}$

35. $\dfrac{x^4}{x^{-3}} = x^{4-(-3)} = x^{4+3} = x^7$

37. $(2 \times 10^{-4})(4 \times 10^5) = 8 \times 10^1 = 80$

39. $20ab^2 - 16ab^2 + 6ab^2 = 10ab^2$

41. $2x^2(3x^2 + 3x - 1) = 2x^2(3x^2) + 2x^2(3x) - 2x^2(1) = 6x^4 + 6x^3 - 2x^2$

43. $(3y - 5)^2 = (3y)^2 - 2(3y)(5) + (5)^2 = 9y^2 - 30y + 25$

45. $(2a^2 + 7)(2a^2 - 7) = (2a^2)^2 - (7)^2 = 4a^4 - 49$

Chapter 5 Review

1. $10x - 20 = 10(x - 2)$

3. $5x - 5y = 5(x - y)$

5. $8x + 4 = 4(2x + 1)$

7. $24y^2 - 40y + 48 = 8(3y^2 - 5y + 6)$

9. $49a^3 - 14b^3 = 7(7a^3 - 2b^3)$

11. $xy + bx + ay + ab = x(y + b) + a(y + b) = (x + a)(y + b)$

13. $2xy + 10x - 3y - 15 = 2x(y + 5) - 3(y + 5) = (y + 5)(2x - 3)$

15. $y^2 + 9y + 14 = (y + 7)(y + 2)$

17. $a^2 - 14a + 48 = (a - 8)(a - 6)$

19. $y^2 + 20y + 99 = (y + 9)(y + 11)$

21. $2x^2 + 13x + 15 = (2x + 3)(x + 5)$

23. $5y^2 + 11y + 6 = (5y + 6)(y + 1)$

25. $6r^2 + 5rt - 6t^2 = (3r - 2t)(2r + 3t)$

27. $n^2 - 81 = (n + 9)(n - 9)$

29. $x^2 + 49$: Prime - Cannot be factored

31. $64a^2 - 121b^2 = (8a + 11b)(8a - 11b)$

33. $y^2 + 20y + 100 = (y + 10)(y + 10) = (y + 10)^2$

35. $64t^2 + 16t + 1 = (8t + 1)(8t + 1) = (8t + 1)^2$

37. $4r^2 - 12rt + 9t^2 = (2r - 3t)(2r - 3t) = (2r - 3t)^2$

39. $2x^2 + 20x + 48 = 2(x^2 + 10x + 24) = 2(x + 4)(x + 6)$

41. $3m^3 - 18m^2 - 21m = 3m(m^2 - 6m - 7) = 3m(m + 1)(m - 7)$

43. $8x^2 + 16x + 6 = 2(4x^2 + 8x + 3) = 2(2x + 1)(2x + 3)$

45. $20m^3 - 34m^2 + 6m = 2m(10m^2 - 17m + 3) = 2m(2m - 3)(5m - 1)$

47. $4x^2 + 40x + 100 = 4(x^2 + 10x + 25) = 4(x + 5)(x + 5) = 4(x + 5)^2$

49. $5x^2 - 45 = 5(x^2 - 9) = 5(x + 3)(x - 3)$

51. $6a^3b + 33a^2b^2 + 15ab^3 = 3ab(2a^2 + 11ab + 5b^2) = 3ab(2a + b)(a + 5b)$

53. $4y^6 + 9y^4 = y^4(4y^2 + 9)$

55. $30a^4b + 35a^3b^2 - 15a^2b^3 = 5a^2b(6a^2 + 7ab - 3b^2) = 5a^2b(3a - b)(2a + 3b)$

57. $(x - 5)(x + 2) = 0$

$\quad x - 5 = 0 \qquad x + 2 = 0$

$\quad\quad x = 5 \qquad\quad x = -2$

59. $\quad m^2 + 3m = 10$

$\quad m^2 + 3m - 10 = 0$

$\quad (m + 5)(m - 2) = 0$

$\quad\quad m + 5 = 0 \qquad m - 2 = 0$

$\quad\quad\quad m = -5 \qquad\quad m = 2$

61. $m^2 - 9m = 0$

$m(m-9) = 0$

$m = 0 \qquad m - 9 = 0$

$m = 0 \qquad m = 9$

63. $9x^4 + 9x^3 = 10x^2$

$9x^4 + 9x^3 - 10x^2 = 0$

$x^2(9x^2 + 9x - 10) = 0$

$x^2(3x + 5)(3x - 2) = 0$

$x = 0, \dfrac{2}{3}, -\dfrac{5}{3}$

65. Let $x =$ the first integer

Let $x + 1 =$ the next consecutive integer

$x(x+1) = 110$

$x^2 + x = 110$

$x^2 + x - 110 = 0$

$(x+11)(x-10) = 0$

$x = -11, 10$

$x + 1 = -10, 11$

The two numbers are either 10 and 11 or −11 and −10.

67. Let x and $20 - x =$ the two numbers.

$x(20 - x) = 75$

$20x - x^2 = 75$

$0 = x^2 - 20x + 75$

$0 = (x-15)(x-5)$

$x = 15, 5$

$20 - x = 5, 15$

The two numbers are 5 and 15.

69. Let $b =$ the base

Let $8b =$ the height

$\dfrac{1}{2}b(8b) = 16$

$4b^2 = 16$

$b^2 - 16 = 0$

$4(b^2 - 4) = 0$

$4(b+2)(b-2) = 0$

$b = 2$

Cumulative Review: Chapters 1-5

1. $-|-9| = -9$

3. $20 - (-9) = 20 + 9 = 29$

5. $\dfrac{9(-2)}{-2} = \dfrac{-18}{-2} = 9$

7. $\dfrac{-3(4-7)-5(7-2)}{-5-2-1} = \dfrac{-3(-3)-5(5)}{-8} = \dfrac{9-25}{-8} = \dfrac{-16}{-8} = 2$

9. $6 - 2(4a+2) - 5 = 6 - 8a - 4 - 5 = -8a - 3$

11. $(9xy)^0 = 1$

13. $\dfrac{50x^8 y^8}{25x^4 y^2} + \dfrac{28x^7 y^7}{14x^3 y} = 2x^4 y^6 + 2x^4 y^6 = 4x^4 y^6$

15. $3x = -18$

$\dfrac{1}{3}(3x) = \dfrac{1}{3}(-18)$

$x = -6$

17. $-\dfrac{x}{3} = 7$

$-3\left(-\dfrac{x}{3}\right) = -3(7)$

$x = -21$

19. Setting each factor equal to 0:

$4m = 0 \qquad m - 7 = 0 \qquad 2m - 7 = 0$

$m = 0 \qquad\quad m = 7 \qquad\quad 2m = 7$

$m = \dfrac{7}{2}$

The solutions are 0, 7, and $\dfrac{7}{2}$.

21. $8x^2 = 10x + 3$

$8x^2 - 10x - 3 = 0$

$(4x+1)(2x-3) = 0$

$4x + 1 = 0 \ \text{ or } \ 2x - 3 = 0$

$x = -\dfrac{1}{4} \qquad x = \dfrac{3}{2}$

23. $7x + 1 \geq 2x - 5$

$5x + 1 \geq -5$

$5x \geq -6$

$x \geq -\dfrac{6}{5}$

25. $y = 2$

See the graph in the back of the textbook.

27. $10x - 3y = 30$

When $x = 0$, $10(0) - 3y = 30$

$-3y = 30$

$y = -10$

When $y = 0$, $10x - 3(0) = 30$

$10x = 30$

$x = 3$

y-intercept: $(0, -10)$

x-intercept: $(3, 0)$

29. $2x + 5y = 10$

$(0, 2)$: $2(0) + 5(2) = 0 + 10 = 10$

$\left(4, \dfrac{2}{5}\right)$: $2(4) + 5\left(\dfrac{2}{5}\right) = 8 + 2 = 10$

$(0, 2)$ and $\left(4, \dfrac{2}{5}\right)$ are solutions.

31. The intersection point is $(-2, 1)$.

See the graph in the back of the textbook.

33. Multiplying the first equation by -3
and the second equation by 7:

$-15x - 21y = 54$

$56x + 21y = 28$

Adding the two equations:

$41x = 82$

$x = 2$

Substituting into the second equation

$8(2) + 3y = 4$

$16 + 3y = 4$

$3y = -12$

$y = -4$

The solution is $(2, -4)$.

35. Multiplying the first equation by -0.04:

$-0.04x - 0.04y = -200$

$0.04x + 0.06y = 270$

Adding the two equations:

$0.02y = 70$

$y = 3500$

Substituting into the first equation:

$x + 3500 = 5000$

$x = 1500$

The solution is $(1500, 3500)$.

37. $n^2 - 5n - 36 = (n - 9)(n + 4)$

39. $16 - a^2 = (4 + a)(4 - a)$

41. $45x^2y - 30xy^2 + 5y^3 = 5y(9x^2 - 6xy + y^2) = 5y(3x - y)(3x - y) = 5y(3x - y)^2$

43. $3xy + 15x - 2y - 10 = 3x(y + 5) - 2(y + 5) = (y + 5)(3x - 2)$

45. Commutative property of addition

47. $\dfrac{28x^4y^4 - 14x^2y^3 + 21xy^2}{-7xy^2} = \dfrac{28x^4y^4}{-7xy^2} + \dfrac{-14x^2y^3}{-7xy^2} + \dfrac{21xy^2}{-7xy^2} = -4x^3y^2 + 2xy - 3$

49. Let x and $x + 4 =$ the length of each piece

$x + x + 4 = 72$

$2x + 4 = 72$

$2x = 68$

$x = 34$

$x + 4 = 38$

The pieces are 34 inches and 38 inches in length.

Chapter 5 Test

1. $5x - 10 = 5(x - 2)$

2. $18x^2y - 9xy - 36xy^2 = 9xy(2x - 1 - 4y)$

3. $x^2 + 2ax - 3bx - 6ab = (x^2 + 2ax) + (-3bx - 6ab)$
$= x(x + 2a) - 3b(x + 2a)$
$= (x + 2a)(x - 3b)$

4. $xy + 4x - 7y - 28 = (xy + 4x) + (-7y - 28)$
$= x(y + 4) - 7(y + 4)$
$= (x - 7)(y + 4)$

5. $x^2 - 5x + 6 = (x - 2)(x - 3)$

6. $x^2 - x - 6 = (x - 3)(x + 2)$

7. $a^2 - 16 = (a - 4)(a + 4)$

8. $x^2 + 25$ Cannot be factored

9. $x^4 - 81 = (x^2 + 9)(x^2 - 9) = (x^2 + 9)(x - 3)(x + 3)$

10. $27x^2 - 75y^2 = 3(9x^2 - 25y^2) = 3(3x - 5y)(3x + 5y)$

11. $x^3 + 5x^2 - 9x - 45 = x^2(x + 5) - 9(x + 5) = (x + 5)(x^2 - 9) = (x + 5)(x + 3)(x - 3)$

12. $x^2 - bx + 5x - 5b = x(x - b) + 5(x - b) = (x - b)(x + 5)$

13. $4a^2 + 22a + 10 = 2(2a^2 + 11a + 5) = 2(2a + 1)(a + 5)$

14. $3m^2 - 3m - 18 = 3(m^2 - m - 6) = 3(m - 3)(m + 2)$

15. $6y^2 + 7y - 5 = (2y - 1)(3y + 5)$

16. $12x^3 - 14x^2 - 10x = 2x(6x^2 - 7x - 5) = 2x(2x + 1)(3x - 5)$

17. Solve by factoring:
$x^2 + 7x + 12 = 0$
$(x + 3)(x + 4) = 0$

18. Solve by factoring:
$x^2 - 4x + 4 = 0$
$(x - 2)^2 = 0$
$x - 2 = 0$
$x = 2$

19. Solve by factoring:
$x^2 - 36 = 0$
$(x + 6)(x - 6) = 0$
$x = -6,\ 6$

20. Solve by factoring:
$x^2 = x + 20$
$x^2 - x - 20 = 0$
$(x + 4)(x - 5) = 0$
$x = -4,\ 5$

21. Solve by factoring:
$x^2 - 11x = -30$
$x^2 - 11x + 30 = 0$
$(x - 6)(x - 5) = 0$
$x = 5,\ 6$

22. Solve by factoring:
$y^3 = 16y$
$y^3 - 16y = 0$
$y(y^2 - 16) = 0$
$y(y + 4)(y - 4) = 0$
$y = 0, -4,\ 4$

23. Solve by factoring:

$$2a^2 = a + 15$$
$$2a^2 - a - 15 = 0$$
$$(2a + 5)(a - 3) = 0$$
$$a = -\frac{5}{2}, 3$$

24. Solve by factoring:

$$30x^3 - 20x^2 = 10x$$
$$30x^3 - 20x^2 - 10x = 0$$
$$10x(3x^2 - 2x - 1) = 0$$
$$10x(3x + 1)(x - 1) = 0$$
$$x = 0, -\frac{1}{3}, 1$$

25. Let x and $20 - x =$ the two numbers.

$$x(20 - x) = 64$$
$$20x - x^2 = 64$$
$$0 = x^2 - 20x + 64$$
$$0 = (x - 16)(x - 4)$$
$$x = 4, 16$$
$$20 - x = 16, 4$$

The two numbers are 4 and 16.

26. Let x and $x + 2 =$ the two consecutive odd integers.

$$x(x + 2) = x + x + 2 + 7$$
$$x^2 + 2x = 2x + 9$$
$$x^2 - 9 = 0$$
$$(x + 3)(x - 3) = 0$$
$$x = -3, 3$$
$$x + 2 = -1, 5$$

The two integers are either -3 and -1, or 3 and 5.

27.

Let $w =$ the width
Let $3w + 5 =$ the length
$$w(3w + 5) = 42$$
$$3w^2 + 5w - 42 = 0$$
$$(3w + 14)(w - 3) = 0$$
$$w = 3$$
$$w = -\frac{14}{3} \text{ is impossible}$$
$$3w + 5 = 14$$

The width is 3 feet and the length is 14 feet.

28.

Let x and $2x + 2 =$ the two legs
$$x^2 + (2x + 2)^2 = 13^2$$
$$x^2 + 4x^2 + 8x + 4 = 169$$
$$5x^2 + 8x - 165 = 0$$
$$(5x + 33)(x - 5) = 0$$
$$x = 5$$
$$x = -\frac{33}{5} \text{ is impossible}$$
$$2x + 2 = 12$$

The two legs are 5 meters and 12 meters in length.

29.

Setting $C = \$800$:
$$800 = 200 + 500x - 100x^2$$
$$100x^2 - 500x + 600 = 0$$
$$100(x^2 - 5x + 6) = 0$$
$$100(x - 2)(x - 3) = 0$$
$$x = 2, 3$$

The company can manufacture either 200 items or 300 items.

30. The revenue is given by:

$$R = xp = (900 - 100p)p$$
Setting $R = \$1,800$:
$$1800 = (900 - 100p)p$$
$$1800 = 900p - 100p^2$$
$$100p^2 - 900p + 1800 = 0$$
$$100(p^2 - 9p + 18) = 0$$
$$100(p - 6)(p - 3) = 0$$
$$p = 3, 6$$

The manufacturer should sell the items at either \$3 or \$6.

CHAPTER 6

Rational Expressions

6.1 Reducing Rational Expressions to Lowest Terms

1. $\dfrac{5}{5x-10} = \dfrac{5}{5(x-2)} = \dfrac{1}{x-2}$

3. $\dfrac{a-3}{a^2-9} = \dfrac{1(a-3)}{(a+3)(a-3)} = \dfrac{1}{a+3}$

5. $\dfrac{x+5}{x^2-25} = \dfrac{x+5}{(x+5)(x-5)} = \dfrac{1}{x-5}$

7. $\dfrac{2x^2-8}{4} = \dfrac{2(x^2-4)}{4} = \dfrac{2(x+2)(x-2)}{4} = \dfrac{(x+2)(x-2)}{2}$

9. $\dfrac{2x-10}{3x-6} = \dfrac{2(x-5)}{3(x-2)}$

11. $\dfrac{10a+20}{5a+10} = \dfrac{10(a+2)}{5(a+2)} = \dfrac{2}{1} = 2$

13. $\dfrac{5x^2-5}{4x+4} = \dfrac{5(x+1)(x-1)}{4(x+1)} = \dfrac{5(x-1)}{4}$

15. $\dfrac{x-3}{x^2-6x+9} = \dfrac{1(x-3)}{(x-3)^2} = \dfrac{1}{x-3}$

17. $\dfrac{3x+15}{3x^2+24x+45} = \dfrac{3(x+5)}{3(x+5)(x+3)} = \dfrac{1}{x+3}$

19. $\dfrac{a^2-3a}{a^3+24x+15a} = \dfrac{a(a-3)}{a(a^2-8a+15)}$

$\qquad\qquad\qquad = \dfrac{a(a-3)}{a(a-3)(a-5)} = \dfrac{1}{a-5}$

21. $\dfrac{3x-2}{9x^2-4} = \dfrac{3x-2}{(3x+2)(3x-2)} = \dfrac{1}{3x+2}$

23. $\dfrac{x^2+8x+15}{x^2+5x+6} = \dfrac{(x+5)(x+3)}{(x+2)(x+3)} = \dfrac{x+5}{x+2}$

25. $\dfrac{2m^3-2m^2-12m}{m^2-5m+6} = \dfrac{2m(m^2-m-6)}{m^2-5m+6} = \dfrac{2m(m-3)(m+2)}{(m-2)(m-3)} = \dfrac{2m(m+2)}{m-2}$

27. $\dfrac{x^3+3x^2-4x}{x^3-16x} = \dfrac{x(x^2+3x-4)}{x(x^2-16)} = \dfrac{x(x+4)(x-1)}{x(x+4)(x-4)} = \dfrac{x-1}{x-4}$

29. $\dfrac{4x^3-10x^2+6x}{2x^3+x^2-3x} = \dfrac{2x(2x^2-5x+3)}{x(2x^2+x-3)} = \dfrac{2x(2x-3)(x-1)}{x(2x+3)(x-1)} = \dfrac{2(2x-3)}{2x+3}$

31. $\dfrac{4x^2-12x+9}{4x^2-9} = \dfrac{(2x-3)^2}{(2x+3)(2x-3)} = \dfrac{2x-3}{2x+3}$

33. $\dfrac{x+3}{x^4-81} = \dfrac{x+3}{(x^2+9)(x+3)(x-3)} = \dfrac{1}{(x^2+9)(x-3)}$

35. $\dfrac{3x^2+x-10}{x^4-16} = \dfrac{(3x-5)(x+2)}{(x^2+4)(x^2-4)} = \dfrac{(3x-5)(x+2)}{(x^2+4)(x+2)(x-2)} = \dfrac{3x-5}{(x^2+4)(x-2)}$

37. $\dfrac{42x^3 - 20x^2 - 48x}{6x^2 - 5x - 4} = \dfrac{2x(21x^2 - 10x - 24)}{(2x+1)(3x-4)} = \dfrac{2x(7x+6)(3x-4)}{(2x+1)(3x-4)} = \dfrac{2x(7x+6)}{2x+1}$

39. $\dfrac{xy + 3x + 2y + 6}{xy + 3x + 5y + 15} = \dfrac{x(y+3) + 2(y+3)}{x(y+3) + 5(y+3)} = \dfrac{(y+3)(x+2)}{(y+3)(x+5)} = \dfrac{x+2}{x+5}$

41. $\dfrac{x^2 - 3x + ax - 3a}{x^2 - 3x + bx - 3b} = \dfrac{x(x-3) + a(x-3)}{x(x-3) + b(x-3)} = \dfrac{(x-3)(x+a)}{(x-3)(x+b)} = \dfrac{x+a}{x+b}$

43. $\dfrac{xy + bx + ay + ab}{xy + bx + 3y + 3b} = \dfrac{x(y+b) + a(y+b)}{x(y+b) + 3(y+b)} = \dfrac{(y+b)(x+a)}{(y+b)(x+3)} = \dfrac{x+a}{x+3}$

45. $\dfrac{8}{6} = \dfrac{4}{3}$ 47. $\dfrac{200}{250} = \dfrac{4}{5}$ 49. $\dfrac{32}{4} = \dfrac{8}{1}$

51. See the table in the back of the textbook.

53. The average speed is: $\dfrac{122 \text{ miles}}{3 \text{ hours}} = 40.7$ miles/hour

55. The average speed is: $\dfrac{785 \text{ feet}}{20 \text{ minutes}} = 39.25$ feet/minute

57. The average speed is: $\dfrac{518 \text{ feet}}{40 \text{ seconds}} = 12.95$ feet/second

59. Her average speed on level ground is: $\dfrac{20 \text{ minutes}}{2 \text{ miles}} = 10$ minutes/mile, or $\dfrac{2 \text{ miles}}{20 \text{ minutes}} = 0.1$ miles/minute

   Her average speed downhill is: $\dfrac{40 \text{ minutes}}{6 \text{ miles}} = \dfrac{20}{3}$ minutes/mile, or $\dfrac{6 \text{ miles}}{40 \text{ minutes}} = \dfrac{3}{20}$ miles/minute

61. The average fuel consumption is: $\dfrac{168 \text{ miles}}{3.5 \text{ gallons}} = 48$ miles/gallon

63. Substituting $x = 5$ and $y = 4$: $\dfrac{x^2 - y^2}{x - y} = \dfrac{5^2 - 4^2}{5 - 4} = \dfrac{25 - 16}{5 - 4} = \dfrac{9}{1} = 9$. The result is equal to $5 + 4$.

65. See the table in the back of the textbook.

   The entries are all -1 because the numerator and denominator are opposites.

   $\dfrac{x-3}{3-x} = \dfrac{-1(3-x)}{3-x} = -1$

67. See the table in the back of the textbook.

69. $\dfrac{27x^5}{9x^2} - \dfrac{45x^8}{15x^5} = 3x^3 - 3x^3 = 0$

71. $\dfrac{72a^3b^7}{9ab^5} + \dfrac{64a^5b^3}{8a^3b} = 8a^2b^2 + 8a^2b^2 = 16a^2b^2$

73. $\dfrac{38x^7 + 42x^5 - 84x^3}{2x^3} = \dfrac{38x^7}{2x^3} + \dfrac{42x^5}{2x^3} - \dfrac{84x^3}{2x^3} = 19x^4 + 21x^2 - 42$

75. $\dfrac{28a^5b^5 + 36ab^4 - 44a^4b}{4ab} = \dfrac{28a^5b^5}{4ab} + \dfrac{36ab^4}{4ab} - \dfrac{44a^4b}{4ab} = 7a^4b^4 + 9b^3 - 11a^3$

6.2 Multiplication and Division of Rational Expressions

1. $\dfrac{x+y}{3} \cdot \dfrac{6}{x+y} = \dfrac{6(x+y)}{3(x+y)} = 2$

3. $\dfrac{2x+10}{x^2} \cdot \dfrac{x^3}{4x+20} = \dfrac{2(x+5)}{x^2} \cdot \dfrac{x^3}{4(x+5)} = \dfrac{x}{2}$

5. $\dfrac{9}{2a-8} \div \dfrac{3}{a-4} = \dfrac{9}{2a-8} \cdot \dfrac{a-4}{3} = \dfrac{9}{2(a-4)} \cdot \dfrac{a-4}{3} = \dfrac{3}{2}$

7. $\dfrac{x+1}{x^2-9} \div \dfrac{2x+2}{x+3} = \dfrac{x+1}{x^2-9} \cdot \dfrac{x+3}{2x+2} = \dfrac{x+1}{(x+3)(x-3)} \cdot \dfrac{x+3}{2(x+1)} = \dfrac{1}{2(x-3)}$

9. $\dfrac{a^2+5a}{7a} \cdot \dfrac{4a^2}{a^2+4a} = \dfrac{a(a+5)}{7a} \cdot \dfrac{4a^2}{a(a+4)} = \dfrac{4a(a+5)}{7(a+4)}$

11. $\dfrac{y^2-5y+6}{2y+4} \div \dfrac{2y-6}{y+2} = \dfrac{y^2-5y+6}{2y+4} \cdot \dfrac{y+2}{2y-6} = \dfrac{(y-2)(y-3)}{2(y+2)} \cdot \dfrac{y+2}{2(y-3)} = \dfrac{y-2}{4}$

13. $\dfrac{2x-8}{x^2-4} \cdot \dfrac{x^2+6x+8}{x-4} = \dfrac{2(x-4)}{(x+2)(x-2)} \cdot \dfrac{(x+2)(x+4)}{(x-4)} = \dfrac{2(x+4)}{x-2}$

15. $\dfrac{x-1}{x^2-x-6} \cdot \dfrac{x^2+5x+6}{x^2-1} = \dfrac{x-1}{(x-3)(x+2)} \cdot \dfrac{(x+2)(x+3)}{(x+1)(x-1)} = \dfrac{x+3}{(x-3)(x+1)}$

17. $\dfrac{a^2+10a+25}{a+5} \div \dfrac{a^2-25}{a-5} = \dfrac{(a+5)(a+5)}{a+5} \cdot \dfrac{a-5}{(a+5)(a-5)} = 1$

19. $\dfrac{y^3-5y^2}{y^4+3y^3+2y^2} \div \dfrac{y^2-5y+6}{y^2-2y-3} = \dfrac{y^3-5y^2}{y^4+3y^3+2y^2} \cdot \dfrac{y^2-2y-3}{y^2-5y+6}$

$$= \dfrac{y^2(y-5)}{y^2(y+2)(y+1)} \cdot \dfrac{(y-3)(y+1)}{(y-2)(y-3)}$$

$$= \dfrac{y-5}{(y+2)(y-2)}$$

21. $\dfrac{2x^2+17x+21}{x^2+2x-35} \cdot \dfrac{x^2-25}{2x^2-7x-15} = \dfrac{(2x+3)(x+7)}{(x+7)(x-5)} \cdot \dfrac{(x+5)(x-5)}{(2x+3)(x-5)} = \dfrac{x+5}{x-5}$

23. $\dfrac{2x^2+10x+12}{4x^2+24x+32} \cdot \dfrac{2x^2+18x+40}{x^2+8x+15} = \dfrac{2(x^2+5x+6)}{4(x^2+6x+8)} \cdot \dfrac{2(x^2+9x+20)}{x^2+8x+15}$

$$= \dfrac{2(x+2)(x+3)}{4(x+4)(x+2)} \cdot \dfrac{2(x+5)(x+4)}{(x+5)(x+3)}$$

$$= 1$$

25. $\dfrac{2a^2+7a+3}{a^2-16} \div \dfrac{4a^2+8a+3}{2a^2-5a-12} = \dfrac{(2a+1)(a+3)}{(a-4)(a+4)} \cdot \dfrac{(2a+3)(a-4)}{(2a+1)(2a+3)} = \dfrac{a+3}{a+4}$

27. $\dfrac{4y^2-12y+9}{y^2-36} \div \dfrac{2y^2-5y+3}{y^2+5y-6} = \dfrac{4y^2-12y+9}{y^2-36} \cdot \dfrac{y^2+5y-6}{2y^2-5y+3}$

$$= \dfrac{(2y-3)^2}{(y+6)(y-6)} \cdot \dfrac{(y+6)(y-1)}{(2y-3)(y-1)}$$

$$= \dfrac{2y-3}{y-6}$$

29. $\dfrac{x^2-1}{6x^2+42x+60} \cdot \dfrac{7x^2+17x+6}{x+1} \cdot \dfrac{6x+30}{7x^2-11x-6} = \dfrac{(x+1)(x-1)}{6(x+2)(x+5)} \cdot \dfrac{(7x+3)(x+2)}{x+1} \cdot \dfrac{6(x+5)}{(7x+3)(x-2)}$

$$= \dfrac{x-1}{x-2}$$

31. $\dfrac{18x^3+21x^2-60x}{21x^2-25x-4} \cdot \dfrac{28x^2-17x-3}{16x^3+28x^2-30x} = \dfrac{3x(6x^2+7x-20)}{21x^2-25x-4} \cdot \dfrac{28x^2-17x-3}{2x(8x^2+14x-15)}$

$$= \dfrac{3x(3x-4)(2x+5)}{(7x+1)(3x-4)} \cdot \dfrac{(7x+1)(4x-3)}{2x(4x-3)(2x+5)}$$

$$= \dfrac{3}{2}$$

33. $(x^2-9)\left(\dfrac{2}{x+3}\right) = \dfrac{(x+3)(x-3)}{1} \cdot \dfrac{2}{x+3} = 2(x-3)$

35. $a(a+5)(a-5)\left(\dfrac{2}{a^2-25}\right) = \dfrac{a(a+5)(a-5)}{1} \cdot \dfrac{2}{(a+5)(a-5)} = 2a$

37. $(x^2-x-6)\left(\dfrac{x+1}{x-3}\right) = \dfrac{x^2-x-6}{1} \cdot \dfrac{x+1}{x-3} = \dfrac{(x-3)(x+2)(x+1)}{x-3} = (x+2)(x+1)$

39. $(x^2-4x-5)\left(\dfrac{-2x}{x+1}\right) = \dfrac{(x-5)(x+1)}{1} \cdot \dfrac{-2x}{x+1} = -2x(x-5)$

41. $\dfrac{x^2-9}{x^2-3x} \cdot \dfrac{2x+10}{xy+5x+3y+15} = \dfrac{(x+3)(x-3)}{x(x-3)} \cdot \dfrac{2(x+5)}{x(y+5)+3(y+5)} = \dfrac{2(x+3)(-3)(x+5)}{x(x-3)(x+3)(y+5)} = \dfrac{2(x+5)}{x(y+5)}$

43. $\dfrac{2x^2+4x}{x^2-y^2} \cdot \dfrac{x^2+3x+xy+3y}{x^2+5x+6} = \dfrac{2x(x+2)}{(x+y)(x-y)} \cdot \dfrac{x(x+3)+y(x+3)}{(x+2)(x+3)} = \dfrac{2x}{x-y}$

45. $\dfrac{x^3-3x^2+4x-12}{x^4-16} \cdot \dfrac{3x^2+5x-2}{3x^2-10x+3} = \dfrac{x^2(x-3)+4(x-3)}{(x^2+4)(x^2-4)} \cdot \dfrac{(x+2)(3x-1)}{(3x-1)(x-3)}$

$= \dfrac{(x-3)(x^2+4)}{(x^2+4)(x+2)(x-2)} \cdot \dfrac{(x+2)(3x-1)}{(3x-1)(x-3)}$

$= \dfrac{1}{x-2}$

47. $\left(1-\dfrac{1}{2}\right)\left(1-\dfrac{1}{3}\right)\left(1-\dfrac{1}{4}\right)\left(1-\dfrac{1}{5}\right) = \left(\dfrac{2}{2}-\dfrac{1}{2}\right)\left(\dfrac{3}{3}-\dfrac{1}{3}\right)\left(\dfrac{4}{4}-\dfrac{1}{4}\right)\left(\dfrac{5}{5}-\dfrac{1}{5}\right) = \dfrac{1}{2} \cdot \dfrac{2}{3} \cdot \dfrac{3}{4} \cdot \dfrac{4}{5} = \dfrac{1}{5}$

49. $\left(1-\dfrac{1}{2}\right)\left(1-\dfrac{1}{3}\right)\left(1-\dfrac{1}{4}\right)\dots\left(1-\dfrac{1}{99}\right)\left(1-\dfrac{1}{100}\right) = \left(\dfrac{1}{2}\right)\left(\dfrac{2}{3}\right)\left(\dfrac{3}{4}\right)\dots\left(\dfrac{98}{99}\right)\left(\dfrac{99}{100}\right) = \dfrac{1}{100}$

51. Since 5,280 feet = 1 mile, the height is: $\dfrac{14,494 \text{ feet}}{5,280 \text{ feet/mile}} \approx 2.7 \text{ miles}$

53. $\dfrac{1088 \text{ feet}}{1 \text{ second}} \cdot \dfrac{1 \text{ mile}}{5280 \text{ feet}} \cdot \dfrac{60 \text{ seconds}}{1 \text{ minute}} \cdot \dfrac{60 \text{ minutes}}{1 \text{ hour}} \approx 742 \text{ miles/hour}$

55. $\dfrac{785 \text{ feet}}{20 \text{ minutes}} \cdot \dfrac{60 \text{ minutes}}{1 \text{ hour}} \cdot \dfrac{1 \text{ mile}}{5280 \text{ feet}} \approx 0.45 \text{ miles/hour}$

57. $\dfrac{518 \text{ feet}}{40 \text{ seconds}} \cdot \dfrac{60 \text{ seconds}}{1 \text{ minute}} \cdot \dfrac{60 \text{ minutes}}{1 \text{ hour}} \cdot \dfrac{1 \text{ mile}}{5280 \text{ feet}} \approx 8.8 \text{ miles/hour}$

59. Her average speed on level ground is: $\dfrac{2 \text{ miles}}{1/3 \text{ hour}} = 6$ miles/hour

Her average speed downhill is: $\dfrac{6 \text{ miles}}{2/3 \text{ hour}} = 9$ miles/hour

61. $\dfrac{1}{2} + \dfrac{5}{2} = \dfrac{6}{2} = 3$

63. $2 + \dfrac{3}{4} = \dfrac{2 \cdot 4}{1 \cdot 4} + \dfrac{3}{4} = \dfrac{8}{4} + \dfrac{3}{4} = \dfrac{11}{4}$

65. $\dfrac{1}{10} + \dfrac{3}{14} = \dfrac{1 \cdot 7}{10 \cdot 7} + \dfrac{3 \cdot 5}{14 \cdot 5} = \dfrac{7}{70} + \dfrac{15}{70} = \dfrac{22}{70} = \dfrac{11}{35}$

67. $\dfrac{10x^4}{2x^2} + \dfrac{12x^6}{3x^4} = 5x^2 + 4x^2 = 9x^2$

69. $\dfrac{12a^2b^5}{3ab^3} + \dfrac{14a^4b^7}{7a^3b^5} = 4ab^2 + 2ab^2 = 6ab^2$

6.3 Addition and Subtraction of Rational Expressions

1. $\dfrac{3}{x} + \dfrac{4}{x} = \dfrac{7}{x}$

3. $\dfrac{9}{a} - \dfrac{5}{a} = \dfrac{4}{a}$

5. $\dfrac{1}{x+1} + \dfrac{x}{x+1} = \dfrac{x+1}{x+1} = 1$

7. $\dfrac{y^2}{y-1} - \dfrac{1}{y-1} = \dfrac{y^2-1}{y-1} = \dfrac{(y+1)(y-1)}{y-1} = y+1$

9. $\dfrac{x^2}{x+2} + \dfrac{4x+4}{x+2} = \dfrac{x^2+4x+4}{x+2} = \dfrac{(x+2)^2}{x+2} = x+2$

11. $\dfrac{x^2}{x-2} - \dfrac{4x-4}{x-2} = \dfrac{x^2-4x+4}{x-2} = \dfrac{(x-2)^2}{x-2} = x-2$

13. $\dfrac{x+2}{x+6} - \dfrac{x-4}{x+6} = \dfrac{x+2-x+4}{x+6} = \dfrac{6}{x+6}$

15. $\dfrac{y}{2} - \dfrac{2}{y} = \dfrac{y \cdot y}{2 \cdot y} - \dfrac{2 \cdot 2}{y \cdot 2} = \dfrac{y^2}{2y} - \dfrac{4}{2y} = \dfrac{y^2-4}{2y} = \dfrac{(y+2)(y-2)}{2y}$

17. $\dfrac{1}{2} + \dfrac{a}{3} = \dfrac{1 \cdot 3}{2 \cdot 3} + \dfrac{a \cdot 2}{3 \cdot 2} = \dfrac{3}{6} + \dfrac{2a}{6} = \dfrac{2a+3}{6}$

19. $\dfrac{x}{x+1} + \dfrac{3}{4} = \dfrac{x \cdot 4}{(x+1) \cdot 4} + \dfrac{3 \cdot (x+1)}{4 \cdot (x+1)} = \dfrac{4x}{4(x+1)} + \dfrac{3x+3}{4(x+1)} = \dfrac{4x+3x+3}{4(x+1)} = \dfrac{7x+3}{4(x+1)}$

21. $\dfrac{x+1}{x-2} - \dfrac{4x+7}{5x-10} = \dfrac{x+1}{x-2} - \dfrac{4x+7}{5(x-2)} = \dfrac{5x+5}{5(x-2)} - \dfrac{4x+7}{5(x-2)} = \dfrac{5x+5-4x-7}{5(x-2)} = \dfrac{x-2}{5(x-2)} = \dfrac{1}{5}$

23. $\dfrac{4x-2}{3x+12} - \dfrac{x-2}{x+4} = \dfrac{4x-2}{3(x+4)} - \dfrac{(x-2) \cdot 3}{(x+4) \cdot 3} = \dfrac{4x-2}{3(x+4)} - \dfrac{3x-6}{3(x+4)} = \dfrac{4x-2-3x+6}{3(x+4)} = \dfrac{x+4}{3(x+4)} = \dfrac{1}{3}$

25. $\dfrac{6}{x(x-2)} + \dfrac{3}{x} = \dfrac{6}{x(x-2)} + \dfrac{3(x-2)}{x(x-2)} = \dfrac{6}{x(x-2)} + \dfrac{3x-6}{x(x-2)} = \dfrac{6+3x-6}{x(x-2)} = \dfrac{3x}{x(x-2)} = \dfrac{3}{x-2}$

27. $\dfrac{4}{a} - \dfrac{12}{a^2+3a} = \dfrac{4(a+3)}{a(a+3)} - \dfrac{12}{a(a+3)} = \dfrac{4a+12}{a(a+3)} - \dfrac{12}{a(a+3)} = \dfrac{4a+12-12}{a(a+3)} = \dfrac{4a}{a(a+3)} = \dfrac{4}{a+3}$

29. $\dfrac{2}{x+5} - \dfrac{10}{x^2-25} = \dfrac{2}{x+5} - \dfrac{10}{(x+5)(x-5)}$

$= \dfrac{2(x-5)}{(x+5)(x-5)} - \dfrac{10}{(x+5)(x-5)}$

$= \dfrac{2x-10-10}{(x+5)(x-5)}$

$= \dfrac{2x-20}{(x+5)(x-5)}$

$= \dfrac{2(x-10)}{(x+5)(x-5)}$

31. $\dfrac{x-4}{x-3} + \dfrac{6}{x^2-9} = \dfrac{(x-4)(x+3)}{(x-3)(x+3)} + \dfrac{6}{(x+3)(x-3)}$

$= \dfrac{x^2-x-12}{(x+3)(x-3)} + \dfrac{6}{(x+3)(x-3)}$

$= \dfrac{x^2-x-12+6}{(x+3)(x-3)}$

$= \dfrac{x^2-x-6}{(x+3)(x-3)}$

$= \dfrac{(x-3)(x+2)}{(x+3)(x-3)}$

$= \dfrac{x+2}{x+3}$

33. $\dfrac{a-4}{a-3} + \dfrac{5}{a^2-a-6} = \dfrac{a-4}{a-3} + \dfrac{5}{(a+2)(a-3)}$

$= \dfrac{a+2}{a+2}\left(\dfrac{a-4}{a-3}\right) + \dfrac{5}{(a+2)(a-3)}$

$= \dfrac{a^2-2a-8}{(a+2)(a-3)} + \dfrac{5}{(a+2)(a-3)}$

$= \dfrac{a^2-2a-3}{(a+2)(a-3)}$

$= \dfrac{(a+1)(a-3)}{(a+2)(a-3)}$

$= \dfrac{a+1}{a+2}$

35. $\dfrac{8}{x^2-16} - \dfrac{7}{x^2-x-12} = \dfrac{8}{(x+4)(x-4)} - \dfrac{7}{(x-4)(x+3)}$

$= \dfrac{8(x+3)}{(x+4)(x-4)(x+3)} - \dfrac{7(x+4)}{(x+4)(x-4)(x+3)}$

$= \dfrac{8x+24}{(x+4)(x-4)(x+3)} - \dfrac{7x+28}{(x+4)(x-4)(x+3)}$

$= \dfrac{8x+24-7x-28}{(x+4)(x-4)(x+3)}$

$= \dfrac{x-4}{(x+4)(x-4)(x+3)}$

$= \dfrac{1}{(x+4)(x+3)}$

37. $\dfrac{4y}{y^2+6y+5}-\dfrac{3y}{y^2+5y+4}=\dfrac{4y}{(y+5)(y+1)}-\dfrac{3y}{(y+4)(y+1)}$

$$=\dfrac{4y(y+4)}{(y+5)(y+1)(y+4)}-\dfrac{3y(y+5)}{(y+1)(y+4)(y+5)}$$

$$=\dfrac{4y^2+16y-3y^2-15y}{(y+5)(y+1)(y+4)}$$

$$=\dfrac{y^2+y}{(y+5)(y+1)(y+4)}$$

$$=\dfrac{y(y+1)}{(y+5)(y+1)(y+4)}$$

$$=\dfrac{y}{(y+5)(y+4)}$$

39. $\dfrac{4x+1}{x^2+5x+4}-\dfrac{x+3}{x^2+4x+3}=\dfrac{4x+1}{(x+4)(x+1)}-\dfrac{x+3}{(x+3)(x+1)}$

$$=\dfrac{(4x+1)(x+3)}{(x+4)(x+1)(x+3)}-\dfrac{(x+3)(x+4)}{(x+4)(x+1)(x+3)}$$

$$=\dfrac{4x^2+13x+3}{(x+4)(x+1)(x+3)}-\dfrac{x^2+7x+12}{(x+4)(x+1)(x+3)}$$

$$=\dfrac{4x^2+13x+3-x^2-7x-12}{(x+4)(x+1)(x+3)}$$

$$=\dfrac{3x^2+6x-9}{(x+4)(x+1)(x+3)}$$

$$=\dfrac{3(x+3)(x-1)}{(x+4)(x+1)(x+3)}$$

$$=\dfrac{3(x-1)}{(x+4)(x+1)}$$

41. $\dfrac{1}{x}+\dfrac{x}{3x+9}-\dfrac{3}{x^2+3x}=\dfrac{1}{x}+\dfrac{x}{3(x+3)}-\dfrac{3}{x(x+3)}$

$$=\dfrac{3(x+3)}{3(x+3)}\left(\dfrac{1}{x}\right)+\dfrac{x}{x}\left(\dfrac{x}{3(x+3)}\right)-\dfrac{3}{3}\left(\dfrac{3}{x(x+3)}\right)$$

$$=\dfrac{3x+9}{3x(x+3)}+\dfrac{x^2}{3x(x+3)}-\dfrac{9}{3x(x+3)}$$

$$=\dfrac{x^2+3x}{3x(x+3)}$$

$$=\dfrac{x(x+3)}{3x(x+3)}$$

$$=\dfrac{1}{3}$$

43. See the table in the back of the textbook.   45. See the table in the back of the textbook.

47. $1 + \dfrac{1}{x+2} = \dfrac{1(x+2)}{1(x+2)} + \dfrac{1}{x+2} = \dfrac{x+2}{x+2} + \dfrac{1}{x+2} = \dfrac{x+2+1}{x+2} = \dfrac{x+3}{x+2}$

49. $1 - \dfrac{1}{x+3} = \dfrac{1(x+3)}{1(x+3)} - \dfrac{1}{x+3} = \dfrac{x+3}{x+3} - \dfrac{1}{x+3} = \dfrac{x+3-1}{x+3} = \dfrac{x+2}{x+3}$

51. $x + 2\left(\dfrac{1}{x}\right) = x + \dfrac{2}{x} = \dfrac{x \cdot x}{1 \cdot x} + \dfrac{2}{x} = \dfrac{x^2}{x} + \dfrac{2}{x} = \dfrac{x^2 + 2}{x}$

53. $\dfrac{1}{x} + \dfrac{1}{2x} = \dfrac{2 \cdot 1}{2x} + \dfrac{1}{2x} = \dfrac{3}{2x}$

55. $2x + 3(x-3) = 6$
$$2x + 3x - 9 = 6$$
$$5x - 9 = 6$$
$$5x = 15$$
$$x = 3$$

57. $x - 3(x+3) = x - 3$
$$x - 3x - 9 = x - 3$$
$$-2x - 9 = x - 3$$
$$-9 = 3x - 3$$
$$-6 = 3x$$
$$-2 = x$$

59. $7 - 2(3x+1) = 4x + 3$
$$7 - 6x - 2 = 4x + 3$$
$$-6x + 5 = 4x + 3$$
$$-10x + 5 = 3$$
$$-10x = -2$$
$$x = \dfrac{1}{5}$$

61. $x^2 + 5x + 6 = 0$
$$(x+2)(x+3) = 0$$
$$x + 2 = 0 \quad \text{or} \quad x + 3 = 0$$
$$x = -2 \qquad\qquad x = -3$$

63. $x^2 - x = 6$
$$x^2 - x - 6 = 0$$
$$(x-3)(x+2) = 0$$
$$x = -2,\ 3$$

65. $x^2 - 5x = 0$
$$x(x-5) = 0$$
$$x = 0 \quad \text{or} \quad x - 5 = 0$$
$$x = 0 \qquad\qquad x = 5$$

6.4 Equations Involving Rational Expressions

1. $$\frac{x}{3} + \frac{1}{2} = -\frac{1}{2}$$

$$6\left(\frac{x}{3}\right) + 6\left(\frac{1}{2}\right) = 6\left(-\frac{1}{2}\right)$$

$$2x + 3 = -3$$

$$2x = -6$$

$$x = -3$$

The solution is $x = -3$.

3. $$\frac{4}{a} = \frac{1}{5}$$

$$5a\left(\frac{4}{a}\right) = 5a\left(\frac{1}{5}\right)$$

$$20 = a$$

The solution is $a = 20$.

5. $$\frac{3}{x} + 1 = \frac{2}{x}$$

$$x\left(\frac{3}{x}\right) + x(1) = x\left(\frac{2}{x}\right)$$

$$3 + x = 2$$

$$x = -1$$

The solution is $x = -1$.

7. $$\left(\frac{3}{a} - \frac{2}{a}\right) = \frac{1}{5}$$

$$5a\left(\frac{3}{a} - \frac{2}{a}\right) = 5a\left(\frac{1}{5}\right)$$

$$15 - 10 = a$$

$$a = 5$$

The solution is $a = 5$.

9. $$\frac{3}{x} + 2 = \frac{1}{2}$$

$$2x\left(\frac{3}{x}\right) + 2x(2) = 2x\left(\frac{1}{2}\right)$$

$$6 + 4x = x$$

$$6 = -3x$$

$$-2 = x$$

The solution is $x = -2$.

11. $$\frac{1}{y} - \frac{1}{2} = -\frac{1}{4}$$

$$4y\left(\frac{1}{y} - \frac{1}{2}\right) = 4y\left(-\frac{1}{4}\right)$$

$$4 - 2y = -y$$

$$4 = y$$

The solution is $y = 4$.

13. $$1 - \frac{8}{x} = \frac{-15}{x^2}$$

$$x^2(1) - x^2\left(\frac{8}{x}\right) = x^2\left(\frac{-15}{x^2}\right)$$

$$x^2 - 8x = -15$$

$$x^2 - 8x + 15 = 0$$

$$(x - 3)(x - 5) = 0$$

$$x = 3, 5$$

The solutions are 3 and 5.

15. $$\frac{x}{2} - \frac{4}{x} = -\frac{7}{2}$$

$$2x\left(\frac{x}{2} - \frac{4}{x}\right) = 2x\left(-\frac{7}{2}\right)$$

$$x^2 - 8 = -7x$$

$$x^2 + 7x - 8 = 0$$

$$(x + 8)(x - 1) = 0$$

$$x = -8, 1$$

The solutions are -8, and 1.

17.
$$\frac{x-3}{2} + \frac{2x}{3} = \frac{5}{6}$$
$$6\left(\frac{x-3}{2}\right) + 6\left(\frac{2x}{3}\right) = 6\left(\frac{5}{6}\right)$$
$$3(x-3) + 2(2x) = 5$$
$$3x - 9 + 4x = 5$$
$$7x - 9 = 5$$
$$7x = 14$$
$$x = \frac{14}{7}$$
$$x = 2$$

The solution is $x = 2$.

19.
$$\frac{x+1}{3} + \frac{x-3}{4} = \frac{1}{6}$$
$$12\left(\frac{x+1}{3} + \frac{x-3}{4}\right) = 12\left(\frac{1}{6}\right)$$
$$4(x+1) + 3(x-3) = 2$$
$$4x + 4 + 3x - 9 = 2$$
$$7x - 5 = 2$$
$$7x = 7$$
$$x = 1$$

The solution is $x = 1$.

21.
$$\frac{6}{x+2} = \frac{3}{5}$$
$$5(x+2)\left(\frac{6}{x+2}\right) = 5(x+2)\left(\frac{3}{5}\right)$$
$$30 = 3(x+2)$$
$$30 = 3x + 6$$
$$24 = 3x$$
$$8 = x$$

The solution is $x = 8$.

23.
$$\frac{3}{y-2} = \frac{2}{y-3}$$
$$(y-2)(y-3) \cdot \frac{3}{y-2} = (y-2)(y-3) \cdot \frac{2}{y-3}$$
$$3(y-3) = 2(y-2)$$
$$3y - 9 = 2y - 4$$
$$y = 5$$

The solution is $y = 5$.

25.
$$\frac{x}{x-2} + \frac{2}{3} = \frac{2}{x-2}$$
$$3(x-2)\left(\frac{x}{x-2}\right) + 3(x-2)\left(\frac{2}{3}\right) = 3(x-2)\left(\frac{2}{x-2}\right)$$
$$3x + 2(x-2) = 6$$
$$3x + 2x - 4 = 6$$
$$5x = 10$$
$$x = 2$$

27.
$$\frac{x}{x-2} + \frac{3}{2} = \frac{9}{2(x-2)}$$
$$2(x-2)\left(\frac{x}{x-2} + \frac{3}{2}\right) = 2(x-2) \cdot \frac{9}{2(x-2)}$$
$$2x + 3(x-2) = 9$$
$$2x + 3x - 6 = 9$$
$$5x - 6 = 9$$
$$5x = 15$$
$$x = 3$$

29.
$$\frac{5}{x+2}+\frac{1}{x+3}=\frac{-1}{x^2+5x+6}$$

$$(x+2)(x+3)\frac{5}{x+2}+(x+2)(x+3)\frac{1}{x+3}=(x+2)(x+3)\frac{-1}{(x+2)(x+3)}$$

$$5(x+3)+x+2=-1$$
$$5x+15+x+2=-1$$
$$6x+17=-1$$
$$6x=-18$$
$$x=-3$$

Since $x=-3$ does not check in the original equation, there is no solution.

31.
$$\frac{8}{x^2-4}+\frac{3}{x+2}=\frac{1}{x-2}$$

$$(x+2)(x-2)\left(\frac{8}{x^2-4}+\frac{3}{x+2}\right)=(x+2)(x-2)\cdot\frac{1}{x-2}$$

$$8+3(x-2)=1(x+2)$$
$$8+3x-6=x+2$$
$$2x=0$$
$$x=0$$

The solution is $x=0$.

33.
$$\frac{a}{2}+\frac{3}{a-3}=\frac{a}{a-3}$$

$$2(a-3)\frac{a}{2}+2(a-3)\left(\frac{3}{a-3}\right)=2(a-3)\frac{a}{a-3}$$

$$a(a-3)+6=2a$$
$$a^2-3a+6=2a$$
$$a^2-5a+6=0$$
$$(a-2)(a-3)=0$$
$$a=2,\,3$$

Since $a=3$ does not check in the original equation, the solution is $a=2$.

35.
$$\frac{6}{(y+2)(y-2)}=\frac{4}{y(y+2)}$$

$$y(y+2)(y-2)\cdot\frac{6}{(y+2)(y-2)}=y(y+2)(y-2)\cdot\frac{4}{y(y+2)}$$

$$6y=4(y-2)$$
$$6y=4y-8$$
$$2y=-8$$
$$y=-4$$

The solution is $y=-4$.

37.
$$\frac{2}{a^2-9}=\frac{3}{a^2+a-12}$$
$$\frac{2}{(a+3)(a-3)}=\frac{3}{(a+4)(a-3)}$$
$$(a+3)(a-3)(a+4)\cdot\frac{2}{(a+3)(a-3)}=(a+3)(a-3)(a+4)\cdot\frac{3}{(a+4)(a-3)}$$
$$2(a+4)=3(a+3)$$
$$2a+8=3a+9$$
$$8=a+9$$
$$-1=a$$

The solution is $a=-1$.

39.
$$\frac{3x}{x-5}-\frac{2x}{x+1}=\frac{-42}{(x-5)(x+1)}$$
$$(x-5)(x+1)\left(\frac{3x}{x-5}-\frac{2x}{x+1}\right)=(x-5)(x+1)\cdot\frac{-42}{(x-5)(x+1)}$$
$$3x(x+1)-2x(x-5)=-42$$
$$3x^2+3x-2x^2+10x=-42$$
$$x^2+13x+42=0$$
$$(x+7)(x+6)=0$$
$$x=-7,\,-6$$

The solutions are -7 and -6.

41.
$$\frac{2x}{x+2}=\frac{x}{x+3}-\frac{3}{x^2+5x+6}$$
$$(x+2)(x+3)\frac{2x}{x+2}=(x+2)(x+3)\frac{x}{x+3}-(x+2)(x+3)\frac{3}{(x+2)(x+3)}$$
$$2x(x+3)=x(x+2)-3$$
$$2x^2+6x=x^2+2x-3$$
$$x^2+4x+3=0$$
$$(x+3)(x+1)=0$$
$$x=-3,\,-1$$

Since $x=-3$ does not check in the original equation, the solution is $x=-1$.

43.
$$x+\frac{4}{x}=5$$
$$x(x)+x\left(\frac{4}{x}\right)=x(5)$$
$$x^2+4=5x$$
$$x^2-5x+4=0$$
$$(x-1)(x-4)=0$$
$$x=1,\,4;\text{ agrees}$$

45.
Let $x=$ the number
$$2(x-3)-5=3$$
$$2x-6-5=3$$
$$2x-11=3$$
$$2x=14$$
$$x=7$$

47. $\quad$ Let $w =$ the width

Let $2w + 5 =$ the length

$2w + 2(2w + 5) = 34$

$2w + 4w + 10 = 34$

$6w + 10 = 34$

$6w = 24$

$w = 4$

$2w + 5 = 13$

The length is 13 inches and the width is 4 inches.

49. Let x and $x + 2 =$ the two integers

$x(x + 2) = 48$

$x^2 + 2x = 48$

$x^2 + 2x - 48 = 0$

$(x + 8)(x - 6) = 0$

$x = -8, \ 6$

$x + 2 = -6, 8$

The two integers are either -8 and -6, or 6 and 8.

51. $\quad$ Let x and $x + 2 =$ the two legs

$x^2 + (x + 2)^2 = 10^2$

$x^2 + x^2 + 4x + 4 = 100$

$2x^2 + 4x - 96 = 0$

$x^2 + 2x - 48 = 0$

$(x + 8)(x - 6) = 0$

$x = 6$

$x + 2 = 8$

$x = -8$ is impossible

The legs are 6 inches and 8 inches.

6.5 Applications

1. $\quad$ Let x and $3x =$ the two numbers

$\dfrac{1}{x} + \dfrac{1}{3x} = \dfrac{16}{3}$

$3x\left(\dfrac{1}{x}\right) + 3x\left(\dfrac{1}{3x}\right) = 3x\left(\dfrac{16}{3}\right)$

$3 + 1 = 16x$

$4 = 16x$

$\dfrac{4}{16} = x$

$\dfrac{1}{4} = x$

$\dfrac{3}{4} = 3x$

The numbers are 1/4 and 3/4.

3. $\quad$ Let $x =$ the number

$x + \dfrac{1}{x} = \dfrac{13}{6}$

$6x\left(x + \dfrac{1}{x}\right) = 6x\left(\dfrac{13}{6}\right)$

$6x^2 + 6 = 13x$

$6x^2 - 13x + 6 = 0$

$(3x - 2)(2x - 3) = 0$

$x = \dfrac{2}{3}, \ \dfrac{3}{2}$

The number is 2/3 or 3/2.

5.   Let x = the number

$$\frac{7+x}{9+x} = \frac{5}{7}$$

$$7(9+x)\frac{7+x}{9+x} = 7(9+x)\frac{5}{7}$$

$$7(7+x) = 5(9+x)$$

$$49+7x = 45+5x$$

$$49+2x = 45$$

$$2x = -4$$

$$x = -2$$

The number is –2.

7.   Let x and $x+2$ = the two integers

$$\frac{1}{x} + \frac{1}{x+2} = \frac{5}{12}$$

$$12x(x+2)\left(\frac{1}{x} + \frac{1}{x+2}\right) = 12x(x+2)\left(\frac{5}{12}\right)$$

$$12(x+2)+12x = 5x(x+2)$$

$$12x+24+12x = 5x^2 +10x$$

$$0 = 5x^2 -14x -24$$

$$(5x+6)(x-4) = 0$$

$$x = 4$$

$$x = -\frac{6}{5} \text{ is impossible}$$

$$x+2 = 6$$

The integers are 4 and 6.

9. Let x = the rate of the boat in still water.

	d	r	t
Upstream	26	$x-3$	$\dfrac{26}{x-3}$
Downstream	38	$x+3$	$\dfrac{38}{x+3}$

$$\frac{26}{x-3} = \frac{38}{x+3}$$

$$(x+3)(x-3)\frac{26}{x-3} = (x+3)(x-3)\frac{38}{x+3}$$

$$(x+3)26 = (x-3)38$$

$$26x+78 = 38x -114$$

$$78 = 12x -114$$

$$192 = 12x$$

$$16 = x$$

The speed is the boat in still water is 16 mph.

11. Let $x =$ the plane speed in still air.

	d	r	t
Against Wind	140	$x-20$	$\dfrac{140}{x-20}$
With Wind	160	$x+20$	$\dfrac{160}{x+20}$

$$\frac{140}{x-20} = \frac{160}{x+20}$$

$$(x+20)(x-20)\cdot\frac{140}{x-20} = (x+20)(x-20)\cdot\frac{160}{x+20}$$

$$140(x+20) = 160(x-20)$$

$$140x+2800 = 160x-3200$$

$$-20x+2800 = -3200$$

$$-20x = -6000$$

$$x = 300$$

The plane speed in still air is 300 mph.

13. Let x and $x+20 =$ the rates of each plane

	d	r	t
Plane 1	285	$x+20$	$\dfrac{285}{x+20}$
Plane 2	255	x	$\dfrac{255}{x}$

$$\frac{285}{x+20} = \frac{255}{x}$$

$$x(x+20)\frac{285}{x+20} = x(x+20)\frac{255}{x}$$

$$285x = 255(x+20)$$

$$285x = 255x+5100$$

$$30x = 5100$$

$$x = 170$$

$$x+20 = 190$$

The plane speeds are 170 mph and 190 mph.

15. Let x = her rate downhill.

	d	r	t
Level Ground	2	$x-3$	$\dfrac{2}{x-3}$
Downhill	6	x	$\dfrac{6}{x}$

$$\frac{2}{x-3}+\frac{6}{x}=1$$

$$x(x-3)\left(\frac{2}{x-3}+\frac{6}{x}\right)=x(x-3)\cdot 1$$

$$2x+6(x-3)=x(x-3)$$

$$2x+6x-18=x^2-3x$$

$$8x-18=x^2-3x$$

$$0=x^2-11x+18$$

$$0=(x-2)(x-9)$$

$$x=9\quad (x=2\text{ is impossible})$$

Tina runs 9 mph on the downhill part of the course.

17. Let x = her rate on level ground.

	d	r	t
Level Ground	4	x	$\dfrac{4}{x}$
Downhill	5	$x+2$	$\dfrac{5}{x+2}$

$$\frac{4}{x}+\frac{5}{x+2}=1$$

$$x(x+2)\left(\frac{4}{x}+\frac{5}{x+2}\right)=x(x+2)\cdot 1$$

$$4(x+2)+5x=x(x+2)$$

$$4x+8+5x=x^2+2x$$

$$9x+8=x^2+2x$$

$$0=x^2-7x-8$$

$$0=(x-8)(x+1)$$

$$x=8\quad (x=-1\text{ is impossible})$$

Jerri jogs 8 mph on level ground.

19. Let t = the time to fill the pool with both pipes left open.

$$\frac{1}{12} - \frac{1}{15} = \frac{1}{t}$$
$$60t\left(\frac{1}{12} - \frac{1}{15}\right) = 60t \cdot \frac{1}{t}$$
$$5t - 4t = 60$$
$$t = 60$$

It will take 60 hours to fill the pool with both pipes left open.

21. Let t = the time to fill the bathtub with both faucets open.

$$\frac{1}{10} + \frac{1}{12} = \frac{1}{t}$$
$$60t\left(\frac{1}{10} + \frac{1}{12}\right) = 60t \cdot \frac{1}{t}$$
$$6t + 5t = 60$$
$$11t = 60$$
$$t = \frac{60}{11} = 5\frac{5}{11}$$

It will take 5 5/11 minutes to fill the tub with both faucets open.

23. Let t = the time to fill the sink with both the faucet and the drain left open.

$$\frac{1}{3} - \frac{1}{4} = \frac{1}{t}$$
$$12t\left(\frac{1}{3} - \frac{1}{4}\right) = 12t \cdot \frac{1}{t}$$
$$4t - 3t = 12$$
$$t = 12$$

It will take 12 minutes for the sink to overflow with both the faucet and drain left open.

25. See the graph in the back of the textbook.

27. $y = \dfrac{-4}{x}$

See the graph in the back of the textbook.

29. $y = \dfrac{8}{x}$

See the graph in the back of the textbook.

31. See the graph in the back of the textbook.

33. $15a^3b^3 - 20a^2b - 35ab^2 = 5ab\left(3a^2b^2 - 4a - 7b\right)$

35. $x^2 - 4x - 12 = \left(x - 6\right)\left(x + 2\right)$

37. $x^4 - 16 = \left(x^2 + 4\right)\left(x^2 - 4\right) = \left(x^2 + 4\right)\left(x + 2\right)\left(x - 2\right)$

39. $5x^3 - 25x^2 - 30x = 5x(x^2 - 5x - 6) = 5x(x-6)(x+1)$

41. Solve the equation by factoring:

$$x^2 - 6x = 0$$
$$x(x-6) = 0$$
$$x = 0, 6$$

43. Solving the equation by factoring:

$$x(x+2) = 80$$
$$x^2 + 2x = 80$$
$$x^2 + 2x - 80 = 0$$
$$(x+10)(x-8) = 0$$
$$x = -10, 8$$

45. Let x and $x+3$ =the two legs.

$$x^2 + (x+3)^2 = 15^2$$
$$x^2 + x^2 + 6x + 9 = 225$$
$$2x^2 + 6x - 216 = 0$$
$$x^2 + 3x - 108 = 0$$
$$(x+12)(x-9) = 0$$
$$x = 9 \quad (x = -12 \text{ is impossible})$$
$$x + 3 = 12$$

The two legs are 9 inches and 12 inches.

6.6 Complex Fractions

1. $\dfrac{\frac{3}{4}}{\frac{1}{8}} = \dfrac{8 \cdot \frac{3}{4}}{8 \cdot \frac{1}{8}} = \dfrac{6}{1} = 6$

3. $\dfrac{\frac{2}{3}}{4} = \dfrac{\frac{2}{3} \cdot 3}{4 \cdot 3} = \dfrac{2}{12} = \dfrac{1}{6}$

5. $\dfrac{\frac{x^2}{y}}{\frac{x}{y^3}} = \dfrac{y^3 \cdot \frac{x^2}{y}}{y^3 \cdot \frac{x}{y^3}} = \dfrac{x^2 y^2}{x} = xy^2$

7. $\dfrac{\frac{4x^3}{y^6}}{\frac{8x^2}{y^7}} = \dfrac{\frac{4x^3}{y^6} \cdot y^7}{\frac{8x^2}{y^7} \cdot y^7} = \dfrac{4x^3 y}{8x^2} = \dfrac{xy}{2}$

9. $\dfrac{y+\frac{1}{x}}{x+\frac{1}{y}} = \dfrac{xy\left(y+\frac{1}{x}\right)}{xy\left(x+\frac{1}{y}\right)}$

$= \dfrac{xy^2+y}{x^2y+x}$

$= \dfrac{y(xy+1)}{x(xy+1)}$

$= \dfrac{y}{x}$

11. $\dfrac{1+\frac{1}{a}}{1-\frac{1}{a}} = \dfrac{\left(1+\frac{1}{a}\right)\cdot a}{\left(1-\frac{1}{a}\right)\cdot a} = \dfrac{a+1}{a-1}$

13. $\dfrac{\frac{x+1}{x^2-9}}{\frac{2}{x+3}} = \dfrac{x+1}{x^2-9}\cdot\dfrac{x+3}{2}$

$= \dfrac{x+1}{(x+3)(x-3)}\cdot\dfrac{x+3}{2}$

$= \dfrac{x+1}{2(x-3)}$

15. $\dfrac{\frac{1}{a+2}}{\frac{1}{a^2-a-6}} = \dfrac{\frac{1}{a+2}(a-3)(a+2)}{\frac{1}{(a-3)(a+2)}(a-3)(a+2)}$

$= \dfrac{a-3}{1} = a-3$

17. $\dfrac{1-\frac{9}{y^2}}{1-\frac{1}{y}-\frac{6}{y^2}} = \dfrac{\left(1-\frac{9}{y^2}\right)y^2}{\left(1-\frac{1}{y}-\frac{6}{y^2}\right)y^2} = \dfrac{y^2-9}{y^2-y-6}$

$= \dfrac{(y+3)(y-3)}{(y+2)(y-3)}$

$= \dfrac{y+3}{y+2}$

19. $\dfrac{\frac{1}{y}+\frac{1}{x}}{\frac{1}{xy}} = \dfrac{\left(\frac{1}{y}+\frac{1}{x}\right)xy}{\left(\frac{1}{xy}\right)xy} = \dfrac{x+y}{1} = x+y$

21. $\dfrac{1-\frac{1}{a^2}}{1-\frac{1}{a}} = \dfrac{\left(1-\frac{1}{a^2}\right)a^2}{\left(1-\frac{1}{a}\right)a^2} = \dfrac{a^2-1}{a^2-a} = \dfrac{(a+1)(a-1)}{a(a-1)}$

$= \dfrac{a+1}{a}$

23. $\dfrac{\frac{1}{10x}-\frac{y}{10x^2}}{\frac{1}{10}-\frac{y}{10x}} = \dfrac{\left(\frac{1}{10x}-\frac{y}{10x^2}\right)10x^2}{\left(\frac{1}{10}-\frac{y}{10x}\right)10x^2} = \dfrac{x-y}{x^2-xy}$

$= \dfrac{1(x-y)}{x(x-y)} = \dfrac{1}{x}$

25. $\dfrac{\dfrac{1}{a+1}+2}{\dfrac{1}{a+1}+3} = \dfrac{\left(\dfrac{1}{a+1}+2\right)(a+1)}{\left(\dfrac{1}{a+1}+3\right)(a+1)} = \dfrac{1+2(a+1)}{1+3(a+1)}$

$$= \dfrac{1+2a+2}{1+3a+3} = \dfrac{2a+3}{3a+4}$$

27. Simplify each parenthesis first:

$1 - \dfrac{1}{x} = \dfrac{x}{x} - \dfrac{1}{x} = \dfrac{x-1}{x}$

$1 - \dfrac{1}{x+1} = \dfrac{x+1}{x+1} - \dfrac{1}{x+1} = \dfrac{x}{x+1}$

$1 - \dfrac{1}{x+2} = \dfrac{x+2}{x+2} - \dfrac{1}{x+2} = \dfrac{x+1}{x+2}$

Multiply

$\left(1 - \dfrac{1}{x}\right)\left(1 - \dfrac{1}{x+1}\right)\left(1 - \dfrac{1}{x+2}\right)$

$= \dfrac{x-1}{x} \cdot \dfrac{x}{x+1} \cdot \dfrac{x+1}{x+2} = \dfrac{x-1}{x+2}$

29. Simplify each parenthesis first:

$1 + \dfrac{1}{x+3} = \dfrac{x+3}{x+3} + \dfrac{1}{x+3} = \dfrac{x+4}{x+3}$

$1 + \dfrac{1}{x+2} = \dfrac{x+2}{x+2} + \dfrac{1}{x+2} = \dfrac{x+3}{x+2}$

$1 + \dfrac{1}{x+1} = \dfrac{x+1}{x+1} + \dfrac{1}{x+1} = \dfrac{x+2}{x+1}$

Multiply

$\left(1 + \dfrac{1}{x+3}\right)\left(1 + \dfrac{1}{x+2}\right)\left(1 + \dfrac{1}{x+1}\right)$

$= \dfrac{x+4}{x+3} \cdot \dfrac{x+3}{x+2} \cdot \dfrac{x+2}{x+1} = \dfrac{x+4}{x+1}$

31. $2 + \dfrac{1}{2+1} = 2 + \dfrac{1}{3} = \dfrac{6}{3} + \dfrac{1}{3} = \dfrac{7}{3}$

$2 + \dfrac{1}{2+\dfrac{1}{2+1}} = 2 + \dfrac{1}{\dfrac{7}{3}} = 2 + \dfrac{3}{7} = \dfrac{14}{7} + \dfrac{3}{7} = \dfrac{17}{7}$

$2 + \dfrac{1}{2+\dfrac{1}{2+\dfrac{1}{2+1}}} = 2 + \dfrac{1}{\dfrac{17}{7}} = 2 + \dfrac{7}{17} = \dfrac{34}{17} + \dfrac{7}{17} = \dfrac{41}{17}$

33. See the table in the back of the textbook.

35. See the table in the back of the textbook.

37. $\quad 2x + 3 < 5$

$\quad 2x + 3 - 3 < 5 - 3$

$\quad\quad\quad 2x < 2$

$\quad \dfrac{1}{2}(2x) < \dfrac{1}{2}(2)$

$\quad\quad\quad x < 1$

39. $\quad -3x \le 21$

$\quad -\dfrac{1}{3}(-3x) \ge -\dfrac{1}{3}(21)$

$\quad\quad\quad x \ge -7$

41. $\quad -2x + 8 > -4$

$\quad -2x + 8 - 8 > -4 - 8$

$\quad\quad\quad -2x > -12$

$\quad -\dfrac{1}{2}(-2x) < -\dfrac{1}{2}(-12)$

$\quad\quad\quad x < 6$

43. $4 - 2(x+1) \ge -2$

$\quad 4 - 2x - 2 \ge -2$

$\quad\quad -2x + 2 \ge -2$

$\quad -2x + 2 - 2 \ge -2 - 2$

$\quad\quad\quad -2x \ge -4$

$\quad -\dfrac{1}{2}(-2x) \le -\dfrac{1}{2}(-4)$

$\quad\quad\quad x \le 2$

6.7 Proportions

1. $\dfrac{x}{2} = \dfrac{6}{12}$

   $12x = 12$

   $x = 1$

3. $\dfrac{2}{5} = \dfrac{4}{x}$

   $2x = 20$

   $x = 10$

5. $\dfrac{10}{20} = \dfrac{20}{x}$

   $10x = 400$

   $x = 40$

7. $\dfrac{a}{3} = \dfrac{5}{12}$

   $12a = 15$

   $a = \dfrac{15}{12} = \dfrac{5}{4}$

9. $\dfrac{2}{x} = \dfrac{6}{7}$

   $14 = 6x$

   $\dfrac{14}{6} = x$

   $\dfrac{7}{3} = x$

11. $\dfrac{x+1}{3} = \dfrac{4}{x}$

    $x^2 + x = 12$

    $x^2 + x - 12 = 0$

    $(x+4)(x-3) = 0$

    $x = -4,\ 3$

    The solutions are −4 and 3.

13. $\dfrac{x}{2} = \dfrac{8}{x}$

    $x^2 = 16$

    $x^2 - 16 = 0$

    $(x+4)(x-4) = 0$

    $x = -4,\ 4$

    The solutions are −4 and 4.

15. $\dfrac{4}{a+2} = \dfrac{a}{2}$

    $a^2 + 2a = 8$

    $a^2 + 2a - 8 = 0$

    $(a+4)(a-2) = 0$

    $a = -4,\ 2$

    The solutions are −4 and 2.

17. $\dfrac{1}{x} = \dfrac{x-5}{6}$

    $6 = x^2 - 5x$

    $x^2 - 5x - 6 = 0$

    $(x+1)(x-6) = 0$

    $x = -1,\ 6$

19. Compare hits to games

    $\dfrac{6}{18} = \dfrac{x}{45}$

    $18x = 270$

    $x = 15$

    He will get 15 hits in 45 games.

21. Compare ml alcohol to ml water.

    $\dfrac{12}{16} = \dfrac{x}{28}$

    $16x = 336$

    $x = 21$

    The solution will have 21 ml of alcohol.

23. Compare grams of fat to total grams.

    $\dfrac{13}{100} = \dfrac{x}{350}$

    $100x = 4550$

    $x = 45.5$

    There are 45.5 grams of fat in 350 grams of ice cream.

25. Compare inches on the map to actual miles.

$$\frac{3.5}{100} = \frac{x}{420}$$

$$100x = 1470$$

$$x = 14.7$$

They are 14.7 inches apart on the map.

27. Compare miles to hours.

$$\frac{245}{5} = \frac{x}{7}$$

$$5x = 1715$$

$$x = 343$$

He will travel 343 miles.

29. $\dfrac{x^2 - x - 6}{x^2 - 9} = \dfrac{(x-3)(x+2)}{(x+3)(x-3)} = \dfrac{x+2}{x+3}$

31. $\dfrac{x^2 - 25}{x+4} \cdot \dfrac{2x+8}{x^2 - 9x + 20} = \dfrac{(x+5)(x-5)}{x+4} \cdot \dfrac{2(x+4)}{(x-5)(x-4)}$

$$= \frac{2(x+5)(x-5)(x+4)}{(x+4)(x-5)(x-4)} = \frac{2(x+5)}{x-4}$$

33. $\dfrac{x}{x^2 - 16} + \dfrac{4}{x^2 - 16} = \dfrac{x+4}{x^2 - 16} = \dfrac{1(x+4)}{(x+4)(x-4)} = \dfrac{1}{x-4}$

Chapter 6 Review

1. $\dfrac{7}{14x - 28} = \dfrac{7}{14(x-2)} = \dfrac{1}{2(x-2)} \quad x \neq 2$

3. $\dfrac{8x - 4}{4x + 12} = \dfrac{4(2x-1)}{4(x+3)} = \dfrac{2x-1}{x+3} \quad x \neq -3$

5. $\dfrac{3x^3 + 16x^2 - 12x}{2x^3 + 9x^2 - 18x} = \dfrac{x(3x^2 + 16x - 12)}{x(2x^2 + 9x - 18)}$

$$= \frac{x(x+6)(3x-2)}{x(x+6)(2x-3)}$$

$$= \frac{x(x+6)(3x-2)}{x(x+6)(2x-3)}$$

$$= \frac{3x-2}{2x-3}$$

7. $\dfrac{x^2 + 5x - 14}{x+7} = \dfrac{(x+7)(x-2)}{x+7}$

$$= \frac{(x+7)(x-2)}{x+7}$$

$$= x - 2$$

9. $\dfrac{xy+bx+ay+ab}{xy+5x+ay+5a} = \dfrac{(xy+bx)+(ay+ab)}{(xy+5x)+(ay+5a)}$

$\qquad\qquad\qquad\quad = \dfrac{x(y+b)+a(y+b)}{x(y+5)+a(y+5)}$

$\qquad\qquad\qquad\quad = \dfrac{(y+b)(x+a)}{(y+5)(x+a)}$

$\qquad\qquad\qquad\quad = \dfrac{(y+b)(x+a)}{(y+5)(x+a)}$

$\qquad\qquad\qquad\quad = \dfrac{y+b}{y+5}$

11. $\dfrac{x^2+8x+16}{x^2+x-12} \div \dfrac{x^2-16}{x^2-x-6}$

$\quad = \dfrac{x^2+8x+16}{x^2+x-12} \cdot \dfrac{x^2-x-6}{x^2-16}$

$\quad = \dfrac{(x+4)(x+4)}{(x+4)(x-3)} \cdot \dfrac{(x-3)(x+2)}{(x+4)(x-4)}$

$\quad = \dfrac{(x+4)(x+4)(x-3)(x+2)}{(x+4)(x-3)(x+4)(x-4)}$

$\quad = \dfrac{x+2}{x-4}$

13. $\dfrac{3x^2-2x-1}{x^2+6x+8} \div \dfrac{3x^2+13x+4}{x^2+8x+16}$

$\quad = \dfrac{3x^2-2x-1}{x^2+6x+8} \cdot \dfrac{x^2+8x+16}{3x^2+13x+4}$

$\quad = \dfrac{(3x+1)(x-1)}{(x+2)(x+4)} \cdot \dfrac{(x+4)(x+4)}{(3x+1)(x+4)}$

$\quad = \dfrac{(3x+1)(x-1)(x+4)(x+4)}{(x+2)(x+4)(3x+1)(x+4)}$

$\quad = \dfrac{x-1}{x+2}$

15. $\dfrac{x^2}{x-9} - \dfrac{18x-81}{x-9} = \dfrac{x^2-18x+81}{x-9}$

$\qquad\qquad\qquad\qquad = \dfrac{(x-9)(x-9)}{x-9}$

$\qquad\qquad\qquad\qquad = x-9$

17. $\dfrac{x}{x+9} + \dfrac{5}{x} = \dfrac{x\cdot x}{(x+9)x} + \dfrac{5(x+9)}{x(x+9)}$

$\qquad\qquad\quad = \dfrac{x^2+5x+45}{x(x+9)}$

19. $\dfrac{3}{x^2-36} - \dfrac{2}{x^2-4x-12}$

$\quad = \dfrac{3}{(x+6)(x-6)} - \dfrac{2}{(x+2)(x-6)}$

$\quad = \dfrac{3(x+2)}{(x+6)(x-6)(x+2)} - \dfrac{2(x+6)}{(x+2)(x-6)(x+6)}$

$\quad = \dfrac{3x+6-2x-12}{(x+6)(x-6)(x+2)}$

$\quad = \dfrac{x-6}{(x+6)(x-6)(x+2)}$

$\quad = \dfrac{1}{(x+6)(x+2)}$

21. $$\frac{3}{x}+\frac{1}{2}=\frac{5}{x}$$
$$2x\left(\frac{3}{x}\right)+2x\left(\frac{1}{2}\right)=2x\left(\frac{5}{x}\right)$$
$$6+x=10$$
$$x=4$$

23. $$1-\frac{7}{x}=\frac{-6}{x^2}$$
$$x^2\left(1\right)+x^2\left(-\frac{7}{x}\right)=x^2\left(\frac{-6}{x^2}\right)$$
$$x^2-7x=-6$$
$$x^2-7x+6=0$$
$$(x-1)(x-6)=0$$
$$x=1,\ 6$$
The solutions are 1 and 6.

25. $$\frac{2}{y^2-16}=\frac{10}{y^2+4y}$$
$$\frac{2}{(y+4)(y-4)}=\frac{10}{y(y+4)}$$
$$y(y+4)(y-4)\cdot\frac{2}{(y+4)(y-4)}=y(y+4)(y-4)\cdot\frac{10}{y(y+4)}$$
$$2y=10y-40$$
$$40=8y$$
$$5=y$$

27. Let x = speed of the boat in still water

	d	r	t
upstream	48	$x-3$	$\dfrac{48}{x-3}$
downstream	72	$x+3$	$\dfrac{72}{x+3}$

$$\frac{48}{x-3}=\frac{72}{x+3}$$
$$(x+3)(x-3)\cdot\frac{48}{x-3}=(x+3)(x-3)\cdot\frac{72}{x+3}$$
$$(x+3)48=(x-3)72$$
$$48x+144=72x-216$$
$$360=24x$$
$$15=x$$
The speed of the boat in still water is 15 mph.

29. $\dfrac{\frac{x+4}{x^2-16}}{\frac{2}{x-4}} = \dfrac{\frac{x+4}{x^2-16}\cdot\frac{(x+4)(x-4)}{1}}{\frac{2}{x-4}\cdot\frac{(x+4)(x-4)}{1}}$

$= \dfrac{x+4}{2(x+4)}$

$= \dfrac{1}{2}$

31. $\dfrac{\frac{1}{a-2}+4}{\frac{1}{a-2}+1} = \dfrac{(a-2)\left(\frac{1}{a-2}+4\right)}{(a-2)\left(\frac{1}{a-2}+1\right)}$

$= \dfrac{(a-2)\left(\frac{1}{a-2}\right)+(a-2)(4)}{(a-2)\left(\frac{1}{a-2}\right)+(a-2)(1)}$

$= \dfrac{1+4a-8}{1+a-2}$

$= \dfrac{4a-7}{a-1}$

33. $\dfrac{40\text{ sec}}{3\text{ min}} = \dfrac{40}{3(60)} = \dfrac{40}{180} = \dfrac{2}{9}$

35. $\dfrac{a}{3} = \dfrac{12}{a}$

$a^2 = 36$

$a^2 - 36 = 0$

$(a+6)(a-6) = 0$

$a = -6,\ 6$

The solutions are -6 and 6.

Cumulative Review: Chapters 1-6

1. $8 - 11 = 8 + (-11) = -3$

3. $\dfrac{-48}{12} = -4$

5. $5x - 4 - 9x = 5x - 9x - 4 = -4x - 4$

7. $9^{-2} = \dfrac{1}{9^2} = \dfrac{1}{81}$

9. $4^1 + 9^0 + (-7)^0 = 4 + 1 + 1 = 6$

11. $(4a^3 - 10a^2 + 6) - (6a^3 + 5a - 7) = 4a^3 - 10a^2 + 6 - 6a^3 - 5a + 7 = -2a^3 - 10a^2 - 5a + 13$

13. $\dfrac{x^2}{x-7} - \dfrac{14x-49}{x-7} = \dfrac{x^2-14x+49}{x-7} = \dfrac{(x-7)^2}{x-7} = x-7$

15. $\dfrac{\frac{x-2}{x^2+6x+8}}{\frac{4}{x+4}} = \dfrac{\frac{x-2}{(x+4)(x+2)}\cdot(x+4)(x+2)}{\frac{4}{x+4}\cdot(x+4)(x+2)} = \dfrac{x-2}{4(x+2)}$

17. $x - \dfrac{3}{4} = \dfrac{5}{6}$

$x = \dfrac{3}{4} + \dfrac{5}{6}$

$x = \dfrac{9}{12} + \dfrac{10}{12}$

$x = \dfrac{19}{12}$

19. $98r^2 - 18 = 0$

$2\left(49r^2 - 9\right) = 0$

$2\left(7r + 3\right)\left(7r - 3\right) = 0$

$r = -\dfrac{3}{7}, \dfrac{3}{7}$

21. Multiple each side of the equation by $3x$:

$$3x\left(\dfrac{5}{x} - \dfrac{1}{3}\right) = 3x\left(\dfrac{3}{x}\right)$$

$15 - x = 9$

$-x = -6$

$x = 6$

The solution is 6.

23. $3\left(x - 3\right) \cdot \dfrac{x}{3} = 3\left(x - 3\right) \cdot \dfrac{6}{x - 3}$

$x\left(x - 3\right) = 18$

$x^2 - 3x = 18$

$x^2 - 3x - 18 = 0$

$\left(x - 6\right)\left(x + 3\right) = 0$

$x = 6, -3$

The solutions are -3 and 6.

25. Multiplying the first equation by -5
and the second equation by 3:

$-45x - 70y = 20$

$45x - 24y = 27$

Adding the two equations:

$-94y = 47$

$y = -\dfrac{1}{2}$

Substituting into the first equation

$9x + 14\left(-\dfrac{1}{2}\right) = -4$

$9x - 7 = -4$

$9x = 3$

$x = \dfrac{1}{3}$

The solution is $\left(\dfrac{1}{3}, -\dfrac{1}{2}\right)$.

27. To clear each equation of fractions, multiply the
first equation by 6 and the second equation by 12:

$$6\left(\dfrac{1}{2}x + \dfrac{1}{3}y\right) = 6(-1) \qquad 12\left(\dfrac{1}{3}x\right) = 12\left(\dfrac{1}{4}y + 5\right)$$

$3x + 2y = -6 \qquad\qquad 4x = 3y + 60$

$\qquad\qquad\qquad\qquad\qquad 4x - 3y = 60$

$3x + 2y = -6$

$4x - 3y = 60$

Multiply the first equation by 3 and the second
equation by 2:

$9x + 6y = -18$

$8x - 6y = 120$

Add the two equations:

$17x = 102$

$x = 6$

Substitute into $3x + 2y = -6$:

$3(6) + 2y = -6$

$18 + 2y = -6$

$2y = -24$

$y = -12$

The solution is $(6, -12)$.

29. $y = -\dfrac{3}{2}x$

   See the graph in the back of the textbook.

31. $xy + 5x + ay + 5a = x(y+5) + a(y+5) = (y+5)(x+a)$

33. $20y^2 - 27y + 9 = (5y - 3)(4y - 3)$

35. $16x^2 + 72xy + 81y^2 = (4x + 9y)(4x + 9y) = (4x + 9y)^2$

37. Check each ordered pair:

   $(3, -1):\ 2(3) - 5 = 6 - 5 = 1 \neq -1$

   $(1, -3):\ 2(1) - 5 = 2 - 5 = -3$

   $(-2, 9):\ 2(-2) - 5 = -4 - 5 = -9 \neq 9$

   The ordered pair $(1, -3)$ is a solution to the equation.

39. Let $x = 33\%$ of 220

   $\begin{aligned} x &= 0.33 \cdot 220 \\ &= 72.6 \end{aligned}$

   33% of 220 is 72.6

41. $\dfrac{1}{2},\ -\dfrac{1}{4},\ \dfrac{1}{8},\ \ldots$

   Each succeeding term is $-\dfrac{1}{2}$ times the previous term. The next term is

   $\dfrac{1}{8}\left(-\dfrac{1}{2}\right) = -\dfrac{1}{16}$.

43. Associative property of addition

45. $\dfrac{6x-12}{6x+12} \cdot \dfrac{3x+3}{12x-24} = \dfrac{6(x-2)}{6(x+2)} \cdot \dfrac{3(x+1)}{12(x-2)} = \dfrac{18(x-2)(x+1)}{72(x+2)(x-2)} = \dfrac{x+1}{4(x+2)}$

47. $\dfrac{2xy+10x+3y+15}{3xy+15x+2y+10} = \dfrac{2x(y+5)+3(y+5)}{3x(y+5)+2(y+5)} = \dfrac{(y+5)(2x+3)}{(y+5)(3x+2)} = \dfrac{2x+3}{3x+2}$

49. Let x and $y =$ the two numbers.

   $x + y = 40$

   $x - y = 18$

   Add the two equations:

   $2x = 58$

   $x = 29$

   Substitute into the first equation:

   $29 + y = 40$

   $y = 11$

   The two numbers are 29 and 11.

Chapter 6 Test

1. $\dfrac{x^2-16}{x^2-8x+16}=\dfrac{(x+4)(x-4)}{(x-4)(x-4)}=\dfrac{x+4}{x-4}$

2. $\dfrac{10a+20}{5a^2+20a+20}=\dfrac{10(a+2)}{5(a+2)(a+2)}=\dfrac{2}{a+2}$

3. $\dfrac{xy+7x+5y+35}{x^2+ax+5x+5a}=\dfrac{x(y+7)+5(y+7)}{x(x+a)+5(x+a)}$

   $\qquad\qquad\qquad=\dfrac{(x+5)(y+7)}{(x+5)(x+a)}$

   $\qquad\qquad\qquad=\dfrac{y+7}{x+a}$

4. $\dfrac{3x-12}{4}\cdot\dfrac{8}{2x-8}=\dfrac{3(x-4)}{4}\cdot\dfrac{8}{2(x-4)}$

   $\qquad\qquad\qquad=\dfrac{24(x-4)}{8(x-4)}$

   $\qquad\qquad\qquad=3$

5. $\dfrac{x^2-49}{x+1}\div\dfrac{x+7}{x^2-1}=\dfrac{x^2-49}{x+1}\cdot\dfrac{x^2-1}{x+7}$

   $\qquad\qquad\qquad=\dfrac{(x+7)(x-7)}{x+1}\cdot\dfrac{(x+1)(x-1)}{x+7}$

   $\qquad\qquad\qquad=\dfrac{(x+7)(x-7)(x+1)(x-1)}{(x+1)(x+7)}$

   $\qquad\qquad\qquad=(x-7)(x-1)$

6. $\dfrac{x^2-3x-10}{x^2-8x+15}\div\dfrac{3x^2+2x-8}{x^2+x-12}=\dfrac{x^2-3x-10}{x^2-8x+15}\cdot\dfrac{x^2+x-12}{3x^2+2x-8}$

   $\qquad\qquad\qquad\qquad=\dfrac{(x-5)(x+2)}{(x-3)(x-5)}\cdot\dfrac{(x+4)(x-3)}{(x+2)(3x-4)}$

   $\qquad\qquad\qquad\qquad=\dfrac{x+4}{3x-4}$

7. $(x^2-9)\left(\dfrac{x+2}{x+3}\right)=\dfrac{(x+3)(x-3)}{1}\left(\dfrac{x+2}{x+3}\right)=(x-3)(x+2)$

8. $\dfrac{3}{x-2}-\dfrac{6}{x-2}=\dfrac{3-6}{x-2}=\dfrac{-3}{x-2}$

9. $\dfrac{x}{x^2-9}+\dfrac{4}{4x-12}=\dfrac{x}{(x+3)(x-3)}+\dfrac{4}{4(x-3)}$

   $\qquad\qquad\qquad=\dfrac{x}{(x+3)(x-3)}+\dfrac{1}{x-3}$

   $\qquad\qquad\qquad=\dfrac{x}{(x+3)(x-3)}+\dfrac{1(x+3)}{(x-3)(x+3)}$

   $\qquad\qquad\qquad=\dfrac{2x+3}{(x+3)(x-3)}$

10. $\dfrac{2x}{x^2-1}+\dfrac{x}{x^2-3x+2}$

$=\dfrac{2x}{(x+1)(x-1)}+\dfrac{x}{(x-1)(x-2)}$

$\dfrac{2x(x-2)}{(x+1)(x-1)(x-2)}+\dfrac{x(x+1)}{(x-1)(x-2)(x+1)}$

$=\dfrac{2x^2-4x+x^2+x}{(x+1)(x-1)(x-2)}$

$=\dfrac{3x^2-3x}{(x+1)(x-1)(x-2)}$

$=\dfrac{3x(x-1)}{(x+1)(x-1)(x-2)}$

$=\dfrac{3x}{(x+1)(x-2)}$

11. $\dfrac{7}{5}=\dfrac{x+2}{3}$

$15\left(\dfrac{7}{5}\right)=\left(\dfrac{x+2}{3}\right)15$

$21=(x+2)5$

$21=5x+10$

$11=5x$

$\dfrac{11}{5}=x$

12. $\dfrac{10}{x+4}=\dfrac{6}{x}-\dfrac{4}{x}$

$x(x+4)\left(\dfrac{10}{x+4}\right)=x(x+4)\left(\dfrac{6}{x}\right)-x(x+4)\left(\dfrac{4}{x}\right)$

$10x=6(x+4)-4(x+4)$

$10x=6x+24-4x-16$

$10x=2x+8$

$8x=8$

$x=1$

13. $\dfrac{3}{x-2}-\dfrac{4}{x+1}=\dfrac{5}{x^2-x-2}$

$(x+1)(x-2)\left(\dfrac{3}{x-2}\right)-(x+1)(x-2)\left(\dfrac{4}{x+1}\right)=(x+1)(x-2)\left(\dfrac{5}{(x+1)(x-2)}\right)$

$3(x+1)-4(x-2)=5$

$3x+3-4x+8=5$

$-x+11=5$

$-x=-6$

$x=6$

14. Let $x=$ the speed of the boat in still water.

	d	r	t
Upstream	26	$x-2$	$\dfrac{26}{x-2}$
Downstream	34	$x+2$	$\dfrac{34}{x+2}$

$$\frac{26}{x-2} = \frac{34}{x+2}$$

$$(x+2)(x-2)\left(\frac{26}{x-2}\right) = (x+2)(x-2)\left(\frac{34}{x+2}\right)$$

$$26(x+2) = 34(x-2)$$

$$26x+52 = 34x-68$$

$$52 = 8x-68$$

$$120 = 8x$$

$$15 = x$$

The speed of the boat in still water is 15 mph.

15. Let t = the time to empty the pool with both pipes open.

$$\frac{1}{12} - \frac{1}{15} = \frac{1}{t}$$

$$60t\left(\frac{1}{12} - \frac{1}{15}\right) = 60t \cdot \frac{1}{t}$$

$$5t - 4t = 60$$

$$t = 60$$

It will take 60 hours to empty the pool with both pipes open.

16.

$$\frac{27}{54} = \frac{1}{2} \qquad \frac{\text{solution of alcohol}}{\text{solution of water}}$$

$$27 + 54 = 81 \qquad \text{total volume}$$

$$\frac{27}{81} = \frac{1}{3} \qquad \frac{\text{alcohol}}{\text{total volume}}$$

17.

$$\frac{8}{100} = \frac{x}{1650} \qquad \frac{\text{defective parts}}{\text{parts produced}}$$

$$8(1650) = 100x$$

$$13{,}200 = 100x$$

$$132 = x$$

18.

$$\frac{1+\dfrac{1}{x}}{1-\dfrac{1}{x}} = \frac{x\left(1+\dfrac{1}{x}\right)}{x\left(1-\dfrac{1}{x}\right)}$$

$$= \frac{x(1) + x\left(\dfrac{1}{x}\right)}{x(1) - x\left(\dfrac{1}{x}\right)}$$

$$= \frac{x+1}{x-1}$$

19.

$$\frac{1-\dfrac{16}{x^2}}{1-\dfrac{2}{x}-\dfrac{8}{x^2}} = \frac{x^2\left(1-\dfrac{16}{x^2}\right)}{x^2\left(1-\dfrac{2}{x}-\dfrac{8}{x^2}\right)}$$

$$= \frac{x^2(1) - x^2\left(\dfrac{16}{x^2}\right)}{x^2(1) - x^2\left(\dfrac{2}{x}\right) - x^2\left(\dfrac{8}{x^2}\right)}$$

$$= \frac{x^2 - 16}{x^2 - 2x - 8}$$

$$= \frac{(x+4)(x-4)}{(x-4)(x+2)}$$

$$= \frac{x+4}{x+2}$$

CHAPTER 7
Transitions

7.1 Review of Solving Equations

1. $2x - 4 = 6$
   $2x = 10$
   $x = 5$

3. $-3 - 4x = 15$
   $-4x = 18$
   $x = -\dfrac{18}{4}$
   $x = -\dfrac{9}{2}$

5. $-300y + 100 = 500$
   $-300y = 400$
   $y = -\dfrac{400}{300}$
   $y = -\dfrac{4}{3}$

7. $-\dfrac{3}{5}a + 2 = 8$
   $-\dfrac{3}{5}a = 6$
   $\left(-\dfrac{5}{3}\right)\left(-\dfrac{3}{5}a\right) = -\dfrac{5}{3}(6)$
   $a = -\dfrac{30}{3}$
   $a = -10$

9. $-x = 2$
   $x = -2$

11. $-a = -\dfrac{3}{4}$
    $a = \dfrac{3}{4}$

13. $7y - 4 = 2y + 11$
    $5y - 4 = 11$
    $5y = 15$
    $y = 3$

15. $5(y + 2) - 4(y + 1) = 3$
    $5y + 10 - 4y - 4 = 3$
    $y + 6 = 3$
    $y = -3$

17. $6 - 7(m - 3) = -1$
    $6 - 7m + 21 = -1$
    $-7m + 27 = -1$
    $-7m = -28$
    $m = 4$

19. $5 = 7 - 2(3x - 1) + 4x$
    $5 = 7 - 6x + 2 + 4x$
    $5 = 9 - 2x$
    $-4 = -2x$
    $2 = x$

21. $$\frac{1}{2}x + \frac{1}{4} = \frac{1}{3}x + \frac{5}{4}$$

$$12\left(\frac{1}{2}x\right) + 12\left(\frac{1}{4}\right) = 12\left(\frac{1}{3}x\right) + 12\left(\frac{5}{4}\right)$$

$$6x + 3 = 4x + 15$$
$$2x + 3 = 15$$
$$2x = 12$$
$$x = 6$$

23. $x^2 - 5x - 6 = 0$

$(x-6)(x+1) = 0$

$x - 6 = 0$ or $x + 1 = 0$

$x = 6$ $x = -1$

25. $x^3 - 5x^2 + 6x = 0$

$x(x^2 - 5x + 6) = 0$

$x(x-2)(x-3) = 0$

$x = 0$ or $x - 2 = 0$ or $x - 3 = 0$

$x = 2$ $x = 3$

27. $60x^2 - 130x + 60 = 0$

$10(6x^2 - 13x + 6) = 0$

$10(3x - 2)(2x - 3) = 0$

$3x - 2 = 0$ or $2x - 3 = 0$

$3x = 2$ $2x = 3$

$x = \dfrac{2}{3}$ $x = \dfrac{3}{2}$

29. $$\frac{1}{5}y^2 - 2 = -\frac{3}{10}y$$

$$\frac{1}{5}y^2 + \frac{3}{10}y - 2 = 0$$

$$2y^2 + 3y - 20 = 0$$

$$(y+4)(2y-5) = 0$$

$y + 4 = 0$ or $2y - 5 = 0$

$y = -4$ $y = \dfrac{5}{2}$

31. $$-100x = 10x^2$$

$$-10x^2 - 100x = 0$$

$$10x^2 + 100x = 0$$

$$10x(x+10) = 0$$

$10x = 0$ or $x + 10 = 0$

$x = 0$ $x = -10$

33. $(x+6)(x-2) = -7$

$x^2 + 4x - 12 + 7 = 0$

$x^2 + 4x - 5 = 0$

$(x+5)(x-1) = 0$

$x + 5 = 0$ or $x - 1 = 0$

$x = -5$ $x = 1$

35. $(x+1)^2 = 3x + 7$

$x^2 + 2x + 1 - 3x - 7 = 0$

$x^2 - x - 6 = 0$

$(x+2)(x-3) = 0$

$x + 2 = 0$ or $x - 3 = 0$

$x = -2$ $x = 3$

37. $x^3 + 3x^2 - 4x - 12 = 0$

$x^2(x+3) - 4(x+3) = 0$

$(x+3)(x^2 - 4) = 0$

$(x+3)(x+2)(x-2) = 0$

$x + 3 = 0$ or $x + 2 = 0$ or $x - 2 = 0$

$x = -3$ $x = -2$ $x = 2$

39. $5 - 2x = 3x + 1$

$5 = 5x + 1$

$4 = 5x$

$\dfrac{4}{5} = x$

41. $\dfrac{1}{10}t^2 - \dfrac{5}{2} = 0$

$t^2 - 25 = 0$

$(t+5)(t-5) = 0$

$t+5 = 0$ or $t-5 = 0$

$\quad\quad t = -5 \quad\quad\quad\quad t = 5$

43. $7 + 3(x+2) = 4(x+1)$

$7 + 3x + 6 = 4x + 4$

$3x + 13 = 4x + 4$

$13 = x + 4$

$9 = x$

45. $\quad -\dfrac{2}{5}x + \dfrac{2}{15} = \dfrac{2}{3}$

$15\left(-\dfrac{2}{5}x\right) + 15\left(\dfrac{2}{15}\right) = 15\left(\dfrac{2}{3}\right)$

$-6x + 2 = 10$

$-6x = 8$

$x = -\dfrac{4}{3}$

47. $\quad \dfrac{1}{2}x + \dfrac{1}{3}x + \dfrac{1}{4}x = \dfrac{13}{12}$

$12\left(\dfrac{1}{2}x\right) + 12\left(\dfrac{1}{3}x\right) + 12\left(\dfrac{1}{4}x\right) = 12\left(\dfrac{13}{12}\right)$

$6x + 4x + 3x = 13$

$13x = 13$

$x = 1$

49. $(2r+3)(2r-1) = -(3r+1)$

$4r^2 + 4r - 3 = -3r - 1$

$4r^2 + 7r - 2 = 0$

$(r+2)(4r-1) = 0$

$r+2 = 0$ or $4r-1 = 0$

$\quad r = -2 \quad\quad\quad r = \dfrac{1}{4}$

51. $\quad\quad 9a^3 = 16a$

$9a^3 - 16a = 0$

$a(9a^2 - 16) = 0$

$a(3a+4)(3a-4) = 0$

$a = 0$ or $3a+4 = 0$ or $3a-4 = 0$

$\quad\quad\quad 3a = -4 \quad\quad\quad 3a = 4$

$\quad\quad\quad a = -\dfrac{4}{3} \quad\quad\quad a = \dfrac{4}{3}$

53. $\quad 4x^3 + 12x^2 - 9x - 27 = 0$

$4x^2(x+3) - 9(x+3) = 0$

$(x+3)(4x^2 - 9) = 0$

$(x+3)(2x+3)(2x-3) = 0$

$x+3 = 0$ or $2x+3 = 0$ or $2x-3 = 0$

$x = -3 \quad\quad\quad 2x = -3 \quad\quad\quad 2x = 3$

$\quad\quad\quad\quad x = -\dfrac{3}{2} \quad\quad\quad x = \dfrac{3}{2}$

55. Any method of solution results in a false statement.

57. $3x - 6 = 3(x+4)$

$3x - 6 = 3x + 12$

$-6 = 12$

A false statement, no solution.

59. $4y + 2 - 3y + 5 = 3 + y + 4$

$y + 7 = y + 7$

$7 = 7$

All real numbers are solutions

61. $2(4t-1)+3 = 5t+4+3t$

$8t-2+3 = 5t+4+3t$

$8t+1 = 8t+4$

$1 = 4$

A false statement, no solution.

63. Let $x = $ width and $2x = $ length

$P = 2w+2l$

$2x+2(2x) = 60$

$2x+4x = 60$

$6x = 60$

$x = 10$

$2x = 20$

Width = 10 feet and length = 20 feet

65. Let $x = $ width and $2x = $ length

$P = 2w+2l$

$2x+2(2x) = 48$

$2x+4x = 48$

$6x = 48$

$x = 8$

$2x = 16$

Cost $= 2.25(2)(8)+1.75(2)(16)$

$= \$92$

67. Let $x = $ amount sold and $0.06x = $ sales tax

$250+x+0.06x = 1204$

$1.06x = 954$

$x = 900$

$0.06x = 54$

The sales tax collected is $54

69. $M = 220 - A$

Age (years)	Maximum Heart Rate (beats per minute)
18	202
19	201
20	200
21	199
22	198
23	197

71. $T = R+0.6(M-R)$

$= R+0.6(200-R) = 0.4R+120$

Resting Heart Rate (beats per minute)	Training Heart Rate (beats per minute)
60	144
62	145
64	146
68	147
70	148
72	149

73. $x^2+(x+2)^2 = 34$

$x^2+x^2+4x+4-34 = 0$

$2x^2+4x-30 = 0$

$2(x^2+2x-15) = 0$

$2(x+5)(x-3) = 0$

$x+5 = 0$ or $x-3 = 0$

$x = -5$ $x = 3$

$x+2 = -3$ $x+2 = 5$

The integers are $-5, -3$ or $3, 5$.

75.   Height $= x$

Length $= 24$

Distance $= 7$

$a^2+b^2 = c^2$

$x^2+7^2 = 24^2$

$x^2+49 = 576$

$x^2 = 527$

$x = \sqrt{527} \approx 23$

The ladder reaches up about 23 feet.

77. Let x = width, $3x+2$ = length,
    area = 16 sq. ft.

$$A = LW$$
$$16 = (3x+2)x$$
$$16 = 3x^2 + 2x$$
$$0 = 3x^2 + 2x - 16$$
$$0 = (3x+8)(x-2)$$

$3x+8 = 0$ or $x-2 = 0$
$3x = -8$ $x = 2$ ft(width)

$x = -\dfrac{8}{\cancel{3}}$ $3x+2 = 8$ ft(length)

No solution

79. $6\left(\dfrac{x}{3} - \dfrac{1}{2}\right) = 6\left(\dfrac{5}{2}\right)$
$$2x - 3 = 15$$
$$2x = 18$$
$$x = 9$$

81. $x^2\left(1 - \dfrac{5}{x}\right) = x^2\left(\dfrac{-6}{x^2}\right)$
$$x^2 - 5x = -6$$
$$x^2 - 5x + 6 = 0$$
$$(x-2)(x-3) = 0$$
$$x = 2, \ 3$$

83. $2(a-4)\left(\dfrac{a}{a-4} - \dfrac{a}{2}\right) = 2(a-4)\cdot\dfrac{4}{a-4}$
$$2a - a(a-4) = 8$$
$$2a - a^2 + 4a = 8$$
$$-a^2 + 6a - 8 = 0$$
$$a^2 - 6a + 8 = 0$$
$$(a-4)(a-2) = 0$$
$$a = 2, \ 4$$

Since $a = 4$ does not check in the original
equation, the solution is $a = 2$.

7.2 Equations With Absolute Value

1. $|x| = 4$
    $x = \pm 4$

3. $2 = |a|$
     $\pm 2 = a$

5. $|x| = -3$
    $\varnothing.$

7. $|a| + 2 = 3$
     $|a| = 1$
       $a = \pm 1$

9. $|y| + 4 = 3$
     $|y| = -1$
       $\varnothing$

11. $4 = |x| - 2$
     $6 = |x|$
    $-6 = x$ or $x = 6$

13. $|x-2|=5$

$\quad x-2=5 \quad$ or $\quad x-2=-5$

$\qquad x=7 \qquad\qquad x=-3$

15. $|a-4|=\dfrac{5}{3}$

$\quad a-4=\dfrac{5}{3} \quad$ or $\quad a-4=-\dfrac{5}{3}$

$\qquad a=\dfrac{5}{3}+\dfrac{12}{3} \qquad\qquad a=-\dfrac{5}{3}+\dfrac{12}{3}$

$\qquad a=\dfrac{17}{3} \qquad\qquad\qquad a=\dfrac{7}{3}$

17. $\quad 1=|3-x|$

$\quad 1=3-x \quad$ or $\quad -1=3-x$

$\quad -2=-x \qquad\qquad -4=-x$

$\qquad 2=x \qquad\qquad\qquad 4=x$

19. $\left|\dfrac{3}{5}a+\dfrac{1}{2}\right|=1$

$\quad \dfrac{3}{5}a+\dfrac{1}{2}=1 \quad$ or $\quad \dfrac{3}{5}a+\dfrac{1}{2}=-1$

$\qquad \dfrac{3}{5}a=\dfrac{1}{2} \qquad\qquad \dfrac{3}{5}a=-\dfrac{3}{2}$

$\qquad a=\dfrac{5}{6} \qquad\qquad\qquad a=-\dfrac{5}{2}$

21. $\quad 60=|20x-40|$

$\quad 60=20x-40 \quad$ or $\quad -60=20x-40$

$\quad 100=20x \qquad\qquad -20=20x$

$\qquad 5=x \qquad\qquad\qquad -1=x$

23. $|2x+1|=-3$

$\quad \varnothing$

25. $\left|\dfrac{3}{4}x-6\right|=9$

$\quad \dfrac{3}{4}x-6=9 \quad$ or $\quad \dfrac{3}{4}x-6=-9$

$\qquad \dfrac{3}{4}x=15 \qquad\qquad \dfrac{3}{4}x=-3$

$\qquad x=20 \qquad\qquad\qquad x=-4$

27. $\left|1-\dfrac{1}{2}a\right|=3$

$\quad 1-\dfrac{1}{2}a=3 \quad$ or $\quad 1-\dfrac{1}{2}a=-3$

$\qquad -\dfrac{1}{2}a=2 \qquad\qquad -\dfrac{1}{2}a=-4$

$\qquad a=-4 \qquad\qquad\qquad a=8$

29. $|3x+4|+1=7$

$\quad |3x+4|=6$

$\quad 3x+4=6 \quad$ or $\quad 3x+4=-6$

$\qquad 3x=2 \qquad\qquad 3x=-10$

$\qquad x=\dfrac{2}{3} \qquad\qquad x=-\dfrac{10}{3}$

31. $|3-2y|+4=3$

$\quad |3-2y|=-1$

$\qquad \varnothing$

33. $3 + |4t - 1| = 8$

$\qquad |4t - 1| = 5$

$\qquad 4t - 1 = 5 \quad$ or $\quad 4t - 1 = -5$

$\qquad\qquad 4t = 6 \qquad\qquad 4t = -4$

$\qquad\qquad t = \dfrac{6}{4} \qquad\qquad t = -1$

$\qquad\qquad t = \dfrac{3}{2}$

35. $\left|9 - \dfrac{3}{5}x\right| + 6 = 12$

$\qquad \left|9 - \dfrac{3}{5}x\right| = 6$

$\qquad 9 - \dfrac{3}{5}x = 6 \quad$ or $\quad 9 - \dfrac{3}{5}x = -6$

$\qquad -\dfrac{3}{5}x = -3 \qquad\qquad -\dfrac{3}{5}x = -15$

$\qquad\qquad x = 5 \qquad\qquad\qquad x = 25$

37. $5 = \left|\dfrac{2x}{7} + \dfrac{4}{7}\right| - 3$

$\quad 8 = \left|\dfrac{2x}{7} + \dfrac{4}{7}\right|$

$\quad 8 = \dfrac{2x}{7} + \dfrac{4}{7} \quad$ or $\quad -8 = \dfrac{2x}{7} + \dfrac{4}{7}$

$\quad 56 = 2x + 4 \qquad\qquad -56 = 2x + 4$

$\quad 52 = 2x \qquad\qquad\qquad -60 = 2x$

$\quad 26 = x \qquad\qquad\qquad -30 = x$

39. $\quad 2 = -8 + \left|4 - \dfrac{1}{2}y\right|$

$\quad 10 = \left|4 - \dfrac{1}{2}y\right|$

$\quad 10 = 4 - \dfrac{1}{2}y \quad$ or $\quad -10 = 4 - \dfrac{1}{2}y$

$\quad 6 = -\dfrac{1}{2}y \qquad\qquad -14 = -\dfrac{1}{2}y$

$\quad -12 = y \qquad\qquad\qquad 28 = y$

41. $|3a + 1| = |2a - 4|$

Equals	Opposites
$3a + 1 = 2a - 4 \quad$ or	$3a + 1 = -(2a - 4)$
$a + 1 = -4$	$3a + 1 = -2a + 4$
$a = -5$	$5a + 1 = 4$
	$5a = 3$
	$a = \dfrac{3}{5}$

43. $\left|x - \dfrac{1}{3}\right| = \left|\dfrac{1}{2}x + \dfrac{1}{6}\right|$

Equals	Opposites
$x - \dfrac{1}{3} = \dfrac{1}{2}x + \dfrac{1}{6} \quad$ or	$x - \dfrac{1}{3} = -\left(\dfrac{1}{2}x + \dfrac{1}{6}\right)$
$6x - 2 = 3x + 1$	$6x - 2 = -3x - 1$
$3x - 2 = 1$	$9x - 2 = -1$
$3x = 3$	$9x = 1$
$x = 1$	$x = \dfrac{1}{9}$

45. $|y - 2| = |y + 3|$

Equals	Opposites
$y - 2 = y + 3 \quad$ or	$y - 2 = -(y + 3)$
$-2 = 3$	$y - 2 = -y - 3$
No solution	$2y - 2 = -3$
	$2y = -1$
	$y = -\dfrac{1}{2}$

47. $|3x - 1| = |3x + 1|$

Equals	Opposites
$3x - 1 = 3x + 1 \quad$ or	$3x - 1 = -(3x + 1)$
$-1 = 1$	$3x - 1 = -3x - 1$
No solution	$6x - 1 = -1$
	$6x = 0$
	$x = 0$

49. $|3-m| = |m+4|$

Equals		Opposites
$3-m = m+4$	or	$3-m = -(m+4)$
$3 = 2m+4$		$3-m = -m-4$
$-1 = 2m$		$3 = -4$
$-\dfrac{1}{2} = m$		No solution

51. $|0.03 - 0.01x| = |0.04 + 0.05x|$

Equals

$0.03 - 0.01x = 0.04 + 0.05x$

$3 - x = 4 + 5x$

$-1 = 6x$

$-\dfrac{1}{6} = x$

Opposites

$0.03 - 0.01x = -(0.04 + 0.05x)$

$0.03 - 0.01x = -0.04 - 0.05x$

$3 - x = -4 - 5x$

$4x = -7$

$x = -\dfrac{7}{4}$

The solutions are $-\dfrac{1}{6},\ -\dfrac{7}{4}.$

53. $|x-2| = |2-x|$

Equals		Opposites
$x-2 = 2-x$	or	$x-2 = -(2-x)$
$2x-2 = 2$		$x-2 = -2+x$
$2x = 4$		$-2 = -2$
$x = 2$		All real numbers

The solution set is all real numbers

55. $\left|\dfrac{x}{5} - 1\right| = \left|1 - \dfrac{x}{5}\right|$

Equals		Opposites
$\dfrac{x}{5} - 1 = 1 - \dfrac{x}{5}$	or	$\dfrac{x}{5} - 1 = -\left(1 - \dfrac{x}{5}\right)$
$x - 5 = 5 - x$		$x - 5 = -(5 - x)$
$2x - 5 = 5$		$x - 5 = -5 + x$
$2x = 10$		$-5 = -5$
$x = 5$		All real numbers

57. $d = rt$

$\dfrac{d}{r} = t$

59. Let $x \cdot 60 = 15$

$x = \dfrac{1}{4}$

$x = 0.25 = 25\%$

15 is 25% of 60

61. $x - 5 > 8$

$x > 13$

63. $\dfrac{1}{4}x \ge 1$

$x \ge 4$

65. $4 - 2x < 12$

$-2x < 8$

$\dfrac{-2x}{-2} > \dfrac{8}{-2}$

$x > -4$

7.3 Compound Inequalities and Interval Notation

1-15. See the graph in the back of the textbook.

17. $3x - 1 < 5$ or $5x - 5 > 10$

$\qquad 3x < 6 \qquad\qquad 5x > 15$

$\qquad x < 2 \qquad\qquad x > 3$

$\qquad\qquad (-\infty, 2) \cup (3, \infty)$

See the graph in the back of the textbook.

19. $x - 2 > -5$ and $x + 7 < 13$

$\qquad x > -3 \qquad\qquad x < 6$

$\qquad\qquad (-3, 6)$

See the graph in the back of the textbook.

21. $11x \le 22$ or $12x \ge 36$

$\qquad x \le 2 \qquad\qquad x \ge 3$

$\qquad\qquad (-\infty, 2] \cup [3, \infty)$

See the graph in the back of the textbook.

23. $3x - 5 \le 10$ and $2x + 1 \ge -5$

$\qquad 3x \le 15 \qquad\qquad 2x \ge -6$

$\qquad x \le 5 \qquad\qquad x \ge -3$

$\qquad\qquad [-3, 5]$

See the graph in the back of the textbook.

25. $2x - 3 < 8$ and $3x + 1 > -10$

$\qquad 2x < 11 \qquad\qquad 3x > -11$

$\qquad x < \dfrac{11}{2} \qquad\qquad x > -\dfrac{11}{3}$

$\qquad\qquad \left(-\dfrac{11}{3}, \dfrac{11}{2}\right)$

See the graph in the back of the textbook.

27. $2x - 1 < 3$ and $3x - 2 > 1$

$\qquad 2x < 4 \qquad\qquad 3x > 3$

$\qquad x < 2 \qquad\qquad x > 1$

$\qquad\qquad (1, 2)$

See the graph in the back of the textbook.

29. $-1 \le x - 5 \le 2$

$\qquad 4 \le x \le 7$

See the graph in the back of the textbook.

31. $-4 \le 2x \le 6$

$\qquad -2 \le x \le 3$

See the graph in the back of the textbook.

33. $-3 < 2x + 1 < 5$

$\qquad -4 < 2x < 4$

$\qquad -2 < x < 2$

See the graph in the back of the textbook.

35. $0 \le 3x + 2 \le 7$

$\qquad -2 \le 3x \le 5$

$\qquad -\dfrac{2}{3} \le x \le \dfrac{5}{3}$

See the graph in the back of the textbook.

37. $-7 < 2x + 3 < 11$

$\qquad -10 < 2x < 8$

$\qquad -5 < x < 4$

See the graph in the back of the textbook.

39. $-1 \le 4x + 5 \le 9$

$\qquad -6 \le 4x \le 4$

$\qquad -\dfrac{3}{2} \le x \le 1$

See the graph in the back of the textbook.

41. $-2 < x < 3$

43. $x \le -2$ or $x \ge 3$

45. (a) $2x + x > 10;\ \ x + 10 > 2x;\ \ 2x + 10 > x$

(b)

$$2x + x > 10 \qquad x + 10 > 2x$$
$$3x > 10 \qquad 10 > x$$
$$x > \frac{10}{3} \qquad x < 10$$
$$\frac{10}{3} < x < 10$$

47. See the graph in the back of the textbook.

49. Let $x =$ the number
$$10 < x + 5 < 20$$
$$5 < x < 15$$
The number is between 5 and 15.

51. Let $x =$ the number
$$5 < 2x - 3 < 7$$
$$8 < 2x < 10$$
$$4 < x < 5$$
The number is between 4 and 5.

53. Let $w =$ the width, $w + 4 =$ the length. Using the formula for perimeter:
$$20 < 2w + 2(w + 4) < 30$$
$$20 < 2w + 2w + 8 < 30$$
$$20 < 4w + 8 < 30$$
$$12 < 4w < 22$$
$$3 < w < \frac{11}{2}$$

The width is between 3 inches and $\dfrac{11}{2} = 5\dfrac{1}{2}$ inches.

55. Simplifying the expression: $-|-5| = -(5) = -5$

57. Simplifying the expression: $-3 - 4(-2) = -3 + 8 = 5$

59. Simplifying the expression: $5|3 - 8| - 6|2 - 5| = 5|-5| - 6|-3| = 5(5) - 6(3) = 25 - 18 = 7$

61. Simplifying the expression: $5 - 2[-3(5 - 7) - 8] = 5 - 2[-3(-2) - 8] = 5 - 2(6 - 8) = 5 - 2(-2) = 5 + 4 = 9$

63. $-3 - (-9) = -3 + 9 = 6$

65. Applying the distributive property: $\frac{1}{2}(4x - 6) = \frac{1}{2} \cdot 4x - \frac{1}{2} \cdot 6 = 2x - 3$

67. The integers are: $-3,\ 0,\ 2$

7.4 Inequalities Involving Absolute Value

1. $|x| < 3$

    $-3 < x < 3$

    See the graph in the back of the textbook.

3. $|x| \geq 2$

    $x \leq -2$ or $x \geq 2$

5. $|x| + 2 < 5$

    $|x| < 3$

    $-3 < x < 3$

    See the graph in the back of the textbook.

7. $|t| - 3 > 4$

    $|t| > 7$

    $t < -7$ or $t > 7$

    See the graph in the back of the textbook.

9. $|y| < -5$

    $\varnothing$

11. $|x| \geq -2$

    All real numbers

13. $|x - 3| < 7$

    $-7 < x - 3 < 7$

    $-4 < x < 10$

    See the graph in the back of the textbook.

15. $|a + 5| \geq 4$

    $a + 5 \leq -4$ or $a + 5 \geq 4$

    $a \leq -9$ or $a \geq -1$

    See the graph in the back of the textbook.

17. $|a - 1| < -3$

    $\varnothing$

19. $|2x - 4| < 6$

    $-6 < 2x - 4 < 6$

    $-2 < 2x < 10$

    $-1 < x < 5$

    See the graph in the back of the textbook.

21. $|3y + 9| \geq 6$

    $3y + 9 \leq -6$ or $3y + 9 \geq 6$

    $3y \leq -15$ or $3y \geq -3$

    $y \leq -5$ or $y \geq -1$

    See the graph in the back of the textbook.

23. $|2k + 3| \geq 7$

    $2k + 3 \leq -7$ or $2k + 3 \geq 7$

    $2k \leq -10$ or $2k \geq 4$

    $k \leq -5$ or $k \geq 2$

    See the graph in the back of the textbook.

25. $|x - 3| + 2 < 6$

    $|x - 3| < 4$

    $-4 < x - 3 < 4$

    $-1 < x < 7$

    See the graph in the back of the textbook.

27. $|2a + 1| + 4 \geq 7$

    $|2a + 1| \geq 3$

    $2a + 1 \leq -3$ or $2a + 1 \geq 3$

    $2a \leq -4$ or $2a \geq 2$

    $a \leq -2$ or $a \geq 1$

    See the graph in the back of the textbook.

29. $|3x+5|-8<5$

$\qquad |3x+5|<13$

$\qquad -13<3x+5<13$

$\qquad -18<3x<8$

$\qquad \dfrac{1}{3}(-18)<\dfrac{1}{3}(3x)<\dfrac{1}{3}(8)$

$\qquad -6<x<\dfrac{8}{3}$

See the graph in the back of the textbook.

31. $|5-x|>3$

$\qquad 5-x<-3 \quad \text{or} \quad 5-x>3$

$\qquad -x<-8 \quad \text{or} \quad -x>-2$

$\qquad x>8 \quad \text{or} \quad x<2$

See the graph in the back of the textbook.

33. $\left|3-\dfrac{2}{3}x\right|\geq 5$

$\qquad 3-\dfrac{2}{3}x\leq -5 \quad \text{or} \quad 3-\dfrac{2}{3}x\geq 5$

$\qquad -\dfrac{2}{3}x\leq -8 \qquad\qquad -\dfrac{2}{3}x\geq 2$

$\qquad\qquad x\geq 12 \qquad\qquad\qquad x\leq -3$

See the graph in the back of the textbook.

35. $\left|2-\dfrac{1}{2}x\right|>1$

$\qquad 2-\dfrac{1}{2}x<-1 \quad \text{or} \quad 2-\dfrac{1}{2}x>1$

$\qquad 4-x<-2 \quad \text{or} \quad 4-x>2$

$\qquad -x<-6 \quad \text{or} \quad -x>-2$

$\qquad x>6 \quad \text{or} \quad x<2$

See the graph in the back of the textbook.

37. $|x-1|<0.01$

$\qquad -0.01<x-1<0.01$

$\qquad 0.99<x<1.01$

39. $|2x+1|\geq \dfrac{1}{5}$

$\qquad 2x+1\leq -\dfrac{1}{5} \quad \text{or} \quad 2x+1\geq \dfrac{1}{5}$

$\qquad 10x+5\leq -1 \quad \text{or} \quad 10x+5\geq 1$

$\qquad 10x\leq -6 \quad \text{or} \quad 10x\geq -4$

$\qquad x\leq -\dfrac{6}{10} \quad \text{or} \quad x\geq -\dfrac{4}{10}$

$\qquad x\leq -\dfrac{3}{5} \quad \text{or} \quad x\geq -\dfrac{2}{5}$

41. $\left|\dfrac{3x-2}{5}\right| \le \dfrac{1}{2}$ becomes

$$-\dfrac{1}{2} \le \dfrac{3x-2}{5} \le \dfrac{1}{2}$$

$$10\left(-\dfrac{1}{2}\right) \le 10\left(\dfrac{3x-2}{5}\right) \le 10\left(\dfrac{1}{2}\right)$$

$$-5 \le 2(3x-2) \le 5$$

$$-5 \le 6x-4 \le 5$$

$$-1 \le 6x \le 9$$

$$-\dfrac{1}{6} \le x \le \dfrac{9}{6}$$

$$-\dfrac{1}{6} \le x \le \dfrac{3}{2}$$

43. $\left|2x-\dfrac{1}{5}\right| < 0.3$

$$-0.3 < 2x-\dfrac{1}{5} < 0.3$$

$$-3 < 20x-2 < 3$$

$$-1 < 20x < 5$$

$$-0.05 < x < 0.25$$

45. The continued inequality $-4 \le x \le 4$ as an absolute value becomes

$$|x| \le 4.$$

47. $-1 \le x-5 \le 1$

$$|x-5| \le 1$$

49. $|c-24| \le 12$

$$-12 \le c-24 \le 12$$

$$12 \le c \le 36$$

Minimum number $= 12$, Maximum number $= 36$

51. $|a-s| \le b$

$$|65-s| \le 20$$

53. $16x^4 - 20x^3 + 8x^2 = 4x^2\left(4x^2 - 5x + 2\right)$

55. $2ax - 3a + 8x - 12 = a(2x-3) + 4(2x-3) = (2x-3)(a+4)$

57. $x^2 - 2x - 35 = (x-7)(x+5)$

59. $x^2 - xy - 6y^2 = (x-3y)(x+2y)$

61. $x^2 - 9 = (x+3)(x-3)$

63. $$9x^2 = 12x - 4$$

$$9x^2 - 12x + 4 = 0$$

$$(3x-2)^2 = 0$$

$$3x - 2 = 0$$

$$x = \dfrac{2}{3}$$

7.5 Factoring the Sum and Difference of Two Cubes

1. $x^2 - 6x + 9 = (x-3)(x-3) = (x-3)^2$

3. $a^2 - 12a + 36 = (a-6)(a-6)$
$$= (a-6)^2$$

5. $25 - 10t + t^2 = (5-t)(5-t) = (5-t)^2$

7. $4y^4 - 12y^2 + 9 = (2y^2 - 3)(2y^2 - 3) = (2y^2 - 3)^2$

9. $16a^2 + 40ab + 25b^2 = (4a+5b)(4a+5b)$
$$= (4a+5b)^2$$

11. $\dfrac{1}{25} + \dfrac{1}{10}t^2 + \dfrac{1}{16}t^4 = \left(\dfrac{1}{5} + \dfrac{1}{4}t^2\right)\left(\dfrac{1}{5} + \dfrac{1}{4}t^2\right)$
$$= \left(\dfrac{1}{5} + \dfrac{1}{4}t^2\right)^2$$

13. $(x+2)^2 + 6(x+2) + 9 = \left[(x+2)+3\right]\left[(x+2)+3\right]$
$$= (x+5)(x+5)$$
$$= (x+5)^2$$

15. $49x^2 - 64y^2 = (7x+8y)(7x-8y)$

17. $4a^2 - \dfrac{1}{4} = \left(2a + \dfrac{1}{2}\right)\left(2a - \dfrac{1}{2}\right)$

19. $x^2 - \dfrac{9}{25} = \left(x + \dfrac{3}{5}\right)\left(x - \dfrac{3}{5}\right)$

21. $25 - t^2 = (5-t)(5+t)$

23. $16a^4 - 81 = \left(4a^2 + 9\right)\left(4a^2 - 9\right)$
$$= \left(4a^2 + 9\right)\left(2a+3\right)\left(2a-3\right)$$

25. $x^2 - 10x + 25 - y^2$
$$= \left(x^2 - 10x + 25\right) - y^2$$
$$= (x-5)^2 - y^2$$
$$= \left[(x-5)+y\right]\left[(x-5)-y\right]$$
$$= (x-5+y)(x-5-y)$$

27. $a^2 + 8a + 16 - b^2$
$$= (a+4)^2 - b^2$$
$$= \left[(a+4)+b\right]\left[(a+4)-b\right]$$
$$= (a+4+b)(a+4-b)$$

29. $x^3 + 2x^2 - 25x - 50 = x^2(x+2) - 25(x+2)$
$$= (x+2)\left(x^2 - 25\right)$$
$$= (x+2)(x+5)(x-5)$$

31. $2x^3 + 3x^2 - 8x - 12$
$$= x^2(2x+3) - 4(2x+3)$$
$$= (2x+3)\left(x^2 - 4\right)$$
$$= (2x+3)(x+2)(x-2)$$

33. $4x^3 + 12x^2 - 9x - 27 = 4x^2(x+3) - 9(x+3)$
$$= (x+3)(4x^2 - 9)$$
$$= (x+3)(2x+3)(2x-3)$$

35. $x^3 - y^3 = (x-y)(x^2 + xy + y^2)$

37. $a^3 + 8 = a^3 + 2^3 = (a+2)(a^2 - 2a + 4)$

39. $y^3 - 1 = y^3 - 1^3$
$$= (y-1)(y^2 + y + 1)$$

41. $10r^3 - 1250 = 10(r^3 - 125)$
$$= 10(r^3 - 5^3)$$
$$= 10(r-5)(r^2 + 5r + 25)$$

43. $64 + 27a^3 = 4^3 + (3a)^3 = (4 + 3a)(16 - 12a + 9a^2)$

45. $t^3 + \dfrac{1}{27}$
$$= t^3 + \left(\dfrac{1}{3}\right)^3$$
$$= \left(t + \dfrac{1}{3}\right)\left(t^2 - \dfrac{1}{3}t + \dfrac{1}{9}\right)$$

47. $x^2 - 81 = (x+9)(x-9)$

49. $x^2 + 2x - 15 = (x+5)(x-3)$

51. $x^2y^2 + 2y^2 + x^2 + 2 = y^2(x^2 + 2) + x^2 + 2$
$$= (x^2 + 2)(y^2 + 1)$$

53. $2a^3b + 6a^2b + 2ab = 2ab(a^2 + 3a + 1)$

55. $x^2 + x + 1$ Prime

57. $12a^2 - 75 = 3(4a^2 - 25) = 3(2a+5)(2a-5)$

59. $25 - 10t + t^2 = (5-t)(5-t) = (5-t)^2$

61. $4x^3 + 16xy^2 = 4x(x^2 + 4y^2)$

63. $x^3 + 5x^2 - 9x - 45$
$$= x^2(x+5) - 9(x+5)$$
$$= (x+5)(x^2 - 9)$$
$$= (x+5)(x+3)(x-3)$$

65. $x^2 + 49$ Prime

67. $x^2(x-3)-14x(x-3)+49(x-3)$

$= (x-3)(x^2-14x+49)$

$= (x-3)(x-7)(x-7)$

$= (x-3)(x-7)^2$

69. $8-14x-15x^2 = (4+3x)(2-5x)$

71. $r^2 - \dfrac{1}{25} = \left(r+\dfrac{1}{5}\right)\left(r-\dfrac{1}{5}\right)$

73. $49x^2 + 9y^2$ Prime

75. $100x^2 - 100x - 600 = 100\left(x^2-x-6\right)$

$= 100(x-3)(x+2)$

77. $3x^4 - 14x^2 - 5 = (3x^2+1)(x^2-5)$

79. $24a^5b - 3a^2b$

$= 3a^2b\left(8a^3-1\right)$

$= 3a^2b(2a-1)\left(4a^2+2a+1\right)$

81. $64-r^3 = 4^3 - r^3 = (4-r)(16+4r+r^2)$

83. $20x^4 - 45x^2 = 5x^2\left(4x^2-9\right)$

$= 5x^2(2x+3)(2x-3)$

85. $16x^5 - 44x^4 + 30x^3$

$= 2x^3\left(8x^2-22x+15\right)$

$= 2x^3(4x-5)(2x-3)$

87. $y^6 - 1 = \left(y^3+1\right)\left(y^3-1\right)$

$= (y+1)\left(y^2-y+1\right)(y-1)\left(y^2+y+1\right)$

89. $50-2a^2 = 2\left(25-a^2\right)$ In the form $a^2 - b^2$

$= 2(5+a)(5-a)$

91. $x^2 - 4x + 4 - y^2 = (x-2)^2 - y^2$

$= (x-2+y)(x-2-y)$

93. $9x^2 + 30x + 25$ $b = 30$

$9x^2 - 30x + 25$ $b = -30$

95. $A(r) = 100\left(1 + r + \dfrac{r^2}{4}\right),$ $r = 12\%$

$A(0.12) = 100\left(1 + 0.12 + \dfrac{(0.12)^2}{4}\right)$

$= 100\left(1 + 0.12 + \dfrac{0.0144}{4}\right)$

$= 100(1 + 0.12 + 0.0036)$

$= 100(1.1236)$

$= \$112.36$

97. $V = p^3 - r^3 = (p-r)\left(p^2 + pr + r^2\right)$

99.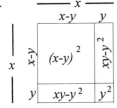

$$x^2 = (x-y)^2 + 2(xy-y^2) + y^2,$$

or

$$x^2 - 2xy + y^2 = (x-y)^2$$

101. $x - y = -2$

$x + y = 4$

See the graph in the back of the textbook.

The solution is $(1, 3)$.

103. $x + y = 4$

$x - y = -2$

. Add the equations

$2x = 2$

$x = 1$

Substitute 1 for x in the first equation.

$1 + y = 4$

$y = 3$

The solution is $(1, 3)$.

105. $x + y = 2$

$y = 2x - 1$

Substitute $2x - 1$ for y in the first equation.

$x + 2x - 1 = 2$

$3x - 1 = 2$

$3x = 3$

$x = 1$

Substitute 1 for x in the second equation.

$y = 2(1) - 1 = 1$

The solution is $(1, 1)$.

107. Let x = number of adult tickets

and y = number of childrens tickets

$x + y = \ 925$ (1)

$2x + y = 1150$ (2)

$-x - y = -925$ (1) times -1

$2x + y = 1150$ (2)

$x = \ 225$ adult tickets

$2(225) + y = 1150$ (2), $x = 225$

$y = \ 700$ children's tickets

109. Let x = gallons of 20% mixture
      and y = gallons of 50% mixture

$$x + y = 9 \qquad (1)$$
$$0.2x + 0.5y = 9(0.3) \qquad (2)$$

$$
\begin{array}{ll}
-2x - 2y = -18 & \quad (1) \text{ times } -2 \\
\underline{2x + 5y = 27} & \quad (2) \text{ times } 10 \\
 3y = 9 & \\
 y = 3 \text{ gals of } 50\% & \\
 x + 3 = 9 & \quad (1), \ y = 3 \\
 x = 6 \text{ gals of } 20\% &
\end{array}
$$

7.6 Review of Systems of Equations in Two Variables

1. $3x - 2y = 6$
    $x - y = 1$
    The solution is $(4, 3)$. See the
    graph in the back of the textbook.

3. $y = \dfrac{3}{5}x - 3$
    $2x - y = -4$
    The solution is $(-5, -6)$. See the
    graph in the back of the textbook.

5. $y = \dfrac{1}{2}x$
    $y = -\dfrac{3}{4}x + 5$
    The solution is $(4, 2)$. See the
    graph in the back of the textbook.

7. $3x + 3y = -2$
    $y = -x + 4$
    Lines are parallel: $\varnothing$
    See the graph in the back of the textbook.

9. $2x - y = 5$
    $y = 2x - 5$
    The lines coincide; any solution to
    one of the equations is a solution
    to the other. See the graph in the
    back of the textbook.

11. $x + y = 5 \qquad (1)$
     $\underline{3x - y = 3 \qquad (2)}$
     $4x = 8$
     $x = 2$
     Substitute 2 for x in (1)
     $2 + y = 5$
     $y = 3$
     The solution is $(2, 3)$.

Problem Set 7.6

13. $3x + y = 4$ (1) $-3x - y = -4$ -1 times (1)

$4x + y = 5$ (2) $\underline{4x + y = 5}$ (2)

$x = 1$

$3(1) + y = 4$ (1), $x = 1$

$y = 1$

The solution is (1, 1).

15. $3x - 2y = 6$ (1)

$6x - 4y = 12$ (2)

Add (2) to -2 times (1)

$-6x + 4y = -12$

$\underline{6x - 4y = 12}$

$0 = 0$

The solution set is $\{(x, y) | 3x - 2y = 6\}$.

17. $x + 2y = 0$ (1) $-2x - 4y = 0$ -2 times (1)

$2x - 6y = 5$ (2) $\underline{2x - 6y = 5}$ (2)

$-10y = 5$

$y = -\dfrac{5}{10} = -\dfrac{1}{2}$

$x + 2\left(-\dfrac{1}{2}\right) = 0$ (1), $y = -\dfrac{1}{2}$

$x - 1 = 0$

$x = 1$

The solution is $(1, -\dfrac{1}{2})$.

19. $2x - 5y = 16$ (1)

$4x - 3y = 11$ (2)

Add (2) to -2 times (1)

$-4x + 10y = -32$

$\underline{4x - 3y = 11}$

$7y = -21$

$y = -3$

Substitute -3 for y in (1)

$2x - 5(-3) = 16$

$x = \dfrac{1}{2}$

The solution is $\left(\dfrac{1}{2}, -3\right)$.

21. $6x + 3y = -1$ (1) $18x + 9y = -3$ 3 times (1)

$9x + 5y = 1$ (2) $\underline{-18x - 10y = -2}$ -2 times (2)

$-y = -5$

$y = 5$

$6x + 3(5) = -1$ (1), $y = 5$

$6x = -16$

$x = -\dfrac{16}{6} = -\dfrac{8}{3}$

The solution is $(-\dfrac{8}{3}, 5)$.

23. $4x + 3y = 14$ (1)

$9x - 2y = 14$ (2)

Add 3 times (2) to 2 times (1)

$8x + 6y = 28$

$\underline{27x - 6y = 42}$

$35x \qquad = 70$

$\quad x \qquad = 2$

Substitute 2 for x in (1)

$4(2) + 3y = 14$

$\qquad y = 2$

The solution is $(2,\ 2)$.

25. $2x - 5y = 3$ (1) $4x - 10y = 6$ 2 times (1)

$-4x + 10y = 3$ (2) $\underline{-4x + 10y = 3}$ (2)

$\qquad\qquad\qquad\qquad\qquad\qquad\qquad 0 = 9$

The lines are parallel.

$\varnothing$

27. $\dfrac{1}{4}x - \dfrac{1}{6}y = -2$ (1)

$-\dfrac{1}{6}x + \dfrac{1}{5}y = 4$ (2)

Add 90 times (2) to 60 times (1)

$15x - 10y = -120$

$\underline{-15x + 18y = 360}$

$\qquad\quad 8y = 240$

$\qquad\quad y = 30$

Substitute 30 for y in (1)

$\dfrac{1}{4}x - \dfrac{1}{6}(30) = -2$

$\qquad \dfrac{1}{4}x = 3$

$\qquad\quad x = 12$

The solution is $(12,\ 30)$.

29. Clear both equations of fractions:

$\dfrac{1}{2}x + \dfrac{1}{3}y = 13$ (1) $3x + 2y = 78$ 6 times (1)

$\dfrac{2}{5}x + \dfrac{1}{4}y = 10$ (2) $8x + 5y = 200$ 20 times (2)

$3x + 2y = 78$ (3) $-15x - 10y = -390$ -5 times (3)

$8x + 5y = 200$ (4) $\underline{16x + 10y = 400}$ 2 times (4)

$\qquad\qquad\qquad\qquad\qquad\qquad\qquad x = 10$

$3(10) + 2y = 78$ (3), $x = 10$

$\qquad 30 + 2y = 78$

$\qquad\qquad 2y = 48$

$\qquad\qquad y = 24$

The solution to the system is $(10,\ 24)$.

31. $\frac{2}{3}x + \frac{2}{5}y = 4$ (1)

$\frac{1}{3}x - \frac{1}{2}y = -\frac{1}{3}$ (2)

Add -30 times (2) to 15 times (1)

$10x + 6y = 60$

$\underline{-10x + 15y = 10}$

$21y = 70$

$y = \frac{10}{3}$

Substitute $\frac{10}{3}$ for y in (1)

$\frac{2}{3}x + \frac{2}{5}\left(\frac{10}{3}\right) = 4$

$\frac{2}{3}x = \frac{8}{3}$

$x = 4$

The solution is $\left(4, \frac{10}{3}\right)$.

33. Substitute $x = 2y + 9$

into $7x - y = 24$

$7(2y + 9) - y = 24$

$14y + 63 - y = 24$

$13y + 63 = 24$

$13y = -39$

$y = -3$

Substitute $y = -3$

into $x = 2y + 9$

$x = 2(-3) + 9$

$x = 3$

The solution to the system is $(3, -3)$.

35. $6x - y = 10$ (1)

$y = -\frac{3}{4}x - 1$ (2)

Substitute $-\frac{3}{4}x - 1$ for y in (1)

$\frac{27}{4}x + 1 = 10$

$\frac{27}{4}x = 9$

$x = \frac{4}{3}$

Substitute $\frac{4}{3}$ for x in (2)

$y = -\frac{3}{4}\left(\frac{4}{3}\right) - 1$

$y = -2$

The solution is $\left(\frac{4}{3}, -2\right)$.

37. Solve $x - y = 4$ for x.

Substitute $x = y + 4$

into $2x - 3y = 6$

$2(y + 4) - 3y = 6$

$2y + 8 - 3y = 6$

$-y + 8 = 6$

$-y = -2$

$y = 2$

Substitute $y = 2$

into $x - y = 4$

$x - 2 = 4$

$x = 6$

The solution to the system is $(6, 2)$.

39. $y = 4x - 4$ (1)

    $y = 3x - 2$ (2)

    Set (1) = (2)

    $4x - 4 = 3x - 2$

         $x = 2$

    Substitute 2 for x in (1)

    $y = 4(2) - 4$

    $y = 4$

    The solution is (2, 4).

41. Solve $2x - y = 5$ for y.

    Substitute $y = 2x - 5$

    into $4x - 2y = 10$

      $4x - 2(2x - 5) = 10$

      $4x - 4x + 10 = 10$

            $0 = 0$

    The lines coincide.

    The solution set is $\{(x,\, y)|2x - y = 5\}$.

43. $\dfrac{1}{3}x - \dfrac{1}{2}y = 0$ (1)

          $x = \dfrac{3}{2}y$ (2)

    Substitute $\dfrac{3}{2}y$ for x in (1)

    $\dfrac{1}{3}\left(\dfrac{3}{2}y\right) - \dfrac{1}{2}y = 0$

      $\dfrac{1}{2}y - \dfrac{1}{2}y = 0$

            $0 = 0$

    The lines coincide.   The solution set is

    $\left\{(x,\, y)\Big|x = \dfrac{3}{2}y\right\}$.

45. $4x - 7y = 3$ (1) $20x - 35y = 15$ 5 times (1)

    $5x + 2y = -3$ (2) $\underline{-20x - 8y = 12}$ -4 times (2)

                              $-43y = 27$

                                  $y = -\dfrac{27}{43}$

    $8x - 14y = 6$ 2 times (1)

    $\underline{35x + 14y = -21}$ 7 times (2)

    $43x \qquad\;\; = -15$

             $x = -\dfrac{15}{43}$

    The solution is $\left(-\dfrac{15}{43},\, -\dfrac{27}{43}\right)$.

47. $9x - 8y = 4$ (1)

    $2x + 3y = 6$ (2)

    Add -9 times (2) to 2 times (1)

    $18x - 16y = 8$

    $\underline{-18x - 27y = -54}$

         $-43y = -46$

             $y = \dfrac{46}{43}$

    Add 8 times (2) to 3 times (1)

    $27x - 24y = 12$

    $\underline{16x + 24y = 48}$

    $43x \qquad = 60$

         $x = \dfrac{60}{43}$

    The solution is $\left(\dfrac{60}{43},\, \dfrac{46}{43}\right)$.

49. $3x - 5y = 2$ $6x - 10y = 4$ 2 times (1)

    $7x + 2y = 1$ $\underline{35x + 10y = 5}$ 5 times (2)

                            $41x \qquad = 9$

                                $x = \dfrac{9}{41}$

    $21x - 35y = 14$ 7 times (1)

    $\underline{-21x - 6y = -3}$ -3 times (2)

        $-41y = 11$

           $y = -\dfrac{11}{41}$

    The solution is $\left(\dfrac{9}{41},\, -\dfrac{11}{41}\right)$.

51. $x + \quad y = 10{,}000$ (1)

     $0.06x + 0.05y = \quad 560$ (2)

     Add 100 times (2) to -6 times (1)

     $-6x - 6y = -60{,}000$

      $\underline{6x + 5y = \quad 56{,}000}$

         $-y = - \quad 4{,}000$

          $y = \quad 4{,}000$

     Substitute 4000 for y in (1)

     $x + 4000 = 10{,}000$

         $x = \quad 6000$

     The solution is $(6000,\ 4000)$.

53. $6x - 9y = 3$ (1) $12x - 18y = \quad 6$ 2 times (1)

     $4x - 6y = c$ (2) $\underline{-12x + 18y = -3c}$ -3 times (2)

                                 $0 = 6 - 3c$

    $6 - 3c = 0$

        $c = 2$

    The solution is $c = 2$.

55. Let x and y = the two numbers

    The second sentence gives us a second equation:

    The linear system that describes the situation is

           $y = 2x + 3$ (1)

        $x + y = 18$ (2)

    Substituting the expression for y from the first equation into the second yields

        $x + (2x + 3) = 18$ (2), $y = 2x + 3$

             $3x + 3 = 18$

               $3x = 15$

                 $x = 5$

    Using $x = 5$ in $y = 2x + 3$ gives the second number:

          $y = 2(5) + 3$ (1), $x = 5$

          $y = 13$

    The two numbers are 5 and 13.

57. $x + y = 300$ (1)
    $2x + 1.5y = 525$ (2)

    $-2x - 2y = -600$ (1) times -2
    $2x + 1.5y = 525$ (2)
    $-0.5y = -75$
    $y = 150$ children's tickets
    $x + 150 = 300$ (1), $y = 150$
    $x = 150$ adult tickets

59. Let x = number of gallons of 20% solution and
    y = number of gallons of 14% solution.
    $x + y = 15$ (1)
    $0.20x + 0.14y = 0.16(15)$ (2)

    $x + y = 15$ (1)
    $20x + 14y = 240$ 100 times (2)
    $-14x - 14y = -210$ -14 times (1)
    $20x + 14y = 240$ (2)
    $6x = 30$
    $x = 5$

    $5 + y = 15$ (1), $x = 5$
    $y = 10$
    5 gallons of 20% solution and 10 gallons of
    14% solution .

61. $3(x - y) = 18$
    $2(x + y) = 24$

    $3x - 3y = 18$ (1)
    $2x + 2y = 24$ (2)

    $6x - 6y = 36$ (1) times 2
    $-6x - 6y = -72$ (2) times -3
    $-12y = -36$
    $y = 3$ mph - speed of the current
    $3x - 3(3) = 18$ (1), $y = 3$
    $x = 9$ mph - speed of the boat

63. Let x = the speed of the plane and
    y = the speed of the air.
    $d = r \cdot t$
    $600 = (x + y) \cdot 2$ (1)
    $600 = (x - y) \cdot 2\frac{1}{2}$ (2)

    $600 = 2x + 2y$ (1)
    $600 = 2\frac{1}{2}x - 2\frac{1}{2}y$ (2)

    $3000 = 10x + 10y$ 5 times (1)
    $2400 = 10x - 10y$ 4 times (2)
    $5400 = 20x$
    $x = 270$
    $600 = 2(270) + 2y$ (1), $x = 270$
    $600 = 540 + 2y$
    $60 = 2y$
    $30 = y$
    The speed of the plane is 270 mph
    and the speed of the wind is 30 mph.

65. $\left(3x^2\right)^3 \left(5y^4\right)^2 = \left(27x^6\right)\left(25y^8\right) = 675x^6 y^8$

67. $\dfrac{x^5}{x^{-2}} = x^{5-(-2)} = x^7$

69. $\left(5\times10^{-8}\right)\left(4\times10^3\right) = 20\times10^{-5} = 2\times10^{-4}$

71. $\left(7a^2 + 8a + 9\right) - \left(5a^2 - 2a + 1\right) = 7a^2 + 8a + 9 - 5a^2 + 2a - 1$
$$= 2a^2 + 10a + 8$$

73. $\left(4a+9\right)\left(4a-9\right) = \left(4a\right)^2 - \left(9\right)^2 = 16a^2 - 81$

75. $6xy^2 \left(4x^2 - 3xy + 2y^2\right) = 24x^3 y^2 - 18x^2 y^3 + 12xy^4$

77.
$$
\begin{array}{r}
x - 2 \\
x - 3 \overline{\smash{\big)}\ x^2 - 5x + 8} \\
\underline{x^2 - 3x} \\
-2x + 8 \\
\underline{-2x + 6} \\
2
\end{array}
$$

The answer is $x - 2 + \dfrac{2}{x-3}$

7.7 Systems of Linear Equations in Three Variables

1. $\quad\begin{aligned} x+y+z &= 4 \quad (1) \\ x-y+2z &= 1 \quad (2) \\ x-y-3z &= -4 \quad (3) \end{aligned}$

$\begin{aligned} x+y+z &= 4 \quad (1) \\ \underline{x-y+2z = 1} \quad &(2) \\ 2x \quad +3z &= 5 \quad (4) \end{aligned}$

$\begin{aligned} x+y+z &= 4 \quad (1) \\ \underline{x-y-3z = -4} \quad &(3) \\ 2x \quad -2z &= 0 \quad (5) \end{aligned}$

$\begin{aligned} 2x+3z &= 5 \quad (4) \\ \underline{-2x+2z = 0} \quad &-1 \text{ times } (5) \\ 5z &= 5 \\ z &= 1 \end{aligned}$

$\begin{aligned} 2x+3(1) &= 5 \quad (4), \ z=1 \\ 2x+3 &= 5 \\ 2x &= 2 \\ x &= 1 \end{aligned}$

$\begin{aligned} 1+y+1 &= 4 \quad (1), \ x=1, \ z=1 \\ y &= 2 \end{aligned}$

The solution is $(1, 2, 1)$.

3. $\quad\begin{aligned} x+y+z &= 6 \quad (1) \\ x-y+2z &= 7 \quad (2) \\ 2x-y-4z &= -9 \quad (3) \end{aligned}$

$\begin{aligned} x+y+z &= 6 \quad (1) \\ \underline{x-y+2z = 7} \quad &(2) \\ 2x \quad +3z &= 13 \quad (4) \end{aligned}$

$\begin{aligned} x+y+z &= 6 \quad (1) \\ \underline{2x-y-4z = -9} \quad &(3) \\ 3x \quad -3z &= -3 \quad (5) \end{aligned}$

$\begin{aligned} 2x+3z &= 13 \quad (4) \\ \underline{3x-3z = -3} \quad &(5) \\ 5x \quad &= 10 \\ x &= 2 \end{aligned}$

$\begin{aligned} 2(2)+3z &= 13 \quad (4), \ x=2 \\ 3z &= 9 \\ z &= 3 \\ 2+y+3 &= 6 \quad (1), \ x=2, \ z=3 \\ y &= 1 \end{aligned}$

The solution is $(2, 1, 3)$.

5. $x + 2y + z = 3$ (1)
    $2x - y + 2z = 6$ (2)
    $3x + y - z = 5$ (3)

$$2x - y + 2z = 6 \quad (2)$$
$$\underline{3x + y - z = 5} \quad (3)$$
$$5x \quad + z = 11 \quad (4)$$

$$x + 2y + z = 3 \quad (1)$$
$$\underline{4x - 2y + 4z = 12} \quad 2 \text{ times } (2)$$
$$5x \quad + 5z = 15 \quad (5)$$

$$5x + z = 11 \quad (4)$$
$$\underline{-5x - 5z = -15} \quad -1 \text{ times } (5)$$
$$-4z = -4$$
$$z = 1$$

$$5x + 1 = 11 \quad (4), \quad z = 1$$
$$5x = 10$$
$$x = 2$$

$$2 + 2y + 1 = 3 \quad (1), \quad x = 2, \ z = 1$$
$$2y + 3 = 3$$
$$2y = 0$$
$$y = 0$$

The solution to is $(2, \ 0, \ 1)$.

7. $2x + 3y - 2z = 4$ (1)
    $x + 3y - 3z = 4$ (2)
    $3x - 6y + z = -3$ (3)

$$2x + 3y - 2z = 4 \quad (1)$$
$$\underline{-x - 3y + 3z = -4} \quad (2) \text{ times } -1$$
$$x \quad + z = 0 \quad (4)$$

$$2x + 6y - 6z = 8 \quad (2) \text{ times } 2$$
$$\underline{3x - 6y + z = -3} \quad (3)$$
$$5x \quad - 5z = 5 \quad (5)$$

$$5x + 5z = 0 \quad (4) \text{ times } 5$$
$$\underline{5x - 5z = 5} \quad (5)$$
$$10x \quad = 5$$
$$x = \frac{1}{2}$$

$$\frac{1}{2} + z = 0 \quad (4), \quad x = \frac{1}{2}$$
$$z = -\frac{1}{2}$$
$$2\left(\frac{1}{2}\right) + 3y - 2\left(-\frac{1}{2}\right) = 4 \quad (1), \quad x = \frac{1}{2}, \ z = -\frac{1}{2}$$
$$3y = 2$$
$$y = \frac{2}{3}$$

The solution is $\left(\dfrac{1}{2}, \dfrac{2}{3}, -\dfrac{1}{2}\right)$.

9.
$$-x+4y-3z = 2 \qquad (1)$$
$$2x-8y+6z = 1 \qquad (2)$$
$$3x-\ y+\ z = 0 \qquad (3)$$

$$-2x+8y-6z = 4 \qquad \text{2 times } (1)$$
$$\underline{\ 2x-8y+6z = 1} \qquad (2)$$
$$0 = 5$$

The equations are inconsistent and there is no solution.

11.
$$\frac{1}{2}x-\ y+z = \ 0 \qquad (1)$$
$$2x+\frac{1}{3}y+z = \ 2 \qquad (2)$$
$$x+\ y+z = -4 \qquad (3)$$

$$\frac{1}{2}x-\ y+z = \ 0 \qquad (1)$$
$$\underline{-2x-\frac{1}{3}y-z = -2} \qquad \text{(2) times } -1$$
$$-\frac{3}{2}x-\frac{4}{3}y\ = -2 \qquad (4)$$

$$2x+\frac{1}{3}y+z = 2 \qquad (2)$$
$$\underline{-x-\ y-z = 4} \qquad \text{(3) times } -1$$
$$x-\frac{2}{3}y\ = 6 \qquad (5)$$

$$-\frac{3}{2}x-\frac{4}{3}y = -2 \qquad (4)$$
$$\underline{-\ 2x+\frac{4}{3}y = -12} \qquad \text{(5) times } -2$$
$$-\frac{7}{2}x\ = -14$$
$$x = \ 4$$

$$4-\frac{2}{3}y = 6 \qquad (5),\ x = 4$$
$$-\frac{2}{3}y = 2$$
$$y = -3$$
$$4+(-3)+z = -4 \qquad (3),\ x = 4,\ y = -3$$
$$z = -5$$

The solution is $(4,\ -3,\ -5)$.

13.
$$2x-\ y-3z = 1 \qquad (1)$$
$$x+2y+4z = 3 \qquad (2)$$
$$4x-2y-6z = 2 \qquad (3)$$

$$x+2y+4z = 3 \qquad (2)$$
$$\underline{4x-2y-6z = 2} \qquad (3)$$
$$5x\ \ -2z = 5 \qquad (4)$$

$$-4x+2y+6z = -2 \qquad -2 \text{ times } (1)$$
$$\underline{\ 4x-2y-6z = \ 2} \qquad (3)$$
$$0 = 0$$

This is a dependent system: no unique solution.

15. $2x - y + 3z = 4$ (1)

    $x + 2y - z = -3$ (2)

    $4x + 3y + 2z = -5$ (3)

$$\begin{array}{ll} 4x - 2y + 6z = 8 & \text{(1) times 2} \\ \underline{x + 2y - z = -3} & \text{(2)} \\ 5x \quad\;\; + 5z = 5 & \text{(4)} \end{array}$$

$$\begin{array}{ll} 6x - 3y + 9z = 12 & \text{(1) times 3} \\ \underline{4x + 3y + 2z = -5} & \text{(3)} \\ 10x \quad\;\; + 11z = 7 & \text{(5)} \end{array}$$

$$\begin{array}{ll} -10x - 10z = -10 & \text{(4) times } -2 \\ \underline{10x + 11z = \quad 7} & \text{(5)} \\ \quad\quad\; z = -3 \end{array}$$

$$\begin{array}{ll} 5x + 5(-3) = 5 & \text{(4), } z = -3 \\ \quad\quad\; x = 4 \end{array}$$

$$\begin{array}{ll} 4 + 2y - (-3) = -3 & \text{(2), } z = -3,\ x = 4 \\ \quad\quad\quad y = -5 \end{array}$$

The solution is $(4, -5, -3)$.

17. $x + y = 9$ (1)

    $y + z = 7$ (2)

    $x - z = 2$ (3)

$$\begin{array}{ll} \quad\; y + z = 7 & \text{(2)} \\ \underline{x \quad\;\; - z = 2} & \text{(3)} \\ x + y \quad\;\; = 9 & \text{(4)} \end{array}$$

$$\begin{array}{ll} \quad x + y = 9 & \text{(1)} \\ \underline{-x - y = -9} & -1 \text{ times } (4) \\ \quad\quad 0 = 0 \end{array}$$

This is a dependent system: no unique solution.

19. $2x + y = 2$ (1)

    $y + z = 3$ (2)

    $4x - z = 0$ (3)

$$\begin{array}{ll} \quad\quad y + z = 3 & \text{(2)} \\ \underline{4x \quad\;\; - z = 0} & \text{(3)} \\ 4x + y \quad\;\; = 3 & \text{(4)} \end{array}$$

$$\begin{array}{ll} \quad 2x + y = 2 & \text{(1)} \\ \underline{-4x - y = -3} & \text{(4) times } -1 \\ \; -2x \quad\;\; = -1 \\ \quad\quad x = \dfrac{1}{2} \end{array}$$

$$\begin{array}{ll} 2\left(\dfrac{1}{2}\right) + y = 2 & \text{(1), } x = \dfrac{1}{2} \\ \quad\quad\;\; y = 1 \\ \quad\; 1 + z = 3 & \text{(2), } y = 1 \\ \quad\quad\;\; z = 2 \end{array}$$

The solution is $\left(\dfrac{1}{2}, 1, 2\right)$.

21. $2x - 3y = 0$ (1)

    $6y - 4z = 1$ (2)

      $x + 2z = 1$ (3)

$$
\begin{array}{ll}
2x - 3y = 0 & (1) \\
\underline{-2x - 4z = -2} & -2 \text{ times } (3) \\
-3y - 4z = -2 & (4)
\end{array}
$$

$$
\begin{array}{ll}
6y - 4z = 1 & (2) \\
\underline{3y + 4z = 2} & -1 \text{ times } (4) \\
9y = 3 & \\
y = \dfrac{3}{9} & \\
y = \dfrac{1}{3} &
\end{array}
$$

$$2x - 3\left(\frac{1}{3}\right) = 0 \quad (1), \; y = \frac{1}{3}$$

$$2x - 1 = 0$$

$$2x = 1$$

$$x = \frac{1}{2}$$

$$6\left(\frac{1}{3}\right) - 4z = 1 \quad (2), \; y = \frac{1}{3}$$

$$2 - 4z = 1$$

$$-4z = -1$$

$$z = \frac{1}{4}$$

The solution is $\left(\dfrac{1}{2}, \dfrac{1}{3}, \dfrac{1}{4}\right)$.

23. $\dfrac{1}{2}x + \dfrac{2}{3}y = \dfrac{5}{2}$ (1)

   $\dfrac{1}{5}x - \dfrac{1}{2}z = -\dfrac{3}{10}$ (2)

   $\dfrac{1}{3}y - \dfrac{1}{4}z = \dfrac{3}{4}$ (3)

$$
\begin{array}{lll}
3x + 4y = 15 & (1) \text{ times } 6 & (4) \\
2x - 5z = -3 & (2) \text{ times } 10 & (5) \\
4y - 3z = 9 & (3) \text{ times } 12 & (6)
\end{array}
$$

$$
\begin{array}{ll}
3x + 4y = 15 & (4) \\
\underline{ -4y + 3z = -9} & (6) \text{ times } -1 \\
3x + 3z = 6 & (7)
\end{array}
$$

$$
\begin{array}{ll}
6x + 6z = 12 & (7) \text{ times } 2 \\
\underline{-6x + 15z = 9} & (5) \text{ times } -3 \\
21z = 21 & \\
z = 1 &
\end{array}
$$

$$3x + 3(1) = 6 \quad (7), \; z = 1$$

$$3x = 3$$

$$x = 1$$

$$3(1) + 4y = 15 \quad (4), \; x = 1$$

$$4y = 12$$

$$y = 3$$

The solution is $(1, 3, 1)$.

25. First eliminate the fractions:

$$\frac{1}{2}x - \frac{1}{4}y + \frac{1}{2}z = -2$$

$$\frac{1}{4}x - \frac{1}{12}y - \frac{1}{3}z = \frac{1}{4}$$

$$\frac{1}{6}x + \frac{1}{3}y - \frac{1}{2}z = \frac{3}{2}$$

$2x - y + 2z = -8$ 4 times (1), (4)

$3x - y - 4z = 3$ 12 times (2), (5)

$x + 2y - 3z = 9$ 6 times (3), (6)

$$2x - y + 2z = -8 \quad (4)$$
$$\underline{-2x - 4y + 6z = -18} \quad -2 \text{ times } (6)$$
$$-5y + 8z = -26 \quad (7)$$

$$3x - y - 4z = 3 \quad (5)$$
$$\underline{-3x - 6y + 9z = -27} \quad -3 \text{ times } (6)$$
$$-7y + 5z = -24 \quad (8)$$

$$35y - 56z = 182 \quad -7 \text{ times } (7)$$
$$\underline{-35y + 25z = -120} \quad 5 \text{ times } (8)$$
$$-31z = 62$$
$$z = -2$$

$$-5y + 8(-2) = -26 \quad (7), \; z = -2$$
$$-5y - 16 = -26$$
$$-5y = -10$$
$$y = 2$$

$$2x - 2 + 2(-2) = -8 \quad (4), \; y = 2, \; z = -2$$
$$2x - 6 = -8$$
$$2x = -2$$
$$x = -1$$

The solution is $(-1, 2, -2)$.

27. $x - y - z = 0$ (1)

     $5x + 20y = 80$ (2)

     $20y - 10z = 50$ (3)

$$5x + 20y = 80 \quad (2)$$
$$\underline{-20y + 10z = -50} \quad (3) \text{ times } -1$$
$$5x + 10z = 30 \quad (4)$$

$$20x - 20y - 20z = 0 \quad (1) \text{ times } 20$$
$$\underline{5x + 20y = 80} \quad (2)$$
$$25x - 20z = 80 \quad (5)$$

$$10x + 20z = 60 \quad (4) \text{ times } 2$$
$$\underline{25x - 20z = 80} \quad (5)$$
$$35x = 140$$
$$x = 4 \text{ amp}$$

$$5(4) + 10z = 30 \quad (4), \; x = 4$$
$$10z = 10$$
$$z = 1 \text{ amp}$$

$$4 - y - 1 = 0 \quad (1), \; x = 4, \; z = 1$$
$$y = 3 \text{ amp}$$

4 amp, 3 amp, 1 amp

29.

Let x = the amount invested at 6% and

y = the amount invested at 9%

$$x + y + z = 2,200 \qquad (1)$$
$$3x = z \qquad (2)$$
$$0.06x + 0.08y + 0.09z = 178 \qquad (3)$$

$$6x + 8y + 9z = 17,800 \qquad \text{100 times } (3) \rightarrow (4)$$

$$x + y + 3x = 2200 \qquad (1), \ z = 3x$$
$$6x + 8y + 9(3x) = 17,800 \qquad (4), \ z = 3x$$

$$4x + y = 2200 \qquad (5)$$
$$33x + 8y = 17,800 \qquad (6)$$

$$-32x - 8y = -17,600 \qquad -8 \text{ times } (5)$$
$$\underline{33x + 8y = \ \ 17,800 \qquad (6)}$$
$$x \qquad = \qquad 200$$

$$3(200) = z \qquad (2), \ x = 200$$
$$600 = z$$

$$200 + y + 600 = 2200 \qquad (1), \ x = 200, \ z = 600$$
$$y = 1400$$

The man invested \$200 at 6%, \$1400 at 8% and \$600 at 9%.

31.

Let x = the number of nickels

y = the number of dimes

z = the number of quarters

$$x + y + z = \ \ 9 \qquad (1)$$
$$x = \ \ y \qquad (2)$$
$$5x + 10y + 25z = 120 \qquad (3)$$

$$y + y + z = 9 \qquad (1), \ x = y$$
$$5y + 10y + 25z = 120 \qquad (3), \ x = y$$

$$2y + z = 9 \qquad (4)$$
$$15y + 25z = 120 \qquad (5)$$

$$-50y - 25z = -225 \qquad -25 \text{ times } (4)$$
$$\underline{15y + 25z = \ \ 120 \qquad (5)}$$
$$-35y \qquad = -105$$
$$y = \ \ 3$$

$$3 + 3 + z = 9 \qquad (1), \ y = z, \ x = 3$$
$$z = 3$$

There are 3 of each type of coin.

33. $h = at^2 + bt + c$

$$a + b + c = 128 \qquad (1)$$
$$9a + 3b + c = 128 \qquad (2)$$
$$25a + 5b + c = 0 \qquad (3)$$

$$\begin{array}{ll} -a - b - c = -128 & (1) \text{ times } -1 \\ \underline{9a + 3b + c = 128} & (2) \\ 8a + 2b = 0 & \\ 4a + b = 0 & (4) \end{array}$$

$$\begin{array}{ll} -9a - 3b - c = -128 & (2) \text{ times } -1 \\ \underline{25a + 5b + c = 0} & (3) \\ 16a + 2b = -128 & \\ 8a + b = -64 & (5) \end{array}$$

$$\begin{array}{ll} -4a - b = 0 & (4) \text{ times } -1 \\ \underline{8a + b = -64} & (5) \\ 4a = -64 & \\ a = -16 & \end{array}$$

$$\begin{array}{ll} 4(-16) + b = 0 & (4), \; a = -16 \\ b = 64 & \end{array}$$

$$\begin{array}{ll} -16 + 64 + c = 128 & (1), \; a = -16, \; b = 64 \\ c = 80 & \\ h = -16t^2 + 64t + 80 & \end{array}$$

35. $\dfrac{x^2 - x - 6}{x^2 - 9} = \dfrac{(x-3)(x+2)}{(x+3)(x-3)} = \dfrac{x+2}{x+3}$

37. $\dfrac{x^2 - 25}{x+4} \cdot \dfrac{2x+8}{x^2 - 9x + 20} = \dfrac{(x+5)(x-5)}{x+4} \cdot \dfrac{2(x+4)}{(x-4)(x-5)} = \dfrac{2(x+5)(x-5)(x+4)}{(x+4)(x-4)(x-5)} = \dfrac{2(x+5)}{x-4}$

39. $\dfrac{x}{x^2 - 16} + \dfrac{4}{x^2 - 16} = \dfrac{x+4}{x^2 - 16} = \dfrac{x+4}{(x+4)(x-4)} = \dfrac{1}{x-4}$

41. $\dfrac{1 - \dfrac{25}{x^2}}{1 - \dfrac{8}{x} + \dfrac{15}{x^2}} \cdot \dfrac{x^2}{x^2} = \dfrac{x^2 - 25}{x^2 - 8x + 15} = \dfrac{(x+5)(x-5)}{(x-5)(x-3)} = \dfrac{x+5}{x-3}$

43. $\dfrac{x}{x^2 - 9} - \dfrac{3}{x-3} = \dfrac{1}{x+3}$

$$(x+3)(x-3)\left(\dfrac{x}{x^2 - 9} - \dfrac{3}{x-3} \right) = (x+3)(x-3) \cdot \dfrac{1}{x+3}$$
$$x - 3(x+3) = x - 3$$
$$x - 3x - 9 = x - 3$$
$$-2x - 9 = x - 3$$
$$-3x = 6$$
$$x = -2$$

The solution is $x = -2$.

45. Let t = the time to fill the pool with both pipes open.

$$\frac{1}{8} - \frac{1}{12} = \frac{1}{t}$$

$$24t\left(\frac{1}{8} - \frac{1}{12}\right) = 24t \cdot \frac{1}{t}$$

$$3t - 2t = 24$$

$$t = 24$$

It will take 24 hours to fill the pool with both pipes left open.

CHAPTER 7 REVIEW

1. $4x - 2 = 7x + 7$

   $-2 = 3x + 7$ Add $-4x$ to both sides

   $-9 = 3x$ Add -7 to both sides

   $-3 = x$ Divide both sides by 3

3. $8 - 3(2t + 1) = 5(t + 2)$

   $8 - 6t - 3 = 5t + 10$ Distributive property

   $-6t + 5 = 5t + 10$ Combine terms

   $5 = 11t + 10$ Add $6t$ to both sides

   $-5 = 11t$ Add -10 to both sides

   $-\dfrac{5}{11} = t$ Divide both sides by 11

5. $\quad 2x^2 - 5x = 12$

   $2x^2 - 5x - 12 = 0$

   $(2x + 3)(x - 4) = 0$

   $2x + 3 = 0 \quad \text{or} \quad x - 4 = 0$

   $\quad 2x = -3 \qquad\qquad x = 4$

   $\quad x = -\dfrac{3}{2}$

7. $(x - 2)(x - 3) = 2$

   $x^2 - 5x + 6 = 2$

   $x^2 - 5x + 4 = 0$

   $(x - 1)(x - 4) = 0$

   $x - 1 = 0 \quad \text{or} \quad x - 4 = 0$

   $\quad x = 1 \qquad\qquad x = 4$

9.   Let x = width, length = $3x$, and perimeter = 32 feet.

   $P = 2l + 2w$

   $32 = 2(3x) + 2x$

   $32 = 6x + 2x$

   $32 = 8x$

   $4 = x$

   $12 = 3x$

   The width is 4 feet & the length is 12 feet.

11. $|x - 3| = 1$

    $x - 3 = 1 \quad \text{or} \quad x - 3 = -1$

    $\quad x = 4 \qquad\qquad x = 2$ Add 3 to both sides

13. $|2y-3| = 5$

    $2y-3 = 5$ or $2y-3 = -5$

       $2y = 8$ $2y = -2$ Add 3 to both sides

        $y = 4$ $y = -1$ Divide both sides by 2

15. $|4x-3| + 2 = 11$

    $|4x-3| = 9$ Add -2 to both sides

    $4x-3 = 9$ or $4x-3 = -9$

        $4x = 12$ $4x = -6$ Add 3 to both sides

           $x = 3$ $x = -\dfrac{6}{4}$ Divide both sides by 4

                        $x = -\dfrac{3}{2}$ Reduce

17. $\dfrac{3}{4}x + 1 \le 10$

    $\dfrac{3}{4}x \le 9$ Add -1 to both sides

    $x \le 12$ Multiply both sides by $\dfrac{4}{3}$

    $(-\infty,\ 12]$

19. $\dfrac{1}{3} \le \dfrac{1}{6}x \le 1$ LCD = 6

    $2 \le x \le 6$ Multiply all members by 6

    $[2,\ 6]$

21. $5t + 1 \le 3t - 2$

    $2t + 1 \le -2$ Add $-3t$ to both sides

    $2t \le -3$ Add -1 to both sides

    $t \le -\dfrac{3}{2}$ Multiply both sides by $\dfrac{1}{2}$

    or

  $-7t \le -21$

    $t \ge 3$ Multiply by $-\dfrac{1}{7}$ and reverse the

                   direction of the inequality symbol

  $\left(-\infty,\ -\dfrac{3}{2}\right] \cup [3,\ \infty)$

23. $|y-2| < 3$

    $-3 < y-2 < 3$

    $-1 < y < 5$ Add 2 to both sides

  See the graph in the back of the textbook.

25. $|2t+1| - 3 < 2$

    $|2t+1| < 5$ Add 3 to both sides

    $-5 < 2t+1 < 5$

    $-6 < 2t < 4$ Add -1 to all three members

    $-3 < t < 2$ Divide all three members by 2

See the graph in the back of the textbook.

27. $|5+8t|+4 \le 1$

$|5+8t| \le -3$

The absolute value will never be negative so the solution is $\varnothing$.

29. $x^4 - 16 = (x^2+4)(x^2-4) = (x^2+4)(x+2)(x-2)$

31. $a^3 - 8 = a^3 - 2^3 = (a-2)(a^2+2a+4)$

33. $3a^3b - 27ab^3 = 3ab(a^2 - 9b^2) = 3ab(a+3b)(a-3b)$

35. $36 - 25a^2 = 6^2 - (5a)^2 = (6+5a)(6-5a)$

37. $6x - 5y = -5$ (1)

   $3x + \ y = \ 1$ (2)

    $6x - 5y = -5$ (1)

   $\underline{-6x - 2y = -2}$ -2 times (2)

       $-7y = -7$

         $y = \ 1$

   $3x + y = 1$

   $3x + 1 = 1$ (2), $y = 1$

     $3x = 0$

      $x = 0$

The solution is $(0, 1)$.

39. $-7x + 4y = -1$ (1)

    $5x - 3y = 0$ (2)

   $-35x + 20y = -5$ 5 times (1)

   $\underline{35x - 21y \ = 0}$ 7 times (2)

        $-y = 5$

          $y = 5$

   $5x - 3(5) = 0$ (2), $y = 5$

       $x = 3$

The solution is $(3, 5)$.

41. $x+y=2$

$y=x-1$

Substitute $y=x-1$ into the first equation

$x+x-1=2$

$2x-1=2$

$2x=3$

$x=\dfrac{3}{2}$

Substitute $x=\dfrac{3}{2}$ into the second equation

$y=\dfrac{3}{2}-1$

$y=\dfrac{1}{2}$

The solution is $\left(\dfrac{3}{2},\dfrac{1}{2}\right)$.

43. $3x+7y=6$

$x=-3y+4$

Substitute $x=-3y+4$ into the first equation

$3(-3y+4)+7y=6$

$-9y+12+7y=6$

$-2y=-6$

$y=3$

Substitute $y=3$ into the second equation

$x=-3(3)+4$

$x=-5$

The solution is $(-5,\,3)$.

45. $x+y+z=6$ (1)

$x-y-3z=-8$ (2)

$x+y-2z=-6$ (3)

$x+y+z=6$ (1)

$\underline{x-y-3z=-8}$ (2)

$2x-2z=-2$ (4)

$x-y-3z=-8$ (2)

$\underline{x+y-2z=-6}$ (3)

$2x-5z=-14$ (5)

$2x-2z=-2$ (4)

$\underline{-2x+5z=14}$ -1 times (5)

$3z=12$

$z=4$

$2x-5(4)=-14$ (5), $z=4$

$x=3$

$3+y+4=6$ (1), $x=3,\ z=4$

$y=-1$

The solution is $(3,\,-1,\,4)$.

47. $5x-2y+z=6$ (1)

$-3x+4y-z=2$ (2)

$6x-8y+2z=-4$ (3)

$5x-2y+z=6$ (1)

$\underline{-3x+4y-z=2}$ (2)

$2x+2y=8$

$-6x+8y-2z=4$ 2 times (2)

$\underline{6x-8y+2z=-4}$ (3)

$0=0$

The system is dependent.

49. $\begin{aligned} 2x - y \quad\quad &= 5 \quad\quad (1) \\ 3x - \quad 2z &= -2 \quad (2) \\ 5y + \quad z &= -1 \quad (3) \end{aligned}$

$$\begin{array}{ll} 3x \quad\quad -2z = -2 & (2) \\ \underline{\quad\quad 10y + 2z = -2} & 2 \text{ times } (3) \\ 3x + 10y \quad\quad = -4 & (4) \end{array}$$

$$\begin{array}{ll} 20x - 10y = 50 & 10 \text{ times } (1) \\ \underline{\quad 3x + 10y = -4} & (4) \\ 23x \quad\quad = 46 & \\ \quad\quad x = 2 & \end{array}$$

$$\begin{aligned} 3(2) - 2z &= -2 \quad (2), \ x = 2 \\ z &= 4 \end{aligned}$$

$$\begin{aligned} 5y + 4 &= -1 \quad (3), \ z = 4 \\ y &= -1 \end{aligned}$$

The solution is $(2, -1, 4)$.

Cumulative Review: Chapters 1-7

1. $15 - 12 \div 4 - 3 \cdot 2 = 15 - 3 - 6$
$$= 6$$

3. $\left(\dfrac{2}{5}\right)^{-2} = \left(\dfrac{5}{2}\right)^2 = \dfrac{5^2}{2^2} = \dfrac{25}{4}$

5. $5 - 3[2x - 4(x-2)] = 5 - 3[2x - 4x + 8]$
$$= 5 - 3[-2x + 8]$$
$$= 5 + 6x - 24$$
$$= 6x - 19$$

7. $(2x+3)(x^2 - 4x + 2) = 2x^3 - 8x^2 + 4x + 3x^2 - 12x + 6$
$$= 2x^3 - 5x^2 - 8x + 6$$

9. $-6 + 2(2x+3) = 0$
$$-6 + 4x + 6 = 0$$
$$4x = 0$$
$$x = 0$$

11. $|2x-3| + 7 = 1$
$$|2x-3| = -6$$
$$\varnothing$$

13. $\quad 8x + 6y = 4 \qquad (1)$
$\quad 12x + 9y = 8 \qquad (2)$

$\quad\quad 24x + 18y = 12 \qquad 3 \text{ times } (1)$
$\quad\underline{-24x - 18y = -16} \qquad -2 \text{ times } (2)$
$\quad\quad\quad\quad 0 \;\; = -4$

The lines are parallel.

15. $2x + y \quad\quad = 8 \qquad (1)$
$\quad\quad 4y - \;\; z = -9 \qquad (2)$
$\;\; 3x \quad\quad - 2z = -6 \qquad (3)$

$\quad -8x - \;\; 4y \quad\quad = -32 \qquad -4 \text{ times } (1)$
$\quad\underline{ 4y - z \;\; = \;\; -9} \qquad (2)$
$\quad -8x \quad\quad\quad - z \;\; = -41 \qquad (4)$

$\quad\quad 3x - 2z = -6 \qquad (3)$
$\quad\underline{16x + 2z = 82} \qquad -2 \text{ times } (4)$
$\quad 19x \quad\quad\quad = 76$
$\quad\quad\quad x = \;\; 4$

$2(4) + y = 8 \qquad (1), \;\; x = 4$
$\quad\quad y = 0$

$4(0) - z = -9 \qquad (2), \;\; y = 0$
$\quad\quad z = 9$

The solution is $(4, 0, 9)$.

17. $|2x - 7| \le 3$

$-3 \le 2x - 7 \le 3$

$4 \le 2x \le 10$

$2 \le x \le 5$

$[2, 5]$

See the graph in the back of the textbook.

19. $-3t \ge 12$

$t \le -4$

$(-\infty, -4]$

21. $-2x > 6$ or $3x - 7 > 2$

$x < -3$ or $3x > 9$

$x < -3$ or $x > 3$

23. $-6\left(\dfrac{7}{54}\right) + \left(-\dfrac{2}{9}\right) = \dfrac{-6 \cdot 7}{9 \cdot 6} + \left(-\dfrac{2}{9}\right)$

$= \dfrac{-7}{9} + \left(-\dfrac{2}{9}\right)$

$= \dfrac{-7 - 2}{9}$

$= \dfrac{-9}{9}$

$= -1$

25. $-1, 0, 2.35, 4$

27. $3 + \dfrac{1}{x} = \dfrac{10}{x^2}$

$3x^2 + x^2\left(\dfrac{1}{x}\right) = x^2\left(\dfrac{10}{x^2}\right)$

$3x^2 + x = 10$

$3x^2 + x - 10 = 0$

$(3x - 5)(x + 2) = 0$

$3x - 5 = 0$ or $x + 2 = 0$

$x = \dfrac{5}{3}$ $x = -2$

29. $\left(5x^{-3}y^2z^{-2}\right)\left(6x^{-5}y^4z^{-3}\right) = 30x^{-8}y^6z^{-5}$

$= \dfrac{30y^6}{x^8z^5}$

31. $\dfrac{\left(7 \times 10^{-5}\right)\left(21 \times 10^{-6}\right)}{3 \times 10^{-12}} = \dfrac{7 \cdot 21 \times 10^{-5-6-(-12)}}{3}$

$= 49 \times 10^1$

$= 4.9 \times 10^1 \times 10^1$

$= 4.9 \times 10^2$

33. $\dfrac{x + 5}{x^2 - 25} = \dfrac{x + 5}{(x + 5)(x - 5)}$

$(x + 5)(x - 5) \ne 0$

$x + 5 \ne 0$ or $x - 5 \ne 0$

$x \ne -5$ $x \ne 5$

35. $y = \dfrac{2}{3}x - 4$

See the graph in the back of the textbook.

37. $16y^2 + 2y + \dfrac{1}{16} = \left(4y + \dfrac{1}{4}\right)^2$

39. $x^3 - 8 = x^3 - 2^3 = (x-2)(x^2 + 2x + 4)$

41. $\begin{array}{r} 2x-3 \\ x-2\overline{)2x^2 - 7x + 9} \\ \underline{2x^2 - 4x} \\ -3x + 9 \\ \underline{-3x + 6} \\ 3 \end{array}$

The answer is $2x - 3 + \dfrac{3}{x-2}$

43. $\dfrac{-2}{x^2 - 2x - 3} + \dfrac{3}{x^2 - 9} = \dfrac{-2}{(x-3)(x+1)} \cdot \dfrac{x+3}{x+3} + \dfrac{3}{(x+3)(x-3)} \cdot \dfrac{x+1}{x+1}$

$= \dfrac{-2(x+3) + 3(x+1)}{(x+3)(x-3)(x+1)}$

$= \dfrac{-2x - 6 + 3x + 3}{(x+3)(x-3)(x+1)}$

$= \dfrac{x-3}{(x+3)(x-3)(x+1)}$

$= \dfrac{1}{(x+3)(x+1)}$

45. Let x = speed of the boat in still water

y = speed of the current

$d = r \cdot t,$

$36 = (x-y)9$

$36 = (x+y)3$

which is equivalent to

$36 = 9x - 9y$ (1)

$36 = 3x + 3y$ (2)

$36 = 9x - 9(4)$ (1), $y = 4$

$72 = 9x$

$8 = x$

The speed of the boat in still water is 8 mph
and the speed of the current is 4 mph.

$\begin{array}{ll} 36 = 9x - 9y & (1) \\ \underline{-108 = -9x - 9y} & -3 \text{ times (2)} \\ -72 = -18y & \\ 4 = y & \end{array}$

Chapter 7 Test

1.
$$5 - \frac{4}{7}a = -11$$
$$-\frac{4}{7}a = -16$$
$$-\frac{7}{4}\left(-\frac{4}{7}a\right) = -\frac{7}{4}(-16)$$
$$a = 28$$

2.
$$3x^2 = 5x + 2$$
$$3x^2 - 5x - 2 = 0$$
$$(3x + 1)(x - 2) = 0$$
$$3x + 1 = 0 \quad \text{or} \quad x - 2 = 0$$
$$x = -\frac{1}{3} \qquad x = 2$$

3.
$$100x^3 = 500x^2$$
$$100x^3 - 500x^2 = 0$$
$$100x^2(x - 5) = 0$$
$$100x^2 = 0 \quad \text{or} \quad x - 5 = 0$$
$$x = 0 \qquad x = 5$$

4. $5(x-1) - 2(2x+3) = 5x - 4$
$$5x - 5 - 4x - 6 = 5x - 4$$
$$x - 11 = 5x - 4$$
$$-11 = 4x - 4$$
$$-7 = 4x$$
$$-\frac{7}{4} = x$$

5. $(x+1)(x+2) = 12$
$$x^2 + 3x + 2 = 12$$
$$x^2 + 3x - 10 = 0$$
$$(x+5)(x-2) = 0$$
$$x + 5 = 0 \quad \text{or} \quad x - 2 = 0$$
$$x = -5 \qquad x = 2$$

6. Let x = the width, $2x$ = the length
and $P = 36$ inches.
$$P = 2l + 2w$$
$$36 = 2(2x) + 2x$$
$$36 = 4x + 2x$$
$$36 = 6x$$
$$6 = x$$
$$12 = 2x$$
The length is 12 inches and the width is 6 inches.

7. When $h = 0$, $h = 32t - 16t^2$
becomes $0 = 32t - 16t^2$
$$16t^2 - 32t = 0$$
$$16t(t - 2) = 0$$
$$16t = 0 \quad \text{or} \quad t - 2 = 0$$
$$t = 0 \qquad t = 2$$
The times are 0 seconds and 2 seconds.

8. $\left|\frac{1}{4}x - 1\right| = \frac{1}{2}$
$$\frac{1}{4}x - 1 = \frac{1}{2} \quad \text{or} \quad \frac{1}{4}x - 1 = -\frac{1}{2}$$
$$x - 4 = 2 \qquad x - 4 = -2$$
$$x = 6 \qquad x = 2$$

9. $|3-2x|+5=2$

$|3-2x|=-3$

$\varnothing$

10. $5-\dfrac{3}{2}x>-1$ or $2x-5\geq7$

$-\dfrac{3}{2}x>-6$ or $2x\geq12$

$x<4$ $x\geq6$

$(-\infty,\,4)\cup[6,\,\infty)$

See the graph in the back of the textbook.

11. $-3\leq5x-1\leq9$

$-2\leq5x\leq10$

$-\dfrac{2}{5}\leq x\leq2$

$\left[-\dfrac{2}{5},\,2\right]$

See the graph in the back of the textbook.

12. $|6x-1|>7$

$6x-1<-7$ or $6x-1>7$

$6x<-6$ $6x>8$

$x<-1$ $x>\dfrac{8}{6}$

$x>\dfrac{4}{3}$

See the graph in the back of the textbook.

13. $|3x-5|-4\leq3$

$|3x-5|\leq7$

$-7\leq3x-5\leq7$

$-2\leq3x\leq12$

$-\dfrac{2}{3}\leq x\leq4$

See the graph in the back of the textbook.

14. $x^2+x-12=(x+4)(x-3)$

15. $16a^4-81y^4=\left(4a^2+9y^2\right)\left(4a^2-9y^2\right)$

$=\left(4a^2+9y^2\right)(2a+3y)(2a-3y)$

16. $t^3+\dfrac{1}{8}=t^3+\left(\dfrac{1}{2}\right)^3=\left(t+\dfrac{1}{2}\right)\left(t^2-\dfrac{1}{2}t+\dfrac{1}{4}\right)$

17. $4a^5b-24a^4b^2-64a^3b^3=4a^3b\left(a^2-6ab-16b^2\right)$

$=4a^3b(a-8b)(a+2b)$

18. $2x-5y=-8$ (1)

$3x+\ y=\ 5$ (2)

$2x-5y=-8$ (1)

$\underline{15x+5y=25}$ 5 times (2)

$17x\ \ \ \ \ =17$

$x\ \ \ \ \ =1$

$2(1)-5y=-8$ (1), $x=1$

$-5y=-10$

$y=2$

The solution is $(1,\,2)$.

19. $\frac{1}{3}x - \frac{1}{6}y = 3$ (1)

    $-\frac{1}{5}x + \frac{1}{4}y = 0$ (2)

    $2x - y = 18$ (3)

    $-4x + 5y = 0$ (4)

    $\frac{1}{3}x - \frac{1}{6}(12) = 3$

         $\frac{1}{3}x = 5$

         $x = 15$

The solution is $(15, 12);$.

20. $2x - 5y = 24$

    $y = 3x - 10$

Substitute $y = 3x - 10$ into the first equation

$2x - 5(3x - 10) = 24$

$2x - 15x + 5 = 24$

$\qquad -13x = -26$

$\qquad\quad x = \frac{26}{13} = 2$

Substitute $x = 2$ into the second equation

$\quad y = 3(2) - 10$

$\quad y = -4$

The solution is $(2, -4)$.

21. $2x - y + z = 9$ (1)

    $x + y - 3z = -2$ (2)

    $3x + y - z = 6$ (3)

Eliminate y

$2x - y + z = 9$ (1)

$\underline{x + y - 3z = -2}$ (2)

$3x \quad\;\; - 2z = 7$ (4)

Eliminate y and z

$2x - y + z = 9$ (1)

$\underline{3x + y - z = 6}$ (3)

$5x \qquad\quad = 15$

$\qquad x = 3$

Substitute $x = 3$ into equation (4)

$3(3) - 2z = 7$

$\qquad z = 1$

Substitute $x = 3$ and $z = 1$ into equation (1)

$2(3) - y + 1 = 9$

$\qquad\quad y = -2$

The solution to the system is $(3, -2, 1)$.

22. $x = 2y - 1$

    $x + y = 14$

Substitute $x = 2y - 1$ into the second equation

$2y - 1 + y = 14$

$\qquad 3y = 15$

$\qquad\; y = 5$

Substitute $y = 5$ into the first equation

$\qquad x = 2(5) - 1 = 9$

The two numbers are 5 and 9.

23. Let x = amount invested at 6% and

y = amount invested at 5%.

The system of equations is

$$x = 2y \qquad (1)$$

$$0.06x + 0.05y = 680 \qquad (2)$$

Substitute $x = 2y$ into the second equation

$$0.06(2y) + 0.05y = 680$$

$$0.12y + 0.05y = 680$$

$$0.17y = 680$$

$$y = 4,000$$

Substitute $y = 4,000$ into the first equation

$$x = 2(4,000) = 8,000$$

John invested $8,000 at 6% and $4,000 at 5%.

24. Let x = the number of nickels,

y = the number of dimes and

z = the number of quarters

$$x + y + z = 15 \qquad (1)$$

$$0.05x + 0.10y + 0.25z = 1.10 \qquad (2)$$

$$x = 4y - 1 \qquad (3)$$

Substitute $x = 4y - 1$ into equation (1)

$$4y - 1 + y + z = 15$$

$$5y + z = 16 \qquad (4)$$

Multiply equation (2) by 100 and then substitute $x = 4y - 1$ into it

$$5(4y - 1) + 10y + 25z = 110$$

$$30y + 25z = 115 \qquad (5)$$

$$\begin{array}{ll} -30y - 6z = -96 & -6 \text{ times (4)} \\ \underline{30y + 25z = 115} & (5) \\ 19z = 19 & \\ z = 1 & \end{array}$$

$$5y + 1 = 16 \quad (4), \ z = 1$$

$$5y = 15$$

$$y = 3$$

$$x = 4(3) - 1 \quad (3), \ y = 3$$

$$x = 11$$

There are 11 nickels, 3 dimes and 1 quarter.

CHAPTER 8

Equations and Inequalities in Two Variables

8.1 The Slope of a Line

1. $(2, 0)$, $(0, -3)$

$$m = \frac{y_2 - y_1}{x_2 - x_1} = \frac{-3 - 0}{0 - 2} = \frac{-3}{-2} = \frac{3}{2}$$

3. $(-4, 1)$, $(-4, 0)$

$$m = \frac{y_2 - y_1}{x_2 - x_1} = \frac{0 - 1}{-4 - (-4)} = -\frac{1}{0}$$

undefined (no slope)

5. $(1, 1)$, $(4, 3)$

$$m = \frac{y_2 - y_1}{x_2 - x_1} = \frac{3 - 1}{4 - 1} = \frac{2}{3}$$

7. $(2, 1)$, $(4, 4)$

$$m = \frac{y_2 - y_1}{x_2 - x_1} = \frac{4 - 1}{4 - 2} = \frac{3}{2}$$

See the graph in the back of the textbook.

9. $(1, 4)$, $(5, 2)$

$$m = \frac{y_2 - y_1}{x_2 - x_1} = \frac{2 - 4}{5 - 1} = \frac{-2}{4} = -\frac{1}{2}$$

See the graph in the back of the textbook.

11. $(1, -3)$, $(4, 2)$

$$m = \frac{y_2 - y_1}{x_2 - x_1} = \frac{2 - (-3)}{4 - 1} = \frac{5}{3}$$

See the graph in the back of the textbook.

13. $(-3, -2)$, $(1, 3)$

$$m = \frac{y_2 - y_1}{x_2 - x_1} = \frac{3 - (-2)}{1 - (-3)} = \frac{5}{4}$$

See the graph in the back of the textbook.

15. $(3, -2)$, $(-3, 2)$

$$m = \frac{y_2 - y_1}{x_2 - x_1} = \frac{-2 - 2}{3 - (-3)} = \frac{-4}{6} = -\frac{2}{3}$$

See the graph in the back of the textbook.

17. $(2, -5)$, $(3, -2)$

$$m = \frac{y_2 - y_1}{x_2 - x_1} = \frac{-2 - (-5)}{3 - 2} = \frac{3}{1} = 3$$

See the graph in the back of the textbook.

19. $2x + 3y = 6$

See the table in the back of the textbook.

$(3, 0)$, $(0, 2)$

$$m = \frac{y_2 - y_1}{x_2 - x_1} = \frac{2 - 0}{0 - 3} = -\frac{2}{3}$$

21. $y = \frac{2}{3}x - 5$

See the table in the back of the textbook

$(0, -5)$, $(3, -3)$

$$m = \frac{y_2 - y_1}{x_2 - x_1} = \frac{-3 - (-5)}{3 - 0} = \frac{2}{3}$$

23. $(3, 0)$, $(0, -2)$

$$m = \frac{y_2 - y_1}{x_2 - x_1} = \frac{-2 - 0}{0 - 3} = \frac{2}{3}$$

See the graph in the back of the textbook.

25. $(4, 0), (0, 2)$

$m = \dfrac{y_2 - y_1}{x_2 - x_1} = \dfrac{0-2}{4-0} = \dfrac{-2}{4} = -\dfrac{1}{2}$

See the graph in the back of the textbook.

27. $(2, 3), (-8, 1)$

$m = \dfrac{y_2 - y_1}{x_2 - x_1} = \dfrac{1-3}{-8-2} = \dfrac{-2}{-10} = \dfrac{1}{5}$

Parallel line: $m_2 = m_1 = -\dfrac{1}{5}$

29. $(5, -6) , (5, 2)$

$m = \dfrac{2-(-6)}{5-5} = \dfrac{8}{0}$ Undefined

Perpendicular line :

$m = -\dfrac{0}{8} = 0$

31. (a) Yes   (b) No

33. $m = \dfrac{2}{3}$ and $h = 8$,

$m = \dfrac{2}{3} = \dfrac{8}{r}$, where r = radius

$\dfrac{2}{3} = \dfrac{8}{r}$

$2r = 24$

Diameter $= 2r = 24$ feet

35. $10 - 0 = 10$ min

37. $(0, -20) , (1, 0)$

$m = \dfrac{y_2 - y_1}{x_2 - x_1}$

$= \dfrac{0-(-20)}{1-0}$

$= 20$

Celsius per minute

39. During the first minute.

41. $(1, 13,200) , (2, 11,950)$

$m = \dfrac{y_2 - y_1}{x_2 - x_1}$

$= \dfrac{11,950-13,200}{2-1}$

$= -1,250$

Dollars per year

43. From 2 to 3 years.

45. (a) Point A is $(3 \text{ mi})(5{,}280 \text{ ft}/\text{mi}) = 15{,}840$ feet.

    (b) Point, $(0, 1{,}106)$ and $m = -\dfrac{7}{100}$

$$y = -\frac{7}{100}x + 1{,}106$$

47.   average yearly salary increase=slope, $m = \dfrac{y_2 - y_1}{x_2 - x_1}$

    (a) $m = \dfrac{40{,}000 - 25{,}000}{1975 - 1967} = \dfrac{15{,}000}{8} = \$1{,}875$

    (b) $m = \dfrac{1{,}300{,}000 - 600{,}000}{1994 - 1990} = \dfrac{700{,}000}{4}$

           $= \$175{,}000$

49. $3x + 2y = 12$

     $3(4) + 2y = 12$ when $x = 4$

       $12 + 2y = 12$

          $2y = 0$

           $y = 0$

51. $3x + 2y = 12$

      $2y = -3x + 12$

        $y = -\dfrac{3}{2}x + 6$

53. $A = P + Prt$

     $A - P = Prt$

     $\dfrac{A - P}{Pr} = t$

55.   Y=2X+B

     See the graph in the back of the textbook.

57. See the graph in the back of the textbook.

59.   Y = AX

     See the graph in the back of the textbook.

8.2 The Equation of a Line

1. $m = 2,\ b = 3$

     $y = mx + b$

     $y = 2x + 3$

3. $m = 1,\ b = -5$

     $y = mx + b$

     $y = 1x - 5$

     $y = x - 5$

5. $m = \dfrac{1}{2},\ b = \dfrac{3}{2}$

     $y = mx + b$

     $y = \dfrac{1}{2}x + \dfrac{3}{2}$

7. $m = 0,\ b = 4$

     $y = mx + b$

     $y = 0x + 4$

     $y = 4$

9. $y = 3x - 2$

Slope $= 3$, y-intercept $= -2$

Perpendicular, slope $= -\dfrac{1}{3}$

See the graph in the back of the textbook.

11. $2x - 3y = 12$

$$y = \frac{2}{3}x - 4$$

Slope $= \dfrac{2}{3}$, y-intercept $= -4$

Perpendicular, slope $= -\dfrac{3}{2}$

See the graph in the back of the textbook.

13. $4x + 5y = 20$

$$5y = -4x + 20$$

$$y = -\frac{4}{5}x + 4$$

Slope $= -\dfrac{4}{5}$, y-intercept $= 4$

Perpendicular slope $= \dfrac{5}{4}$.

See the graph in the back of the textbook.

15. $(0, -4)$, $(4, -2)$

$$m = \frac{-2 - (-4)}{4 - 0}$$

$$= \frac{2}{4} = \frac{1}{2}$$

$$b = -4$$

$$y = \frac{1}{2}x - 4$$

17. $(3, 1)$, $(-3, 5)$

$$m = \frac{5 - 1}{-3 - 3}$$

$$= \frac{4}{-6}$$

$$= -\frac{2}{3}$$

Slope $= -\dfrac{2}{3}$ y-intercept $= 3$.

$$y = -\frac{2}{3}x + 3$$

19. In the following exercises use

$$y - y_1 = m(x - x_1).$$

$(-2, -5)$, $m = 2$

$$y - (-5) = 2[x - (-2)]$$

$$y + 5 = 2x + 4$$

$$y = 2x - 1$$

21. $(-4, 1)$, $m = -\dfrac{1}{2}$

$$y - 1 = -\frac{1}{2}(x + 4)$$

$$y - 1 = -\frac{1}{2}x - 2$$

$$y = -\frac{1}{2}x - 1$$

23. $\left(-\dfrac{1}{3}, 2\right)$, $m = -3$

$$y - 2 = -3\left[x - \left(-\frac{1}{3}\right)\right]$$

$$y - 2 = -3x - 1$$

$$y = -3x + 1$$

25. $(-2, -4)$, $(1, -1)$

$$m = \frac{-1-(-4)}{1-(-2)} = \frac{3}{3} = 1$$

$(1, -1)$, $m = 1$

$$y + 1 = 1(x-1)$$

$$x - y = 2$$

27. $(-1, -5)$, $(2, 1)$

$$m = \frac{1-(-5)}{2-(-1)} = \frac{6}{3} = 2$$

$$y - 1 = 2(x-2)$$

$$y - 1 = 2x - 4$$

$$2x - y = 3$$

29.

$$\left(\frac{1}{3}, -\frac{1}{5}\right), \left(-\frac{1}{3}, -1\right)$$

$$m = \frac{-1-\left(-\frac{1}{5}\right)}{-\frac{1}{3}-\left(\frac{1}{3}\right)} = \frac{-\frac{4}{5}}{-\frac{2}{3}} = -\frac{4}{5} \cdot -\frac{3}{2} = \frac{12}{10} = \frac{6}{5}$$

$$\left(\frac{1}{3}, -\frac{1}{5}\right), m = \frac{6}{5}$$

$$y + \frac{1}{5} = \frac{6}{5}\left(x - \frac{1}{3}\right)$$

$$y + \frac{1}{5} = \frac{6}{5}x - \frac{6}{15}$$

$$y = \frac{6}{5}x - \frac{3}{5} \Rightarrow 6x - 5y = 3$$

31. Points: $(2, 0)$, $(0, -4)$

$$m = \frac{-4-0}{0-2} = 2$$

$$b = -4$$

$$y = 2x - 4$$

33. $(0, 4)$, $(-2, 0)$

Rise $= 4$

Run $= 2$

$$m = \frac{\text{rise}}{\text{run}} = \frac{4}{2} = 2$$

y-intercept $= 4$

$$y = 2x + 4$$

35. $y = -2$

Slope$= 0$, y-intercept $= -2$

See the graph in the back of the textbook.

37. $l_1 : 3x - y = 5$

$$-y = -3x + 5$$

$$y = 3x - 5$$

$$m_1 = 3$$

l_2 : parallel $m_2 = m_1 = 3$; $(-1, 4)$

$$y - 4 = 3(x+1)$$

$$y - 4 = 3x + 3$$

$$y = 3x + 7$$

39.

$$l_1: 2x - 5y = 10$$

$$-5y = -2x + 10$$

$$y = \frac{2}{5}x - 2$$

$$m_1 = \frac{2}{5}$$

l_2: perpendicular, $m_2 = -\frac{1}{m_1} = -\frac{5}{2}$; $(-4, -3)$

$$y - (-3) = -\frac{5}{2}[x - (-4)]$$

$$y + 3 = -\frac{5}{2}x - \frac{20}{2}$$

$$y = -\frac{5}{2}x - 13$$

41. $l_1: \ y = -4x + 2$

$m_1 = -4$

$l_2: \ \text{perpendicular}, \ m_2 = -\dfrac{1}{m_1} = \dfrac{1}{4}; \ (-1, 0)$

$y - 0 = \dfrac{1}{4}\left[x - (-1)\right]$

$y = \dfrac{1}{4}x + \dfrac{1}{4}$

43. $(3, 0), (0, 2)$

$y = \dfrac{2 - 0}{0 - 3} = -\dfrac{2}{3}$

$y - y_1 = m(x - x_1)$

$y - 0 = -\dfrac{2}{3}(x - 3)$

$y = -\dfrac{2}{3}x + 2$

45. a. $(0, 32), (25, 77)$

$m = \dfrac{77 - 32}{25 - 0} = \dfrac{45}{25} = \dfrac{9}{5}, \ b = 32$

$F = mC + b$

$F = \dfrac{9}{5}C + 32$

b. $C = 30$

$F = \dfrac{9}{5}C + 32$

$F = \dfrac{9}{5}(30) + 32 = 86°$

47. (a) $(98, 1), (155, 1.5)$

$m = \dfrac{1.5 - 1}{155 - 98} = 0.00877$

$y - y_1 = m(x - x_1)$

$y - 1 = 0.00877(x - 98)$

$y = 0.00877x - 1.85946$

(b) $40 \le x \le 220$

49. (a) Let $x =$ the number of years since 1984 and $y =$ the number of AIDS cases.

$(0, 3,000), (4, 20,000)$

$m = \dfrac{20,000 - 3,000}{4 - 0} = 4,250$

Using $(x_1, \ y_1) = (0, \ 3,000), \ m = 4,250$

$y - y_1 = m(x - x_1)$ yields

$y - 3,000 = 4,250(x - 0)$

$y = 4,250x + 3,000$

(b) 1986 represents $x = 2$

$y = 4,250(2) + 3,000$

$= 11,500$ cases of AIDS in 1986

51. $\quad x+y+z=6 \quad$ (1)
$\quad\quad 2x-y+z=3 \quad$ (2)
$\quad\quad x+2y-3z=-4 \quad$ (3)

$\quad\quad\quad x+y+z=6$
$\quad\quad\underline{-2x+y-z=-3}$
$\quad\quad -\ x+2y\ \ =3 \quad$ (4)

$\quad\quad 3x+3y+3z=18$
$\quad\quad\underline{x+2y-3z=-4}$
$\quad\quad 4x+5y\quad\quad=14 \quad$ (5)

53. $\ 3x+4y\quad=15 \quad$ (1)
$\quad\ 2x\quad-5z=-3 \quad$ (2)
$\quad\quad\ 4y-3z=\ 9 \quad$ (3)

$\quad\quad 6x+8y\quad\quad=30 \quad$ (1) times 2
$\quad\underline{-6x\quad\quad+15z=\ 9} \quad$ (2) times -3
$\quad\quad\quad 8y+15z=39 \quad$ (4)

$\quad\quad -8y+\ 6z=-18 \quad$ (3) times -2
$\quad\quad\underline{\ 8y+15z=\ \ 39} \quad$ (4)
$\quad\quad\quad\quad 21z=\ \ 21$
$\quad\quad\quad\quad\ z=\ \ 1$

$\quad -4x+8y=12 \quad$ 4 times (4)
$\quad\underline{\ \ 4x+5y=14} \quad$ (5)
$\quad\quad\quad 13y=26$
$\quad\quad\quad\ y=2$

$4x+5(2)=14 \quad$ (5), $\ y=2$
$\quad\quad\ x=1$

$1+2+z=6 \quad$ (1), $\ y=2,\ x=1$
$\quad\quad z=3$
The solution is $(1,\ 2,\ 3)$.

$4y-3(1)=9 \quad$ (3), $z=1$
$\quad\quad\ y=3$

$2x-5(1)=-3 \quad$ (2), $z=1$
$\quad\quad\ x=1$
The solution is $(1,\ 3,\ 1)$.

55. See the graph in the back of the textbook.

57. See the graph in the back of the textbook.

59. $\quad \dfrac{x}{2}+\dfrac{y}{3}=1$
$\quad\quad 3x+2y=6$
$\quad\quad\quad 2y=-3x+6$
$\quad\quad\quad\ y=-\dfrac{3}{2}x+3$
$\quad\quad$ Slope $=-\dfrac{3}{2}$
$\quad y$- intercept $=3$
$\quad x$- intercept $=2$

61. $\quad \dfrac{x}{-2}+\dfrac{y}{3}=1$
$\quad\quad 3x-2y=6$
$\quad\quad\quad 2y=3x+6$
$\quad\quad\quad\ y=\dfrac{3}{2}x+3$
$\quad\quad$ Slope $=\dfrac{3}{2}$
$\quad y$- intercept $=3$
$\quad x$- intercept $=\dfrac{-2}{1}=-2$

63. $\dfrac{x}{a} + \dfrac{y}{b} = 1$

$bx + ay = ab$

$ay = -bx + ab$

$y = -\dfrac{b}{a}x + b$

Slope $= -\dfrac{b}{a}$

y - intercept $= b$

x - intercept $= a$

8.3 Linear Inequalities in Two Variables

1. $x + y < 5$
See the graph in the back of the textbook.

3. $x - y \geq -3$
See the graph in the back of the textbook.

5. $2x + 3y < 6$
See the graph in the back of the textbook.

7. $-x + 2y > -4$
See the graph in the back of the textbook.

9. $2x + y < 5$
See the graph in the back of the textbook.

11. $y < 2x - 1$
See the graph in the back of the textbook.

13. $m = \dfrac{-4}{4} = -1, \ b = 4$

$y > -1x + 4$

$y > -x + 4$

$x + y > 4$

15. $m = \dfrac{2}{4} = \dfrac{1}{2}, \ b = 2$

$y \leq \dfrac{1}{2}x + 2$

17. $x \geq 3$
See the graph in the back of the textbook.

19. $y \leq 4$
See the graph in the back of the textbook.

21. $y < 2x$
See the graph in the back of the textbook.

23. $y \geq \dfrac{1}{2}x$
See the graph in the back of the textbook.

25. $y \geq \dfrac{3}{4}x - 2$
See the graph in the back of the textbook.

27. $\dfrac{x}{3} + \dfrac{y}{2} > 1$
See the graph in the back of the textbook.

29. $x + y \leq 200,\quad x \geq 0,\ \text{and}\ y \geq 0$
See the graph in the back of the textbook.

31. $\dfrac{x}{12} + \dfrac{y}{22} \leq 30$
$11x + 6y \leq 3960,\ x \geq 0,\ y \geq 0$
See the graph in the back of the textbook.

33. $y - 0.08x \leq 0,\quad x \geq 0,\ \text{and}\ y \geq 0$
See the graph in the back of the textbook.

35. $5t - 4 > 3t - 8$
$\quad 5t > 3t - 4$
$\quad 2t > -4$
$\quad\ t > -2$

37. $2x < -8\quad \text{or}\quad x + 2 > 6$
$\ \ x < -4\quad \text{or}\quad x > 4$

39. $-9 < -4 + 5t < 6$
$\quad -5 < 5t < 10$
$\quad -1 < t < 2$

41. $y < |x + 2|$
See the graph in the back of the textbook.

43. $y > |x - 3|$
See the graph in the back of the textbook.

45. $x + 2y \geq 100,\quad x + y \leq 200,\quad x \geq 0,\ \text{and}\ y \geq 0$
See the graph in the back of the textbook.

8.4 Introduction to Functions

1. (a) $y = 8.5x$ for $10 \leq x \leq 40$
   (b) See the table in the back of the textbook.
   (c) See the line graph in the back of the textbook.
   (d) Domain $= \{x | 10 \leq x \leq 40\}$
       Range $= \{y | 85 \leq y \leq 340\}$
   (e) Minimum $= \$85$; Maximum $= \$340$

3. Domain is $\{1, 2, 4\}$ range is $\{1, 3, 5\}$
   This is a function.

5. Domain is $\{-1, 1, 2\}$ range is $\{3, -5\}$
   This is a function.

7. Domain is $\{7, 3\}$ range is $\{-1, 4\}$
   This is not a function.

9. Yes, a function.

11. No, not a function.

13. No, not a function.

15. Yes, a function.

17. Yes, a function.

19. (a) See the table in the back of the textbook.

   (b) Domain $= \{t \mid 0 \le t \le 1\}$; Range $= \{h \mid 0 \le h \le 4\}$

   (c) See the graph in the back of the textbook.

21. Domain $= \{x \mid -5 \le x \le 5\}$

   Range $= \{y \mid 0 \le y \le 5\}$.

23. Domain: $\{x \mid -5 \le x \le 3\}$ Range: $\{y \mid y = 3\}$

25. $y = x^2 - 1$

   Domain = all real numbers

   Range $= \{y \mid y \ge -1\}$

   A function

   See the graph in the back of the textbook.

27. $y = x^2 + 4$

   Domain = all reals

   Range $= \{y \mid y \ge 4\}$

   Function

   See the graph in the back of the textbook.

29. $x = y^2 - 1$

   Domain $= \{x \mid x \ge -1\}$

   Range = all real numbers

   Not a function

   See the graph in the back of the textbook.

31. $x = y^2 + 4$

   Domain $= \{x \mid x \ge 4\}$

   Range = all reals

   Not a function

   See the graph in the back of the textbook.

33. $y = |x - 2|$

   Domain = all real numbers

   Range $= \{y \mid y \ge 0\}$

   A function

   See the graph in the back of the textbook.

35. $A = \pi r^2$, $0 \le r \le 3$ See the graph back of textbook.

37. Let x = width and $x + 2$ = length

   $P = 2w + 2l$

   $P(x) = 2x + 2(x + 2)$

   $P(x) = 2x + 2x + 4$

   $P(x) = 4x + 4$

   The variable x (width) must be positive.

39. $A = x(x + 2)$

   $\quad = x^2 + 2x$

   $\quad x > 0$

41. (a) yes, a function

   (b) Domain $= \{t \mid 0 \le t \le 6\}$,

      Range $= \{h(t) \mid 0 \le h(t) \le 60$

   (c) $t = 3$

   (d) $h(3) = 60$

   (e) $t = 6$

43. (a) Figure 11   (b) Figure 12

   (c) Figure 10   (d) Figure 9

45. $x = 4$,

$y = 3(4) - 2 = 10$

47. $y = 3x - 2$, $x = -4$

$y = 3(-4) - 2 = -14$

49. $x = 2$,

$y = 2^2 - 3 = 1$

51. $x = 0$

$y = (0)^2 - 3 = -3$

8.5 Function Notation

In Exercises 1-9, use $f(x) = 2x - 5$ and $g(x) = x^2 + 3x + 4$.

1. $f(2) = 2(2) - 5 = 4 - 5 = -1$

3. $f(-3) = 2(-3) - 5 = -11$

5. $g(-1) = (-1)^2 + 3(-1) + 4$

$= 1 - 3 + 4 = 2$

7. $g(-3) = (-3)^2 + 3(-3) + 4 = 4$

9. $g(4) = 4^2 + 3(4) + 4 = 16 + 12 + 4 = 32$

$f(4) = 2(4) - 5 = 8 - 5 = 3$

$g(4) + f(4) = 32 + 3 = 35$

11. $f(3) = 2(3) - 5 = 1$

$g(2) = 2^2 + 3(2) + 4 = 14$

$f(3) - g(2) = 1 - 14 = -13$

In Exercises 13-25 use $f(x) = 3x^2 - 4x + 1$ and $g(x) = 2x - 1$.

13. $f(0) = 3(0)^2 - 4(0) + 1 = 1$

15. $g(-4) = 2(-4) - 1 = -9$

17. $f(-1) = 3(-1)^2 - 4(-1) + 1$

$f(-1) = 3 + 4 + 1 = 8$

19. $g(10) = 2(10) - 1 = 19$

21. $f(3) = 3(3)^2 - 4(3) + 1$

$f(3) = 27 - 12 + 1 = 16$

23. $g(\frac{1}{2}) = 2(\frac{1}{2}) - 1 = 0$

25. $f(a) = 3a^2 - 4a + 1$

In Exercises 27-32 use $f = \{(1,4),(-2,0),(3,\frac{1}{2}),(\pi,0)\}$
and $g = \{(1,1),(-2,2),(\frac{1}{2},0)\}$.

27. $f(1) = 4$

29. $g\left(\dfrac{1}{2}\right) = 0$

31. $g(-2) = 2$

33. $f(x) = 2x^2 - 8$
$$f(0) = 2(0)^2 - 8 = 0 - 8 = -8$$

In Exercises 35-45 use $f(x) = 2x^2 - 8$ and $g(x) = \dfrac{1}{2}x + 1$.

35. $g(-4) = \frac{1}{2}(-4) + 1 = -1$

37. $f(x) = 2x^2 - 8$
$$f(a) = 2a^2 - 8$$

39. $f(b) = 2b^2 - 8$

41. $\quad g(2) = 1 + 1 = 2$
$$f[g(2)] = f\left[\frac{1}{2}(2) + 1\right]$$
$$f(2) = 2(2)^2 - 8 = 8 - 8 = 0$$

43. $g[f(-1)] = g\left[2(-1)^2 - 8\right]$
$$= g(-6)$$
$$= \frac{1}{2}(-6) + 1$$
$$= -2$$

45. $g[f(0)] = g\left[2(0)^2 - 8\right]$
$$= g(-8)$$
$$= \frac{1}{2}(-8) + 1$$
$$= -3$$

47. $f(x) = \dfrac{1}{2}x + 2$

See the graph in the back of the textbook.

49. $\quad f(x) = \dfrac{1}{2}x + 2$ and $f(x) = x$ then:
$$\frac{1}{2}x + 2 = x$$
$$2 = \frac{1}{2}x$$
$$4 = x$$

51. $f(x) = x^2$

See the graph in the back of the textbook.

53. If $V(t) = 150 \cdot 2^{t/3}$ for $t \geq 0$ then
$$V(3) = 150 \cdot 2^{3/3} = 150 \cdot 2 = 300$$
The painting is worth $300 in 3 years.
$$V(6) = 150 \cdot 2^{6/3} = 150 \cdot 2^2 = 150 \cdot 4 = 600$$
The painting is worth $600 in 6 years.

55. $P(x) = 2x + 2(2x + 3)$

$6x + 6$, where $x > 0$

57. $A(r) = \pi r^2$
$$A(2) = 3.14(2)^2 = 3.14(4) = 12.56$$
$$A(5) = 3.14(5)^2 = 3.14(25) = 78.5$$
$$A(10) = 3.14(10)^2 = 3.14(100) = 314$$

59. (a) $f(x) = 24x + 33, \quad x = 9$

$\quad f(9) = 24(9) + 33 = 249$ cents

$\qquad = \$2.49$

(b) $f(5)$ represents the charge for a 6-minute call.

$\quad f(5) = 24(5) + 33 = 153$ cents

$\qquad = \$1.53$, the cost of a 6-minute call

(c) $f(x) = 129$

$\quad 129 = 24x + 33$

$\quad 96 = 24x$

$\quad 4 = x$

5 minutes

61. $V(t) = -3,300t + 18,000$

(a) 3years and 9 months=3.75 years: $V(3.75) = -3,300(3.75) + 18,000 = \$5,625$

(b) Salvage means after 5 years: $V(5) = -3,300(5) + 18,000 = \$1,500$

(c) Domain: $\{t \mid 0 \leq t \leq 5\}$

(d) See the graph in the back of the textbook.

(e) Range: $\{V \mid 1,500 \leq V \leq 18,000\}$

(f) $10,000 = -3,300t + 18,000$

$\quad 3,300t = 8,000$

$\qquad t \approx 2.4$ years

63. (a) See the table in the back of the textbook.

    (b) Over 2 ounces, but not over 3 ounces.

    $2 < x \leq 3$

    (c) Domain $\{x | 0 < x \leq 6\}$

    (d) Range $\{C(x) | C(x) = 32, 55, 78, 101, 124, 147\}$

65. $|3x - 5| = 7$

    $3x - 5 = 7$ or $3x - 5 = -7$

       $3x = 12$ $3x = -2$

        $x = 4$ $x = -\dfrac{2}{3}$

67. $|4y + 2| - 8 = -2$

    $|4y + 2| = 6$

    $4y + 2 = 6$ or $4y + 2 = -6$

        $4y = 4$ $4y = -8$

         $y = 1$ $y = -2$

69. $5 + |6t + 2| = 3$

    $|6t + 2| = -2$

    $\varnothing$

71. (a) $f(2) = 2$ (b) $f(-4) = 0$

    (c) $g(0) = 1$ (d) $g(3) = 4$

    (e) $g(2) - f(2) = 3 - 2 = 1$

    (f) $f(1) + g(1) = 1 + 2 = 3$

    (g) $f[g(3)] = f[4] = 0$

    (h) $g[f(3)] = g[1] = 2$

8.6 Algebra with Functions

1. $(f + g)(x) = (4x - 3) + (2x + 5)$

   $= 6x + 2$

3. $(g - f)(x) = (2x + 5) - (4x - 3)$

   $= 2x + 5 - 4x + 3$

   $= -2x + 8$

5. $(fg)(x) = (4x - 3)(2x + 5)$

   $= 8x^2 + 20x - 6x - 15$

   $= 8x^2 + 14x - 15$

7. $(g/f)(x) = \dfrac{2x + 5}{4x - 3}$

9. $(g + f)(x) = g(x) + f(x)$

   $= (x - 2) + (3x - 5)$

   $= 4x - 7$

11. $(g + h)(x) = g(x) + h(x)$

    $= (x - 2) + (3x^2 - 11x + 10)$

    $= 3x^2 - 10x + 8$

13. $(g-f)(x) = g(x) - f(x)$
$$= (x-2) - (3x-5)$$
$$= -2x+3$$

15. $(fg)(x) = f(x)g(x)$
$$= (3x-5)(-2)$$
$$= 3x^2 - 11x + 10$$
$$= h(x)$$

17. $(fh)(x) = f(x)h(x)$
$$= (3x-5)(3x^2 - 11x + 10)$$
$$= 9x^3 - 48x^2 + 85x - 50$$

19. $\left(\dfrac{h}{f}\right)(x) = \dfrac{h(x)}{f(x)}$
$$= \frac{3x^2 - 11x + 10}{3x-5}$$
$$= \frac{(3x-5)(x-2)}{3x-5}$$
$$= x-2$$
$$= g(x)$$

21. $\left(\dfrac{f}{h}\right)(x) = \dfrac{f(x)}{h(x)}$
$$= \frac{3x-5}{3x^2 - 11x + 10}$$
$$= \frac{3x-5}{(3x-5)(x-2)}$$
$$= \frac{1}{x-2}$$
$$= \frac{1}{g(x)}$$

23. $(f+g+h)(x) = f(x) + g(x) + h(x)$
$$= (3x-5) + (x-2) + (3x^2 - 11x + 10)$$
$$= 3x^2 - 7x + 3$$

25. $(h+fg)(x) = h(x) + f(x)g(x)$
$$= (3x^2 - 11x + 10) + (3x-5)(x-2)$$
$$= (3x^2 - 11x + 10) + (3x^2 - 11x + 10)$$
$$= 6x^2 - 22x + 20$$
$$= 2h(x)$$

27. $(f+g)(x) = (2x+1) + (4x+2)$
$$= 6x+3$$
$$(f+g)(2) = 6(2) + 3 = 15$$

29. $(fg)(x) = (2x+1)(4x+2)$

$\qquad = 8x^2 + 4x + 4x + 2$

$\qquad = 8x^2 + 8x + 2$

$\quad (fg)(3) = 8(3)^2 + 8(3) + 2 = 98$

31. $(h/g)(x) = \dfrac{4x^2 + 4x + 1}{4x + 2}$

$\qquad = \dfrac{(2x+1)^2}{2(2x+1)}$

$\qquad = \dfrac{2x+1}{2}, \quad x \neq -\dfrac{1}{2}$

$\quad (h/g)(1) = \dfrac{2(1)+1}{2} = \dfrac{3}{2}$

33. $(fh)(x) = (2x+1)(4x^2 + 4x + 1)$

$\qquad = 8x^3 + 8x^2 + 2x + 4x^2 + 4x + 1$

$\qquad = 8x^3 + 12x^2 + 6x + 1$

$\quad (fh)(0) = 8(0)^3 + 12(0)^2 + 6(0) + 1 = 1$

35. $(f + g + h)(x) = (2x+1) + (4x+2) + (4x^2 + 4x + 1)$

$\qquad = 4x^2 + 10x + 4$

$\quad (f + g + h)(2) = 4(2)^2 + 10(2) + 4 = 40$

37. $(h + fg)(x) = (4x^2 + 4x + 1) + (8x^2 + 8x + 2)$

$\qquad = 12x^2 + 12x + 3$

$\quad (h + fg)(3) = 12(3)^2 + 12(3) + 3 = 147$

39. $(fog)(x) = f[g(x)] = (x+4)^2 = x^2 + 8x + 16$

$\quad (gof)(x) = g[f(x)] = (x^2) + 4 = x^2 + 4$

(a) $(fog)(5) = (5)^2 + 8(5) + 16 = 81$

(b) $(gof)(5) = (5)^2 + 4 = 29$

(c) $(fog)(x) = x^2 + 8x + 16$

(d) $(gof)(x) = x^2 + 4$

41. $(fog)(x) = f[g(x)] = (4x-1)^2 + 3(4x-1) = 16x^2 - 8x + 1 + 12x - 3$

$\qquad = 16x^2 + 4x - 2$

$\quad (gof)(x) = g[f(x)] = 4(x^2 + 3x) - 1 = 4x^2 + 12x - 1$

(a) $(fog)(0) = 16(0)^2 + 4(0) - 2 = -2$

(b) $(gof)(0) = 4(0)^2 + 12(0) - 1 = -1$

(c) $(fog)(x) = 16x^2 + 4x - 2$

(d) $(gof)(x) = 4x^2 + 12x - 1$

43. $(fog)(x) = f[g(x)] = 5\left(\dfrac{x+4}{5}\right) - 4 = x + 4 - 4 = x$

$\quad (gof)(x) = g[f(x)] = \dfrac{(5x-4)+4}{5} = \dfrac{5x}{5} = x$

45. (a) $R(x) = n(x) \cdot p(x) = x(11.5 - 0.05x) = 11.5x - 0.05x^2$

(b) $C(x) = f(x) + v(x) = 200 + 2x$

(c) $P(x) = R(x) - C(x) = (11.5x - 0.05x^2) - (200 + 2x) = -0.05x^2 + 9.5x - 200$

(d) $\overline{C}(x) = \dfrac{C(x)}{n(x)} = \dfrac{200 + 2x}{x} = \dfrac{200}{x} + 2$

47. $T(M) = 62 + 0.6(M - 62) = 62 + 0.6M - 0.6(62) = 0.6M + 24.8$

(a) $M(x) = 220 - x$

(b) $M(24) = 200 - 24 = 196$

(c) $T(196) = 0.6(196) + 24.8 = 142.4 \approx 142$

(d) $T(x) = 0.6(220 - x) + 24.8 = 132 - 0.6x + 24.8 = 156.8 - 0.6x$

$\quad T(36) = 156.8 - 0.6(36) = 135.2 \approx 135$

(e) $T(48) = 156.8 - 0.6(48) = 128$

49. $4x + 3y = 10 \qquad (1)$

$\quad\ 2x + y = 4 \qquad (2)$

$\quad\ 4x + 3y = 10 \qquad (1)$

$\underline{-4x - 2y = -8} \qquad -2 \text{ times } (2)$

$\qquad\quad y = 2$

$4x + 3(2) = 10 \qquad (1), \ y = 2$

$\qquad\ x = 1$

The solution is $(1, 2)$.

51. $4x + 5y = 5 \qquad (1)$

$\quad \dfrac{6}{5}x + y = 2 \qquad (2)$

$\quad\ 4x + 5y = \ \ 5 \qquad (1)$

$\underline{-6x - 5y = -10} \qquad -5 \text{ times } (2)$

$\ -2x \qquad\ \ = -5$

$\qquad x = \dfrac{5}{2}$

$\dfrac{6}{5}\left(\dfrac{5}{2}\right) + y = 2 \qquad (2), \ x = \dfrac{5}{2}$

$\qquad 3 + y = 2$

$\qquad\quad y = -1$

The solution is $\left(\dfrac{5}{2}, -1\right)$.

53. $\quad x + y = 3 \qquad\qquad y = x + 3$

$\quad x + x + 3 = 3 \qquad\ y = 0 + 3$

$\qquad\quad x = 0 \qquad\qquad y = 3$

$\qquad$ The solution set is $(0, 3)$

55. $\qquad 2x - 3y = -6 \qquad (1)$

$\qquad\qquad y = 3x - 5 \qquad (2)$

$\quad 2x - 3(3x - 5) = -6 \qquad (1), \ y = 3x - 5$

$\qquad 2x - 9x + 15 = -6$

$\qquad\qquad\ -7x = -21$

$\qquad\qquad\qquad x = 3$

$\qquad\qquad\qquad y = 3(3) - 5 \qquad (2), \ x = 3$

$\qquad\qquad\qquad y = 4$

$\qquad$ The solution is $(3, 4)$

8.7 Variation

1. $y = Kx$: $y = 10$, $x = 2$
   $10 = K(2) \Rightarrow 1 = K = 5$
   $y = 5x$ Let $x = 6$
   $y = 5 \cdot 6$
   $y = 30$

3. $y = Kx$: $y = -32$, $x = 4$
   $-32 = K(4) \Rightarrow K = -8$
   $y = -8x$ Let $y = -40$
   $-40 = -8x$
   $5 = x$

5. $r = \dfrac{K}{s}$: $r = -3$, $s = 4$
   $-3 = \dfrac{K}{4} \Rightarrow K = -12$
   $r = \dfrac{-12}{s}$ Let $s = 2$
   $r = \dfrac{-12}{2}$
   $r = -6$

7. $r = \dfrac{K}{s}$: $r = 8$, $s = 3$
   $8 = \dfrac{K}{3} \Rightarrow K = 24$
   $r = \dfrac{24}{s}$ Let $r = 48$
   $48 = \dfrac{24}{s}$
   $s = \dfrac{1}{2}$

9. $d = Kr^2$: $d = 10$, $r = 5$
   $10 = K \cdot 25 \Rightarrow K = \dfrac{2}{5}$
   $d = \dfrac{2}{5} r^2$ Let $r = 10$
   $d = \dfrac{2}{5}(10)^2$
   $d = \dfrac{2}{5}(100)$
   $d = 2(20)$
   $d = 40$

11. $d = Kr^2$: $d = 100$, $r = 2$
    $100 = K(4) \Rightarrow K = 25$
    $d = 25r^2$ Let $r = 3$
    $d = 25(9)$
    $d = 225$

13. $y = \dfrac{K}{x^2}:$ $y = 45,$ $x = 3$

$45 = \dfrac{K}{3^2}$

$45 = \dfrac{K}{9} \Rightarrow K = 405$

$y = \dfrac{405}{x^2}$ Let $x = 5$

$y = \dfrac{405}{5^2}$

$y = \dfrac{405}{25}$

$y = \dfrac{81}{5}$

15. $y = \dfrac{K}{x^2}:$ $y = 18,$ $x = 3$

$18 = \dfrac{K}{9} \Rightarrow K = 162$

$y = \dfrac{162}{x^2}$ Let $x = 2$

$y = \dfrac{162}{4}$

$y = 40.5$

17. $z = Kxy^2,$ $z = 54,$ $x = 3,$ $y = 3$

$54 = K(3)(3)^2$

$54 = 27K \Rightarrow K = 2$

$z = 2xy^2,$ $x = 2,$ $y = 4$

$z = 2(2)(4^2)$

$z = 64$

19. $z = Kxy^2,$ $z = 64,$ $x = 1,$ $y = 4$

$64 = K(1)(16) \Rightarrow K = 4$

$z = 4xy^2,$ $z = 32,$ $y = 1$

$32 = 4(x)(1)$

$8 = x$

21. Let $l =$ length the spring stretches and $F =$ force

$l = KF,$ $l = 3,$ $F = 5$

$3 = K \cdot 5 \Rightarrow K = \dfrac{3}{5}$

$l = \dfrac{3}{5}F,$ $l = 10,$

$10 = \dfrac{3}{5}F$

$F = \dfrac{50}{3}$ pounds

23.   (a) Let $t =$ temperature, $p =$ pressure,

$t = Kp$ $t = 200,$ $p = 50$

$200 = K(50) \Rightarrow K = 4$

$t = 4p$

(c) $t = 4p,$ $t = 280$

$280 = 4p$

$p = 70$ lbs/sq in

See the graph in the back of the textbook.

25. Let v = volume, p = pressure,

$$v = \frac{K}{p}, \quad v = 25, \quad p = 36$$

$$25 = \frac{K}{36}$$

$$K = (25)(36) = 900$$

$$v = \frac{900}{p}, \quad v = 75$$

$$75 = \frac{900}{p}$$

$$p = 12 \text{ lbs/sq in}$$

27. (a) Let f = f-stop, d = diameter

$$f = \frac{K}{d}, \quad f = 2, \quad d = 40$$

$$2 = \frac{K}{40} \Rightarrow K = 80$$

$$f = \frac{80}{d}$$

(b) See the graph in the back of the textbook.

$$f = \frac{80}{d}, \quad d = 10$$

$$f = \frac{80}{10}$$

$$= 8$$

29. Let S = surface area, h = height, r = radius

$$S = Khr, \quad S = 94, \quad h = 5, \quad r = 3$$

$$94 = K(5)(3)$$

$$94 = 15K \Rightarrow K = \frac{94}{15}$$

$$S = \frac{95}{15} \text{ hr}, \quad h = 8, \quad r = 2$$

$$S = \frac{94}{15}(8)(2)$$

$$S = \frac{1504}{15} \text{ inches}$$

31. Let R = resistance, L = length, d = diameter

$$R = \frac{KL}{d^2}, \quad L = 100, \quad d = 0.01, \quad R = 10$$

$$10 = \frac{K(100)}{0.0001} \Rightarrow K = 0.00001$$

$$K = \frac{0.00001L}{d^2}, \quad L = 60, \quad d = 0.02$$

$$R = \frac{0.00001(60)}{0.004} = 1.5 \text{ ohms}$$

33. a. Let P = period and l = length;

$$P = K\sqrt{l}, \quad P = 2.1, \quad l = 100$$

$$2.1 = K\sqrt{100}$$

$$2.1 = K(10) \Rightarrow K = 0.21$$

$$P = 0.21\sqrt{l}$$

b. See the graph in the back of the textbook.

c. When $l = 225$, the equation becomes

$$P = 0.21\sqrt{225} = 3.15$$

It takes 3.15 seconds

35. Let f = pitch, λ = wavelength

$$f = \frac{K}{\lambda}, \quad f = 420, \quad \lambda = 2.2$$

$$420 = \frac{K}{2.2} \Rightarrow K = 924$$

$$f = \frac{924}{\lambda}, \quad f = 720$$

$$720 = \frac{924}{\lambda}$$

$$\lambda = \frac{924}{720} = 1.28 \text{ meters}$$

37. $F = G\dfrac{m_1 m_2}{d^2}$

39. $\left|\dfrac{x}{5}+1\right| \geq \dfrac{4}{5}$

$\dfrac{x}{5}+1 \leq -\dfrac{4}{5}$ or $\dfrac{x}{5}+1 \geq \dfrac{4}{5}$

$\dfrac{x}{5} \leq -1\dfrac{4}{5}$ or $\dfrac{x}{5} \geq -\dfrac{1}{5}$

$\dfrac{x}{5} \leq -\dfrac{9}{5}$ or $x \geq -1$

$x \leq -9$ or $x \geq -1$

See the graph in the back of the textbook.

41. **All real numbers**

43. $-8+|3y+5| < 5$

$|3y+5| < 13$

$-13 < 3y+5 < 13$

$-18 < 3y\ \ \ < 8$

$-6 < \ \ y\ \ \ \ < \dfrac{8}{3}$

See the graph in the back of the textbook.

45. (a) See the table in the back of the textbook.

(b) See the graph in the back of the textbook.

(c) Illumination F is inversely proportional to the square of distance h. Area A is directly proportional to the square of distance h.

(d) $A = Kh^2 \Rightarrow \pi(1)^2 = K(2)^2 \Rightarrow K = \dfrac{\pi}{4} \Rightarrow A = \dfrac{\pi}{4}h^2$

$F = \dfrac{K}{h^2} \Rightarrow 900 = \dfrac{K}{(2)^2} \Rightarrow K = 3600 \Rightarrow F = \dfrac{3600}{h^2}$

Chapter 8 Review

1. $(5, 2),\ (3, 6)$

$$m = \frac{y_2 - y_1}{x_2 - x_1}$$

$$m = \frac{6-2}{3-5} = \frac{4}{-2} = -2$$

The slope is -2.

3. $3x + 2y = 6$

See the graph in the back of the textbook.

Use $(0, 3),\ (2, 0)$

$$m = \frac{0-3}{2-0} = -\frac{3}{2}$$

5. $x = 3$

See the graph in the back of the textbook.

It has no slope.

7. $m = -\dfrac{1}{3};(-4,7),(2,x)$

$$m = \frac{y_2 - y_1}{x_2 - x_1}$$

$$-\frac{1}{3} = \frac{x-7}{2-(-4)}$$

$$-\frac{1}{3} = \frac{x-7}{6}$$

$$-2 = x - 7$$

$$5 = x$$

9. $(5, 3y)$ and $(2, y)$

$$m = \frac{y-3y}{2-5} = \frac{-2y}{-3},\ \text{ parallel, } m = 4$$

$$4 = \frac{-2y}{-3}$$

$$-12 = -2y$$

$$6 = y$$

11. $m = -2,\ b = 0$

$$y = mx + b$$

$$y = -2x + 0$$

$$y = -2x$$

13. $2x - 3y = 9$

$$-3y = -2x + 9$$

$$y = \frac{2}{3}x - 3$$

The slope is $\dfrac{2}{3}$ and the y-intercept is -3.

15. $(x_1,\ y_1) = (-3,\ 1)\ ,\ m = -\dfrac{1}{3}$

$$y - y_1 = m(x - x_1)$$

$$y - 1 = -\frac{1}{3}(x+3)$$

$$y - 1 = -\frac{1}{3}x - 1$$

$$y = -\frac{1}{3}x + 0$$

$$y = -\frac{1}{3}x$$

17. $(-3, 7), (4, 7)$

$$m = \frac{7-7}{4-(-3)} = \frac{0}{7} = 0$$

$(x_1, y_1) = (-3, 7)$, $m = 0$

$y - y_1 = m(x - x_1)$

$y - 7 = 0(x + 3)$

$y - 7 = 0$

$y = 7$

19. $l_1 : 2x - y = 4$

$y = 2x - 4$

$m_1 = 2$

$l_2 :$ parallel $(2, -3)$

$m_2 = m_1 = 2$

$y - y_1 = m(x - x_1)$

$y + 3 = 2(x - 2)$

$y + 3 = 2x - 4$

$y = 2x - 4 - 3$

$y = 2x - 7$

21. $y \le 2x - 3$

See the graph in the back of the textbook.

23. Domain $= \{2, 3, 4\}$, Range $= \{4, 3, 2\}$, a function

25. $f(-3) = 0$

27. $f(x) = 2x^2 - 4x + 1$

$f(0) = 2(0)^2 - 4(0) + 1 = 1$

29. $g(x) = 3x + 2$

$g(0) = 3(0) + 2 = 2$

$f[g(0)] = f(2)$

$f(x) = 2x^2 - 4x + 1$

$f(2) = 2(2)^2 - 4(2) + 1$

$f(2) = 8 - 8 + 1 = 1$

$f[g(0)] = 1$

31. $(f + g)(x) = 2x + 1 + x^2 - 4 = x^2 + 2x - 3$

33. $(fg)(x) = (2x + 1)(x^2 - 4) = 2x^3 - 8x + x^2 - 4 = 2x^3 + x^2 - 8x - 4$

$(fg)(1) = 2(1)^3 + (1)^2 - 8(1) - 4 = -9$

35. $(fog)(x) = 2(x^2 - 4) + 1 = 2x^2 - 8 + 1$

$(fog)(-1) = 2(-1)^2 - 7 = -5$

37. $y = Kx:$ $y = 6,$ $x = 2$
$6 = K \cdot 2 \Rightarrow K = 3$
$y = 3x$ Let $x = 8$
$y = 3 \cdot 8$
$y = 24$

39. $y = \dfrac{K}{x^2}:$ $y = 9,$ $x = 2$
$9 = \dfrac{K}{4} \Rightarrow K = 36$
$y = \dfrac{36}{x^2}$ Let $x = 3$
$y = \dfrac{36}{9}$
$y = 4$

41. Let t = tension, d = distance
$t = Kd:$ $d = 2,$ $t = 42$
$42 = K(2) \Rightarrow K = 21$
$t = 21d$ Let $d = 4$
$t = 21(4)$
$t = 84$ pounds

Cumulative Review: Chapters 1-8

1. $11 - (-9) - 7 - (-5) = 11 + 9 + (-7) + 5$
$= 18$

3. $\dfrac{x^{-5}}{x^{-8}} = x^{-5-(-8)} = x^3$

5. $-3(5x + 4) + 12x = -15x - 12 + 12x$
$= -3x - 12$

7. $(x + 3)^2 - (x - 3)^2 = x^2 + 6x + 9 - (x^2 - 6x + 9)$
$= x^2 + 6x + 9 - x^2 + 6x - 9$
$= 12x$

9. $-3\left(\dfrac{5}{12}\right) - \dfrac{3}{4} = \dfrac{-15}{12} + \dfrac{-3}{4}$
$= \dfrac{-5}{4} + \dfrac{-3}{4}$
$= \dfrac{-8}{4}$
$= -2$

11. $8x^3 - 27 = (2x - 3)(4x^2 + 6x + 9)$

13. $(3x - 4) + 2y = (3x + 2y) - 4$
commutative and associative properties

15. $\dfrac{y}{x^2 - y^2} - \dfrac{x}{x^2 - y^2} = \dfrac{-1(x - y)}{(x + y)(x - y)} = -\dfrac{1}{x + y}$

17. $\dfrac{x^4-16}{x^3-8}\cdot\dfrac{x^2+2x+4}{x^2+4}=\dfrac{\left(x^2+4\right)(x+2)(x-2)}{(x-2)\left(x^2+2x+4\right)}\cdot\dfrac{x^2+2x+4}{x^2+4}=x+2$

19.
$$
\begin{array}{r}
a^3-a^2+2a-4+\dfrac{7}{a+2} \\
a+2\overline{\smash{)}a^4+a^3+0a^2+0a-1}
\end{array}
$$

$\quad\quad\quad$ – $\quad$ –

$\quad\quad\underline{a^4+2a^3}$

$\quad\quad\quad -a^3+0a^2$

$\quad\quad\quad$ + $\quad$ +

$\quad\quad\quad\underline{-a^3-2a^2}$

$\quad\quad\quad\quad 2a^2+0a$

$\quad\quad\quad\quad$ – $\quad$ –

$\quad\quad\quad\quad\underline{2a^2+4a}$

$\quad\quad\quad\quad\quad -4a-1$

$\quad\quad\quad\quad\quad$ + $\quad$ +

$\quad\quad\quad\quad\quad\underline{-4a-8}$

$\quad\quad\quad\quad\quad\quad\quad 7$

21. $7y-6=2y+9$

$\quad\quad 5y-6=9$

$\quad\quad\quad 5y=15$

$\quad\quad\quad\; y=3$

23. $|a|-5=7$

$\quad\quad |a|=12$

$\quad\quad a=12\quad$ or $\quad a=-12$

25. $\dfrac{3}{y-2}=\dfrac{2}{y-3}$

$\quad\quad \text{LCD}=(y-2)(y-3)$

$\quad\quad 3(y-3)=2(y-2)$

$\quad\quad\; 3y-9=2y-4$

$\quad\quad\quad\; y-9=-4$

$\quad\quad\quad\quad\; y=5$

27. $5x - 2y = -1$

$\quad\quad\quad y = 3x + 2$

Substitute $y = 3x + 2$ into the first equation

$5x - 2(3x + 2) = -1$

$\quad 5x - 6x - 4 = -1$

$\quad\quad\quad -x = 3$

$\quad\quad\quad x = -3$

Substitute $x = -3$ into the second equation

$\quad y = 3(-3) + 2$

$\quad y = -7$

The solution is $(-3, -7)$.

29. $-5x + 3y = 1$ No change $\rightarrow$ $-5x + 3y = 1$

$\dfrac{5}{3}x - y = 2$ Multiply by 3 $\rightarrow$ $\dfrac{5x - 3y = 6}{}$

$\quad\quad\quad\quad\quad\quad\quad\quad\quad\quad\quad\quad\quad 0 = 6$

A false statement means the lines are parallel.

31. $\quad 7x - 9y = 2 \quad (1)$

$\quad -3x + 11y = 1 \quad (2)$

Add 3 times (1) and 7 times (2)

$\quad\quad 50y = 13$

$\quad\quad\quad y = \dfrac{13}{50}$

Substitute $\dfrac{13}{50}$ for y in (1)

$7x - 9\left(\dfrac{13}{50}\right) = 2$

$\quad 7x - \dfrac{117}{50} = 2$

$\quad\quad\quad 7x = \dfrac{217}{50}$

$\quad\quad\quad\quad x = \dfrac{31}{50}$

The solution is $\left(\dfrac{31}{50}, \dfrac{13}{50}\right)$.

33. $-3(3x - 1) \leq -2(3x - 3)$

$\quad -9x + 3 \leq -6x + 6$

$\quad\quad -3x + 3 \leq 6$

$\quad\quad\quad -3x \leq 3$

$\quad\quad\quad\quad x \geq -1$

See the graph in the back of the textbook.

35. Use $(-3, 0)$ and $(2, -5)$.

$$m = \frac{y_2 - y_1}{x_2 - x_1} = \frac{-5 - 0}{2 - (-3)} = \frac{-5}{5} = -1$$

The slope is -1

37. $f(x) = \frac{1}{2}x - 1$

See the graph in the back of the textbook.

39. $168 = 2 \cdot 2 \cdot 2 \cdot 3 \cdot 7 = 2^3 \cdot 3 \cdot 7$

41. $x^2 + 10x + 25 - y^2 = (x + 5)^2 - y^2$
$$= (x + 5 + y)(x + 5 - y)$$

43. $9,270,000 = 9.27 \times 10^6$

45. $f(x) = x^2 - 2x \qquad\qquad g(x) = x + 5$

$f(-2) = (-2)^2 - 2(-2) = 8 \qquad g(3) = 3 + 5 = 8$

$$f(-2) - g(3) = 8 - 8 = 0$$

47. For a line through $(-2, 6)$ and $(-2, 3)$

$$m = \frac{y_2 - y_1}{x_2 - x_1} = \frac{3 - 6}{-2 - (-2)} = \frac{-3}{0} \text{ is undefined}$$

The line is vertical. $x = -2$

49. Let $x = $ base and $2x - 5 = $ the height

$$A = \frac{1}{2}bh$$

$$75 = \frac{1}{2}x(2x - 5)$$

$$75 = x^2 - \frac{5}{2}x$$

$$0 = 2x^2 - 5x - 150$$

$$0 = (2x + 15)(x - 10)$$

$2x + 15 = 0 \qquad \text{or} \qquad x - 10 = 0$

$\quad 2x = -15 \qquad\qquad\qquad x = 10$

$\quad\quad x = -\frac{15}{2} \qquad\qquad 2x - 5 = 15$

A base cannot be negative, so $x \neq -\dfrac{15}{2}$

The height is 15 feet, base is 10 feet.

Chapter 8 Test

1. $2x + y = 6$

    $y = -2x + 6$

    $x\text{-intercept} = 3$ when $y = 0$

    $y\text{-intercept} = 6$

    $\text{slope} = -2$

    See the graph in the back of the textbook.

2. $y = -2x - 3$

    $\text{slope} = -2$

    $y\text{-intercept} = -3$

    $x\text{-intercept} = -\dfrac{3}{2}$ when $y = 0$

    See the graph in the back of the textbook.

3. $y = \dfrac{3}{2}x + 4$

    $\text{slope} = \dfrac{3}{2}$

    $y\text{-intercept} = 4$

    $x\text{-intercept} = -\dfrac{8}{3}$ when $y = 0$

    See the graph in the back of the textbook.

4. $x = -2$

    $x\text{-intercept} = -2$

    no $y\text{-intercept}$

    no slope

    See the graph in the back of the textbook.

5. $y - y_1 = m(x - x_1)$ when $(x_1,\ y_1) = (-1,\ 3)$ and $m = 2$

    $y - 3 = 2[x - (-1)]$

    $y - 3 = 2x + 2$

    $y = 2x + 5$

6. The slope through $(-3,\ 2)$ and $(4,\ -1)$

    is $m = \dfrac{-1 - 2}{4 - (-3)} = -\dfrac{3}{7}$

    $y - y_1 = m(x - x_1)$ when

    $(x_1,\ y_1) = (-3,\ 2)$ and $m = -\dfrac{3}{7}$

    $y - 2 = -\dfrac{3}{7}[x - (-3)]$

    $y - 2 = -\dfrac{3}{7}x - \dfrac{9}{7}$

    $y = -\dfrac{3}{7}x + \dfrac{5}{7}$

7. $2x - 5y = 10$

$$-5y = -2x + 10$$

$$y = \frac{2}{5}x - 2$$

$$y - y_1 = m(x - x_1)$$

$$y - (-3) = \frac{2}{5}(x - 5) \quad \text{when}$$

$$(x_1, y_1) = (5, -3) \text{ and } m = \frac{2}{5}$$

$$y + 3 = \frac{2}{5}x - 2$$

$$y = \frac{2}{5}x - 5$$

8.

$$y = 3x - 1$$

$$y - y_1 = m(x - x_1) \quad \text{when}$$

$$(x_1, y_1) = (-1, -2) \text{ and } m = -\frac{1}{3}$$

$$y - (-2) = -\frac{1}{3}[x - (-1)]$$

$$y + 2 = -\frac{1}{3}x - \frac{1}{3}$$

$$y = -\frac{1}{3}x - \frac{7}{3}$$

9. $x = 4$, since it is a vertical line.

10. $3x - 4y < 12$

The boundary $3x - 4y = 12$ is a broken line because of the $<$ symbol. Substituting (0, 0) gives a true statement so the shading is above the boundary since (0, 0) is above the boundary. See the graph in the back of the textbook.

11. $y \leq -x + 2$

The boundary $y = -x + 2$ is a solid line because of the $\leq$ symbol. Substituting (0, 0) gives a true statement so the shading is below the boundary since (0, 0) is below the boundary. See the graph in the back of the textbook.

12. Domain $= \{-2, -3\}$, Range $= \{0, 1\}$, not a function

13. Domain $=$ all real numbers, Range $= \{y | y \geq -9\}$, a function

14. $f(x) = x - 2 \qquad g(x) = 3x + 4$

$f(3) = 3 - 2 = 1 \qquad g(2) = 3(2) + 4 = 10$

$f(3) + g(2) = 1 + 10 = 11$

15. $h(x) = 3x^2 - 2x - 8 \qquad g(x) = 3x + 4$

$h(0) = 3(0)^2 - 2(0) - 8 \qquad g(0) = 3(0) + 4$

$h(0) = -8 \qquad\qquad\qquad g(0) = 4$

$h(0) + g(0) = -8 + 4 = -4$

16. $g(x) = 3x + 4$

$g(2) = 3(2) + 4 = 10$

$f(x) = x - 2$

$f[g(2)] = f(10)$

$\qquad = 10 - 2$

$\qquad = 8$

17. $f(x) = x - 2$

$f(2) = 2 - 2 = 0$

$g(x) = 3x + 4$

$g[f(2)] = g(0)$

$\qquad = 3(0) + 4$

$\qquad = 4$

18. $0 < x < 4$

19. $V(x) = x(8 - 2x)^2$

20. $V(2) = 2(8 - 2 \cdot 2)^2$

    $V(2) = 2(8 - 4)^2$

    $V(2) = 2(4)^2$

    $V(2) = 2(16)$

    $V(2) = 32 \text{ in}^3$

    32 in^3 is the volume of the box if a square with 2−in. sides is cut from each corner.

21. $y = Kx^2$ $\qquad y = 50,\ x = 5$

    $50 = K(25)$

    $2 = K$

    $y = 2x^2$ $\qquad x = 3$

    $y = 2(9)$

    $y = 18$

22. $z = Kxy^3$ $\qquad z = 15,\ x = 5,\ y = 2$

    $15 = K(5)(2)^3$

    $15 = 40K$

    $K = \dfrac{15}{40} = \dfrac{3}{8}$

    $z = \dfrac{3}{8}xy^3$ $\qquad x = 2,\ y = 3$

    $z = \dfrac{3}{8}(2)(3)^3$

    $z = \dfrac{81}{4}$

23. Let L = load, w = width, d = depth, l = length

    $L = 800,\ l = 10,\ w = 3,\ d = 4$

    $L = \dfrac{Kwd^2}{l}$

    $800 = \dfrac{K(3)(16)}{10}$

    $800 = \dfrac{48K}{10}$

    $\dfrac{8000}{48} = K$

    $\dfrac{500}{3} = K$

    $L = \dfrac{\dfrac{500}{3}wd^2}{l}$ $\qquad l = 12,\ w = 3,\ d = 4$

    $L = \dfrac{\dfrac{500}{3}(3)(16)}{12}$

    $L = \dfrac{8000}{12}$

    $L = \dfrac{2000}{3}$ lb.

CHAPTER 9

Rational Exponents and Roots

9.1 Rational Exponents

1. $\sqrt{144} = 12$

3. $\sqrt{-144}$, not a real number

5. $-\sqrt{49} = -7$

7. $\sqrt[3]{-27} = -3$

9. $\sqrt[4]{16} = 2$

11. $\sqrt[4]{-16}$, not a real number

13. $\sqrt{0.04} = 0.2$

15. $\sqrt[3]{0.008} = 0.2$

17. $\sqrt{36a^8} = 6a^4$

19. $\sqrt[3]{27a^{12}} = 3a^4$

21. $\sqrt[3]{x^3 y^6} = xy^2$

23. $\sqrt[5]{32x^{10}y^5} = 2x^2 y$

25. $\sqrt[4]{16a^{12}b^{20}} = 2a^3 b^5$

27. $36^{1/2} = \sqrt{36} = 6$

29. $-9^{1/2} = -\sqrt{9} = -3$

31. $8^{1/3} = \sqrt[3]{8} = 2$

33. $(-8)^{1/3} = \sqrt[3]{-8} = -2$

35. $32^{1/5} = \sqrt[5]{32} = 2$

37. $\left(\dfrac{81}{25}\right)^{1/2} = \sqrt{\dfrac{81}{25}} = \dfrac{9}{5}$

39. $\left(\dfrac{64}{125}\right)^{1/3} = \sqrt[3]{\dfrac{64}{125}} = \dfrac{4}{5}$

41. $27^{2/3} = \left(27^{1/3}\right)^2 = 3^2 = 9$

43. $25^{3/2} = \left(25^{1/2}\right)^3 = 5^3 = 125$

45. $16^{3/4} = \left(16^{1/4}\right)^3 = 2^3 = 8$

47. $27^{-1/3} = \dfrac{1}{27^{1/3}} = \dfrac{1}{3}$

49. $81^{-3/4} = \left(81^{1/4}\right)^{-3} = 3^{-3} = \dfrac{1}{3^3} = \dfrac{1}{27}$

51. $\left(\dfrac{25}{36}\right)^{-1/2} = \left(\dfrac{36}{25}\right)^{1/2} = \dfrac{6}{5}$

53. $\left(\dfrac{81}{16}\right)^{-3/4} = \left[\left(\dfrac{81}{16}\right)^{1/4}\right]^{-3} = \left(\dfrac{3}{2}\right)^{-3} = \left(\dfrac{2}{3}\right)^{3} = \dfrac{8}{27}$

55. $16^{1/2} + 27^{1/3} = 4 + 3 = 7$

57. $8^{-2/3} + 4^{-1/2} = \left(8^{1/3}\right)^{-2} + \left(4^{1/2}\right)^{-1}$

$= 2^{-2} + 2^{-1}$

$= \dfrac{1}{2^2} + \dfrac{1}{2^1}$

$= \dfrac{1}{4} + \dfrac{1}{2}$

$= \dfrac{3}{4}$

59. $x^{3/5} \cdot x^{1/5} = x^{4/5}$

61. $\left(a^{3/4}\right)^{4/3} = a^{(3/4)(4/3)}$

$= a$

63. $\dfrac{x^{1/5}}{x^{3/5}} = \dfrac{1}{x^{2/5}}$

65. $\dfrac{x^{5/6}}{x^{2/3}} = x^{5/6 - 2/3}$

$= x^{5/6 - 4/6}$

$= x^{1/6}$

67. $\left(x^{3/5} y^{5/6} z^{1/3}\right)^{3/5} = x^{9/25} y^{15/30} z^{3/15}$

$= x^{9/25} y^{1/2} z^{1/5}$

69. $\dfrac{a^{3/4} b^2}{a^{7/8} b^{1/4}} = a^{3/4 - 7/8} \cdot b^{2 - 1/4}$

$= a^{6/8 - 7/8} \cdot b^{8/4 - 1/4}$

$= a^{-1/8} \cdot b^{7/4}$

$= \dfrac{b^{7/4}}{a^{1/8}}$

71. $\dfrac{\left(y^{2/3}\right)^{3/4}}{\left(y^{1/3}\right)^{3/5}} = \dfrac{y^{1/2}}{y^{1/5}}$

$= y^{5/10 - 2/10}$

$= y^{3/10}$

73. $\left(\dfrac{a^{-1/4}}{b^{1/2}}\right)^8 = \dfrac{\left(a^{-1/4}\right)^8}{\left(b^{1/2}\right)^8}$

$= \dfrac{a^{-2}}{b^4}$

$= \dfrac{1}{a^2 b^4}$

75. $\dfrac{\left(r^{-2} s^{1/3}\right)^6}{r^8 s^{3/2}} = \dfrac{r^{-12} s^2}{r^8 s^{3/2}}$

$= \dfrac{s^{1/2}}{r^{20}}$

77. $\dfrac{\left(25a^6b^4\right)^{1/2}}{\left(8a^{-9}b^3\right)^{-1/3}} = \dfrac{\left(25\right)^{1/2}\left(a^6\right)^{1/2}\left(b^4\right)^{1/2}}{\left(8\right)^{-1/3}\left(a^{-9}\right)^{-1/3}\left(b^3\right)^{-1/3}}$

$= \dfrac{5a^3b^2}{2^{-1}a^3b^{-1}}$

$= \dfrac{5\cdot 2a^3b^2b}{a^3}$

$= 10b^3$

79. $\left(a^{1/2}+b^{1/2}\right)^2 = a+b \qquad a=9 \text{ and } b=4$

$\left(9^{1/2}+4^{1/2}\right)^2 = (9+4)$

$(3+2)^2 = 13$

$25 = 13 \quad \text{False}$

Therefore, $(a^{1/2}+b^{1/2})^2 \neq a+b$

81. $\sqrt{\sqrt{a}} = \sqrt[4]{a} \qquad (a\geq 0)$

$\left(a^{1/2}\right)^{1/2} = a^{1/4}$

$a^{1/4} = a^{1/4}$

83. $v(r) = \left(\dfrac{5r}{2}\right)^{1/2} \qquad r=250$

$v(250) = \left(\dfrac{5(250)}{2}\right)^{1/2}$

$= \left(\dfrac{1250}{2}\right)^{1/2}$

$= (625)^{1/2}$

$= 25 \text{ miles / hour}$

85. $\dfrac{1+\sqrt{5}}{2} = 1.618$ to the nearest thousandth

87. $\dfrac{3}{2}, \dfrac{5}{3}, \dfrac{8}{5}, \dfrac{13}{8}$

Numerator and denominator are consecutive members of the Fibonacci sequence.

89. (a) $s = 60+2\cdot150+60 = 420$

(b) $d^2 = s^2+s^2 = \sqrt{2s^2} = \sqrt{2}s$

$= \sqrt{2}(420) = 594 \text{ pm}$

(c) $d = 594 \text{ pm}\left(\dfrac{1\text{ m}}{10^{12}\text{ pm}}\right) = 5.94\times10^{-10} \text{ m}$

91. (a) $CH^2 = 1^2+1^2 = 2$

$CH = \sqrt{2} \text{ inches}$

(b) $CF^2 = 1^2+1^2+1^2 = 3$

$CF = \sqrt{3} \text{ inches}$

93. (a) $y=x$, graph B

(b) $y=x^2$, graph A

(c) $y=x^{2/3}$, graph C

(d) $\quad x^2 = x$

$x^2-x = 0$

$x(x-1) = 0$

$x = 0, 1$

They intersect at $(0,\,0)$, $(1,\,1)$

95. $x^2(x^4-x) = x^2(x^4)-x^2(x) = x^6-x^3$

97. $(x-3)(x+5) = x^2 + 2x - 15$

99. $\left(x^2 - 5\right)^2 = x^4 - (2)5x^2 + 5^2$
$$= x^4 - 10x^2 + 25$$

101. $(x-3)\left(x^2 + 3x + 9\right) = x^2 - 27$

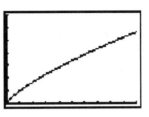

Use the following graph of $y = x^{3/4}$ in Problems 103 and 105.

103. $2^{3/4} = 1.7$

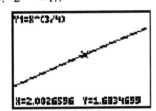

105. $10^{3/4} = 5.6$

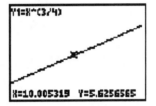

107. $Y_1 = x^{3/4}$
$Y_2 = x^{4/3}$
They intersect at $(1, 1)$ and $(0, 0)$.

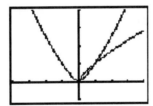

109. Use $A \cdot 2^{-t/5,600} = 3 \cdot 2^{-t/5,600}$

   (a) $t = 5,000$
   $3 \cdot 2^{-5000/5600} = 1.62$ micrograms

   (b) $t = 10,000$
   $3 \cdot 2^{-10,000/5600} = 0.87$ micrograms

   (c) $t = 56,000$
   $3 \cdot 2^{-56,000/5600} = 0.00293$ micrograms

   d). $t = 112,000$
   $3 \cdot 2^{-112,000/5600} = 0.0000029$ micrograms

9.2 More Expressions Involving Rational Exponents

1. $x^{2/3}\left(x^{1/3}+x^{4/3}\right)=x^{2/3}x^{1/3}+x^{2/3}x^{4/3}$

$\qquad = x^{3/3}+x^{6/3}$

$\qquad = x+x^2$

3. $a^{1/2}\left(a^{3/2}-a^{1/2}\right)=a^{4/2}-a^{2/2}$

$\qquad = a^2-a$

5. $2x^{1/3}\left(3x^{8/3}-4x^{5/3}+5x^{2/3}\right)=2x^{1/3}3x^{8/3}-2x^{1/3}4x^{5/3}+2x^{1/3}5x^{2/3}$

$\qquad = 6x^{9/3}-8x^{6/3}+10x^{3/3}$

$\qquad = 6x^3-8x^2+10x$

7. $4x^{1/2}y^{3/5}\left(3x^{3/2}y^{-3/5}-9x^{-1/2}y^{7/5}\right)$

$\qquad = 12x^{4/2}y^0-36x^0y^{10/5}$

$\qquad = 12x^2-36y^2$

9. $\left(x^{2/3}-4\right)\left(x^{2/3}+2\right)$

$\qquad = x^{2/3}x^{2/3}+2x^{2/3}-4x^{2/3}-8$

$\qquad = x^{4/3}-2x^{2/3}-8$

11. $\left(a^{1/2}-3\right)\left(a^{1/2}-7\right)=a^{1/2}a^{1/2}-7a^{1/2}-3a^{1/2}+21$

$\qquad = a-10a^{1/2}+21$

13. $\left(4y^{1/3}-3\right)\left(5y^{1/3}+2\right)=4y^{1/3}5y^{1/3}+4y^{1/3}\cdot2-3\cdot5y^{1/3}-3\cdot2$

$\qquad = 20y^{2/3}+8y^{1/3}-15y^{1/3}-6$

$\qquad = 20y^{2/3}-7y^{1/3}-6$

15. $\left(5x^{2/3}+3y^{1/2}\right)\left(2x^{2/3}+3y^{1/2}\right)$

$\qquad = 10x^{4/3}+15x^{2/3}y^{1/2}+6x^{2/3}y^{1/2}+9y$

$\qquad = 10x^{4/3}+21x^{2/3}y^{1/2}+9y$

17. $\left(t^{1/2}+5\right)^2=\left(t^{1/2}+5\right)\left(t^{1/2}+5\right)$

$\qquad = t^{1/2}t^{1/2}+5t^{1/2}+5t^{1/2}+25$

$\qquad = t+10t^{1/2}+25$

19. $\left(x^{3/2}+4\right)^2=\left(x^{3/2}+4\right)\left(x^{3/2}+4\right)$

$\qquad = x^{6/2}+4x^{3/2}+4x^{3/2}+16$

$\qquad = x^3+8x^{3/2}+16$

21. $\left(a^{1/2}-b^{1/2}\right)^2$

$\qquad =\left(a^{1/2}-b^{1/2}\right)\left(a^{1/2}-b^{1/2}\right)$

$\qquad = a^{1/2}a^{1/2}-a^{1/2}b^{1/2}-a^{1/2}b^{1/2}+b^{1/2}b^{1/2}$

$\qquad = a-2a^{1/2}b^{1/2}+b$

23. $\left(2x^{1/2}-3y^{1/2}\right)^2=\left(2x^{1/2}-3y^{1/2}\right)\left(2x^{1/2}-3y^{1/2}\right)$

$\qquad = 4x-6x^{1/2}y^{1/2}-6x^{1/2}y^{1/2}+9y$

$\qquad = 4x-12x^{1/2}y^{1/2}+9y$

25. $\left(a^{1/2}-3^{1/2}\right)\left(a^{1/2}+3^{1/2}\right)=\left(a^{1/2}\right)^2-\left(3^{1/2}\right)^2=a-3$

27. $\left(x^{3/2}+y^{3/2}\right)\left(x^{3/2}-y^{3/2}\right)=\left(x^{3/2}\right)^2-\left(y^{3/2}\right)^2$

$\qquad = x^3-y^3$

29. $\left(t^{1/2} - 2^{3/2}\right)\left(t^{1/2} + 2^{3/2}\right) = \left(t^{1/2}\right)^2 - \left(2^{3/2}\right)^2 = t - 2^3 = t - 8$

31. $\left(2x^{3/2} + 3^{1/2}\right)\left(2x^{3/2} - 3^{1/2}\right) = \left(2x^{3/2}\right)^2 - \left(3^{1/2}\right)^2$
$$= 4x^3 - 3$$

33.
$$
\begin{array}{r}
x^{2/3} - x^{1/3}y^{1/3} + y^{2/3} \\
\underline{x^{1/3} + y^{1/3}} \\
x^{3/3} - x^{2/3}y^{1/3} + x^{1/3}y^{2/3} \\
\underline{x^{2/3}y^{1/3} - x^{1/3}y^{2/3} + y^{3/3}} \\
x \qquad\qquad\qquad\quad + y
\end{array}
$$

35.
$$
\begin{array}{r}
a^{2/3} + 2a^{1/3} + 4 \\
\underline{a^{1/3} - 2} \\
-2a^{2/3} - 4a^{1/3} - 8 \\
\underline{a + 2a^{2/3} + 4a^{1/3}} \\
a \qquad\qquad\quad - 8
\end{array}
$$

The product is $a - 8$.

37.
$$
\begin{array}{r}
4x^{2/3} - 2x^{1/3} + 1 \\
\underline{2x^{1/3} + 1} \\
8x^{3/3} - 4x^{2/3} + 2x^{1/3} \\
\underline{4x^{2/3} - 2x^{1/3} + 1} \\
8x \qquad\qquad\quad + 1
\end{array}
$$

39. $\left(t^{1/4} - 1\right)\left(t^{1/4} + 1\right)\left(t^{1/2} + 1\right) = \left(t^{1/2} - 1\right)\left(t^{1/2} + 1\right)$
$$= t - 1$$

41. $\dfrac{18x^{3/4} + 27x^{1/4}}{9x^{1/4}} = \dfrac{18x^{3/4}}{9x^{1/4}} + \dfrac{27x^{1/4}}{9x^{1/4}} = 2x^{1/2} + 3$

43. $\dfrac{12x^{2/3}y^{1/3} - 16x^{1/3}y^{2/3}}{4x^{1/3}y^{1/3}} = \dfrac{12x^{2/3}y^{1/3}}{4x^{1/3}y^{1/3}} - \dfrac{16x^{1/3}y^{2/3}}{4x^{1/3}y^{1/3}}$
$$= 3x^{1/3} - 4y^{1/3}$$

45. $\dfrac{21a^{7/5}b^{3/5} - 14a^{2/5}b^{8/5}}{7a^{2/5}b^{3/5}} = \dfrac{21a^{7/5}b^{3/5}}{7a^{2/5}b^{3/5}} - \dfrac{14a^{2/5}b^{8/5}}{7a^{2/5}b^{3/5}} = 3a - 2b$

47. $12(x-2)^{3/2} - 9(x-2)^{1/2} = 3(x-2)^{1/2}\left[4(x-2) - 3\right]$
$$= 3(x-2)^{1/2}(4x - 11)$$

49. $5(x-3)^{12/5} - 15(x-3)^{7/5}$
$$= 5(x-3)^{7/5}\left[(x-3)^{5/5} - 3\right]$$
$$= 5(x-3)^{7/5}\left[(x-3)^1 - 3\right]$$
$$= 5(x-3)^{7/5}(x-6)$$

51. $9x(x+1)^{3/2} + 6(x+1)^{1/2}$
$$= 3(x+1)^{1/2}\left[3x(x+1) + 2\right]$$
$$= 3(x+1)^{1/2}\left(3x^2 + 3x + 2\right)$$

53. $x^{2/3} - 5x^{1/3} + 6 = \left(x^{1/3} - 2\right)\left(x^{1/3} - 3\right)$

55. $a^{2/5} - 2a^{1/5} - 8 = \left(a^{1/5} + 2\right)\left(a^{1/5} - 4\right)$

57. $2y^{2/3} - 5y^{1/3} - 3 = \left(2y^{1/3} + 1\right)\left(y^{1/3} - 3\right)$

59. $9t^{2/5} - 25 = \left(3t^{1/5} + 5\right)\left(3t^{1/5} - 5\right)$

61. $4x^{2/7} + 20x^{1/7} + 25 = \left(2x^{1/7} + 5\right)\left(2x^{1/7} + 5\right) = \left(2x^{1/7} + 5\right)^2$

63. $\dfrac{3}{x^{1/2}} + x^{1/2} = \dfrac{3}{x^{1/2}} + \dfrac{x}{x^{1/2}}$

$$= \dfrac{3+x}{x^{1/2}}$$

65. $x^{2/3} + \dfrac{5}{x^{1/3}} = \dfrac{x^{3/3}}{x^{1/3}} + \dfrac{5}{x^{1/3}} = \dfrac{x+5}{x^{1/3}}$

67. $\dfrac{3x^2}{\left(x^3+1\right)^{1/2}} + \left(x^3+1\right)^{1/2} = \dfrac{3x^2}{\left(x^3+1\right)^{1/2}} + \dfrac{\left(x^3+1\right)}{\left(x^3+1\right)^{1/2}}$

$$= \dfrac{x^3 + 3x^2 + 1}{\left(x^3+1\right)^{1/2}}$$

69. $\dfrac{x^2}{\left(x^2+4\right)^{1/2}} - \left(x^2+4\right)^{1/2} = \dfrac{x^2}{\left(x^2+4\right)^{1/2}} - \dfrac{x^2+4}{\left(x^2+4\right)^{1/2}} = \dfrac{-4}{\left(x^2+4\right)^{1/2}}$

71. $6^{0.25} = 2$

73. $9^{1.5} = 27$

75. $\left(\dfrac{1}{2}\right)^{1/5} = 0.871$

77. $A = 900$, $P = 500$, and $t = 4$,

$$r = \left(\dfrac{A}{P}\right)^{1/t} - 1$$

$$r = \left(\dfrac{900}{500}\right)^{1/4} - 1$$

$$= (1.8)^{0.25} - 1$$

$$= 1.158 - 1$$

$$= 0.158$$

$$= 15.8\%$$

79. $A = \$80,000$, $P = \$60,000$ and $t = 5$

$$r = \left(\dfrac{A}{P}\right)^{1/t} - 1$$

$$r = \left(\dfrac{80,000}{60,000}\right)^{1/5} - 1$$

$$r = \left(\dfrac{4}{3}\right)^{1/5} - 1$$

$$r = 1.059 - 1$$

$$r = 0.059$$

$$r = 5.9\%$$

81. Given $(-4, -1)$ and $(-2, 5)$, we have

$$m = \dfrac{y_2 - y_1}{x_2 - x_1}$$

$$= \dfrac{5 - (-1)}{-2 - (-4)}$$

$$= \dfrac{6}{2}$$

$$= 3$$

83. $2x - 3y = 6$

$\qquad -3y = -2x + 6$

$\qquad y = \dfrac{-2x}{-3} + \dfrac{6}{-3}$

$\qquad y = \dfrac{2}{3}x - 2$

$\qquad m = \dfrac{2}{3},\ b = -2$

85. Given $m = \dfrac{2}{3}$, and the point $(-6,\ 2)$

$\qquad y - y_1 = m(x - x_1)$ becomes

$\qquad y - 2 = \dfrac{2}{3}\left[x - (-6)\right]$

$\qquad y - 2 = \dfrac{2}{3}x + 4$

$\qquad y = \dfrac{2}{3}x + 6$

9.3 Simplified Form for Radicals

1. $\sqrt{8} = \sqrt{4 \cdot 2} = \sqrt{4}\ \sqrt{2} = 2\sqrt{2}$

3. $\sqrt{98} = \sqrt{49 \cdot 2} = 7\sqrt{2}$

5. $\sqrt{288} = \sqrt{144 \cdot 2} = \sqrt{144} \cdot \sqrt{2} = 12\sqrt{2}$

7. $\sqrt{80} = \sqrt{16 \cdot 5} = 4\sqrt{5}$

9. $\sqrt{48} = \sqrt{16 \cdot 3} = \sqrt{16} \cdot \sqrt{3} = 4\sqrt{3}$

11. $\sqrt{675} = \sqrt{225 \cdot 3} = 15\sqrt{3}$

13. $\sqrt[3]{54} = \sqrt[3]{27 \cdot 2} = \sqrt[3]{27} \cdot \sqrt[3]{2} = 3\sqrt[3]{2}$

15. $\sqrt[3]{128} = \sqrt[3]{64 \cdot 2} = 4\sqrt[3]{2}$

17. $\sqrt[3]{432} = \sqrt[3]{216 \cdot 2} = \sqrt[3]{216} \cdot \sqrt[3]{2} = 6\sqrt[3]{2}$

19. $\sqrt[5]{64} = \sqrt[5]{32 \cdot 2} = 2\sqrt[5]{2}$

21. $\sqrt{18x^3} = \sqrt{9x^2 \cdot 2x} = \sqrt{9x^2}\ \sqrt{2x} = 3x\sqrt{2x}$

23. $\sqrt[4]{32y^7} = \sqrt[4]{16y^4 \cdot 2y^3} = 2y\ \sqrt[4]{2y^3}$

25. $\sqrt[3]{40x^4y^7} = \sqrt[3]{8x^3y^6 \cdot 5xy} = \sqrt[3]{8x^3y^6}\ \sqrt[3]{5xy} = 2xy^2\ \sqrt[3]{5xy}$

27. $\sqrt{48a^2b^3c^4} = \sqrt{16a^2b^2c^4 \cdot 3b} = 4abc^2\ \sqrt{3b}$

29. $\sqrt[3]{48a^2b^3c^4} = \sqrt[3]{8b^3c^3 \cdot 6a^2c} = \sqrt[3]{8b^3c^3}\ \sqrt[3]{6a^2c} = 2bc\sqrt[3]{6a^2c}$

31. $\sqrt[5]{64x^8y^{12}} = \sqrt[5]{32x^5y^{10} \cdot 2x^3y^2}$

$\qquad\qquad = 2xy^2\ \sqrt[5]{2x^3y^2}$

33. $\sqrt[5]{243x^7y^{10}z^5} = \sqrt[5]{243x^5y^{10}z^5 \cdot x^2} = \sqrt[5]{243x^5y^{10}z^5} \cdot \sqrt[5]{x^2} = 3xy^2z\sqrt[5]{x^2}$

35. $\sqrt{b^2 - 4ac} = \sqrt{(-6)^2 - 4(2)(3)}$

$= \sqrt{36 - 24}$

$= \sqrt{12}$

$= \sqrt{4 \cdot 3}$

$= 2\sqrt{3}$

37. $a = 1, \, b = 2 \, , c = 6$

$\sqrt{b^2 - 4ac} = \sqrt{2^2 - 4(1)(6)} = \sqrt{4 - 24} = \sqrt{-20},$

not a real number.

39. $\sqrt{b^2 - 4ac} = \sqrt{\left(-\dfrac{1}{2}\right)^2 - 4\left(\dfrac{1}{2}\right)\left(-\dfrac{5}{4}\right)}$

$= \sqrt{\dfrac{1}{4} + \dfrac{5}{2}}$

$= \sqrt{\dfrac{11}{4}}$

$= \dfrac{\sqrt{11}}{2}$

41. $\dfrac{2}{\sqrt{3}} = \dfrac{2}{\sqrt{3}} \cdot \dfrac{\sqrt{3}}{\sqrt{3}} = \dfrac{2\sqrt{3}}{\sqrt{3^2}} = \dfrac{2\sqrt{3}}{3}$

43. $\dfrac{5}{\sqrt{6}} = \dfrac{5}{\sqrt{6}} \cdot \dfrac{\sqrt{6}}{\sqrt{6}} = \dfrac{5\sqrt{6}}{6}$

45. $\sqrt{\dfrac{1}{2}} = \dfrac{\sqrt{1}}{\sqrt{2}} = \dfrac{1}{\sqrt{2}} \cdot \dfrac{\sqrt{2}}{\sqrt{2}} = \dfrac{\sqrt{2}}{\sqrt{2^2}} = \dfrac{\sqrt{2}}{2}$

47. $\sqrt{\dfrac{1}{5}} = \dfrac{\sqrt{1}}{\sqrt{5}} \cdot \dfrac{\sqrt{5}}{\sqrt{5}} = \dfrac{\sqrt{5}}{5}$

49. $\dfrac{4}{\sqrt[3]{2}} = \dfrac{4}{\sqrt[3]{2}} \cdot \dfrac{\sqrt[3]{2^2}}{\sqrt[3]{2^2}} = \dfrac{4\sqrt[3]{2^2}}{\sqrt[3]{2^3}} = \dfrac{4\sqrt[3]{4}}{2} = 2\sqrt[3]{4}$

51. $\dfrac{2}{\sqrt[3]{9}} = \dfrac{2}{\sqrt[3]{9}} \cdot \dfrac{\sqrt[3]{3}}{\sqrt[3]{3}} = \dfrac{2\sqrt[3]{3}}{3}$

53. $\sqrt[4]{\dfrac{3}{2x^2}} = \dfrac{\sqrt[4]{3}}{\sqrt[4]{2x^2}} \cdot \dfrac{\sqrt[4]{2^3 x^2}}{\sqrt[4]{2^3 x^2}} = \dfrac{\sqrt[4]{3 \cdot 2^3 x^2}}{\sqrt[4]{2^4 x^4}} = \dfrac{\sqrt[4]{24x^2}}{2x}$

55. $\sqrt[4]{\dfrac{8}{y}} = \dfrac{\sqrt[4]{8}}{\sqrt[4]{y}} \cdot \dfrac{\sqrt[4]{y^3}}{\sqrt[4]{y^3}} = \dfrac{\sqrt[4]{8y^3}}{y}$

57. $\sqrt[3]{\dfrac{4x}{3y}} = \dfrac{\sqrt[3]{4x}}{\sqrt[3]{3y}} \cdot \dfrac{\sqrt[3]{3^2 y^2}}{\sqrt[3]{3^2 y^2}} = \dfrac{\sqrt[3]{4 \cdot 3^2 xy^2}}{\sqrt[3]{3^3 y^3}} = \dfrac{\sqrt[3]{36xy^2}}{3y}$

59. $\sqrt[3]{\dfrac{2x}{9y}} = \dfrac{\sqrt[3]{2x}}{\sqrt[3]{9y}} \cdot \dfrac{\sqrt[3]{3y^2}}{\sqrt[3]{3y^2}} = \dfrac{\sqrt[3]{6xy^2}}{3y}$

61. $\sqrt[4]{\dfrac{1}{8x^3}} = \dfrac{\sqrt[4]{1}}{\sqrt[4]{2^3 x^3}} \cdot \dfrac{\sqrt[4]{2x}}{\sqrt[4]{2x}} = \dfrac{\sqrt[4]{2x}}{\sqrt[4]{2^4 x^4}} = \dfrac{\sqrt[4]{2x}}{2x}$

63. $\sqrt{\dfrac{27x^3}{5y}} = \dfrac{\sqrt{9x^2 \cdot 3x}}{\sqrt{5y}} \cdot \dfrac{\sqrt{5y}}{\sqrt{5y}} = \dfrac{3x\sqrt{15xy}}{5y}$

65. $\sqrt{\dfrac{75x^3 y^2}{2z}} = \dfrac{\sqrt{75x^3 y^2}}{\sqrt{2z}}$

$= \dfrac{\sqrt{25x^2 y^2}\,\sqrt{3x}}{\sqrt{2z}}$

$= \dfrac{5xy\sqrt{3x}}{\sqrt{2z}} \cdot \dfrac{\sqrt{2z}}{\sqrt{2z}}$

$= \dfrac{5xy\sqrt{6xz}}{\sqrt{2^2 z^2}}$

$= \dfrac{5xy\sqrt{6xz}}{2z}$

67. $\sqrt[3]{\dfrac{16a^4 b^3}{9c}} = \dfrac{\sqrt[3]{8a^3 b^3 \cdot 2a}}{\sqrt[3]{9c}} \cdot \dfrac{\sqrt[3]{3c^2}}{\sqrt[3]{3c^2}} = \dfrac{2ab\sqrt[3]{6ac^2}}{3c}$

69. $\sqrt[3]{\dfrac{8x^3 y^6}{9z}} = \dfrac{\sqrt[3]{8x^3 y^6}}{\sqrt[3]{9z}}$

$= \dfrac{2xy^2}{\sqrt[3]{9z}}$

$= \dfrac{2xy^2}{\sqrt[3]{3^2 z}} \cdot \dfrac{\sqrt[3]{3z^2}}{\sqrt[3]{3z^2}}$

$= \dfrac{2xy^2 \sqrt[3]{3z^2}}{\sqrt[3]{3^3 z^3}}$

$= \dfrac{2xy^2 \sqrt[3]{3z^2}}{3z}$

71. $\sqrt{25x^2} = 5|x|$

73. $\sqrt{27x^3 y^2} = \sqrt{9x^2 y^2 \cdot 3x} = \sqrt{9x^2 y^2} \cdot \sqrt{3x} = 3|xy|\,\sqrt{3x}$

75. $\sqrt{x^2 - 10x + 25} = \sqrt{(x-5)^2} = |x-5|$

77. $\sqrt{4x^2 + 12x + 9} = \sqrt{(2x+3)^2} = |2x+3|$

79. $\sqrt{4a^4 + 16a^3 + 16a^2} = \sqrt{4a^2 \left(a^2 + 4a + 4\right)}$

$= 2\sqrt{a^2 (a+2)^2}$

$= 2\left|a(a+2)\right|$

81. $\sqrt{4x^3 - 8x^2} = \sqrt{4x^2 (x-2)} = \sqrt{4x^2} \cdot \sqrt{x-2} = 2|x|\,\sqrt{x-2}$

83. $\sqrt{9+16} = \sqrt{9} + \sqrt{16}$

$\sqrt{25} = 3 + 4$

$5 \neq 7$

85. Let $l = 15$ and $w = 10$

$$d = \sqrt{l^2 + w^2}$$

$$d = \sqrt{15^2 + 10^2}$$

$$d = \sqrt{225 + 100}$$

$$d = \sqrt{325}$$

$$d = \sqrt{25 \cdot 13} = \sqrt{25} \cdot \sqrt{13} = 5\sqrt{13} \text{ feet}$$

87. Spirals of roots

89. $f(x) = \sqrt{x^2 + 1}$

1st term $f(1) = \sqrt{1^2 + 1} = \sqrt{2}$

2nd term $f(f(1)) = f(\sqrt{2}) = \sqrt{(\sqrt{2})^2 + 1} = \sqrt{2 + 1} = \sqrt{3}$

3rd term $f(f(f(1))) = f(\sqrt{3}) = \sqrt{(\sqrt{3})^2 + 1} = \sqrt{3 + 1} = \sqrt{4}$

4th term $f(f(f(f(1)))) = f(\sqrt{4}) = \sqrt{(\sqrt{4})^2 + 1} = \sqrt{4 + 1} = \sqrt{5}$

5th term $f(f(f(f(f(1))))) = f(\sqrt{5}) = \sqrt{(\sqrt{5})^2 + 1} = \sqrt{5 + 1} = \sqrt{6}$

6th term $f\left(f(f(f(f(f(1)))))\right) = f(\sqrt{6}) = \sqrt{(\sqrt{6})^2 + 1} = \sqrt{6 + 1} = \sqrt{7}$

10th term $= \sqrt{11}$

100th term $= \sqrt{101}$

91. $f(x) = \dfrac{1}{2}x + 3$

$f(0) = \dfrac{1}{2}(0) + 3 = 3$

93. $g(x) = x^2 - 4$

$g(2) = (2)^2 - 4 = 0$

95. $f(x) = \dfrac{1}{2}x + 3$

$f(-4) = \dfrac{1}{2}(-4) + 3 = 1$

97. From Exercise 93, $g(2) = 0$

$f(x) = \dfrac{1}{2}x + 3$

$f[g(2)] = f(0) = \dfrac{1}{2}(0) + 3 = 3$

Problem Set 9.4

99. $\quad f(x) = \frac{1}{2}x + 3 \quad g(x) = x^2 - 4$

$(f+g)(x) = \frac{1}{2}x + 3 + x^2 - 4$

$(f+g)(4) = \frac{1}{2}(4) + 3 + (4)^2 - 4$

$\qquad = 17$

101. $\dfrac{1}{\sqrt[10]{a^3}} = \dfrac{1}{\sqrt[10]{a^3}} \cdot \dfrac{\sqrt[10]{a^7}}{\sqrt[10]{a^7}} = \dfrac{\sqrt[10]{a^7}}{\sqrt[10]{a^{10}}} = \dfrac{\sqrt[10]{a^7}}{a}$

103. $\dfrac{1}{\sqrt[20]{a^{11}}} = \dfrac{1}{\sqrt[20]{a^{11}}} \cdot \dfrac{\sqrt[20]{a^9}}{\sqrt[20]{a^9}}$

$\qquad = \dfrac{\sqrt[20]{a^9}}{a}$

105. See the graph in the back of the textbook.

107. They are about $\frac{3}{4}$ apart at $x = 2$.

109. When $x = 0$, $\sqrt{0^2 + 1} = 0 + 1$

$\qquad\qquad\qquad\qquad\qquad 1 = 1$

111. (a) $s = (a+b+c)/2$

(b) $A = \sqrt{s(s-a)(s-b)(s-c)}$, $a = 5$, $b = 6$, $c = 7$

$\quad s = (5+6+7)/2 = 9$

$\quad A = \sqrt{9(9-5)(9-6)(9-7)} = \sqrt{216} = 14.70$

9.4 Addition and Subtraction of Radical Expressions

1. $3\sqrt{5} + 4\sqrt{5} = (3+4)\sqrt{5} = 7\sqrt{5}$

3. $3x\sqrt{7} - 4x\sqrt{7} = -x\sqrt{7}$

5. $5\sqrt[3]{10} - 4\sqrt[3]{10} = (5-4)\sqrt[3]{10} = \sqrt[3]{10}$

7. $8\sqrt[5]{6} - 2\sqrt[5]{6} + 3\sqrt[5]{6} = 9\sqrt[5]{6}$

9. $3x\sqrt{2} - 4x\sqrt{2} + x\sqrt{2} = (3x - 4x + x)\sqrt{2} = 0\sqrt{2} = 0$

11. $\sqrt{20} - \sqrt{80} + \sqrt{45} = 2\sqrt{5} - 4\sqrt{5} + 3\sqrt{5}$

$\qquad\qquad\qquad\qquad\quad = \sqrt{5}$

13. $4\sqrt{8} - 2\sqrt{50} - 5\sqrt{72} = 4\sqrt{4}\sqrt{2} - 2\sqrt{25}\sqrt{2} - 5\sqrt{36}\sqrt{2}$

$\qquad\qquad\qquad\qquad\quad = 8\sqrt{2} - 10\sqrt{2} - 30\sqrt{2}$

$\qquad\qquad\qquad\qquad\quad = (8 - 10 - 30)\sqrt{2}$

$\qquad\qquad\qquad\qquad\quad = -32\sqrt{2}$

15. $5x\sqrt{8} + 3\sqrt{32x^2} - 5\sqrt{50x^2}$

$\quad = 10x\sqrt{2} + 12x\sqrt{2} - 25x\sqrt{2}$

$\quad = -3x\sqrt{2}$

17. $5\sqrt[3]{16} - 4\sqrt[3]{54} = 5\sqrt[3]{8}\sqrt[3]{2} - 4\sqrt[3]{27}\sqrt[3]{2}$

$\qquad = 5 \cdot 2\sqrt[3]{2} - 4 \cdot 3\sqrt[3]{2}$

$\qquad = 10\sqrt[3]{2} - 12\sqrt[3]{2}$

$\qquad = -2\sqrt[3]{2}$

19. $\sqrt[3]{x^4 y^2} + 7x\sqrt[3]{xy^2} = x\sqrt[3]{xy^2} + 7x\sqrt[3]{xy^2}$

$\qquad = 8x\sqrt[3]{xy^2}$

21. $5a^2\sqrt{27ab^3} - 6b\sqrt{12a^5 b} = 5a^2\sqrt{9b^2}\sqrt{3ab} - 6b\sqrt{4a^4}\sqrt{3ab}$

$\qquad = 5a^2 \cdot 3b\sqrt{3ab} - 6b \cdot 2a^2\sqrt{3ab}$

$\qquad = 15a^2 b\sqrt{3ab} - 12a^2 b\sqrt{3ab}$

$\qquad = 3a^2 b\sqrt{3ab}$

23. $b\sqrt[3]{24a^5 b} + 3a\sqrt[3]{81a^2 b^4}$

$\qquad = 2ab\sqrt[3]{3a^2 b} + 9ab\sqrt[3]{3a^2 b}$

$\qquad = 11ab\sqrt[3]{3a^2 b}$

25. $5x\sqrt[4]{3y^5} + y\sqrt[4]{243x^4 y} + \sqrt[4]{48x^4 y^5} = 5x\sqrt[4]{y^4}\sqrt[4]{3y} + y\sqrt[4]{81x^4}\sqrt[4]{3y} + \sqrt[4]{16x^4 y^4}\sqrt[4]{3y}$

$\qquad = 5x \cdot y\sqrt[4]{3y} + y \cdot 3x\sqrt[4]{3y} + 2xy\sqrt[4]{3y}$

$\qquad = 5xy\sqrt[4]{3y} + 3xy\sqrt[4]{3y} + 2xy\sqrt[4]{3y}$

$\qquad = 10xy\sqrt[4]{3y}$

27. $\dfrac{\sqrt{2}}{2} + \dfrac{1}{\sqrt{2}} = \dfrac{\sqrt{2}}{2} + \dfrac{1}{\sqrt{2}} \cdot \dfrac{\sqrt{2}}{\sqrt{2}}$

$\qquad = \dfrac{\sqrt{2} + \sqrt{2}}{2}$

$\qquad = \dfrac{2\sqrt{2}}{2}$

$\qquad = \sqrt{2}$

29. $\dfrac{\sqrt{5}}{3} + \dfrac{1}{\sqrt{5}} = \dfrac{\sqrt{5}}{3} + \dfrac{1}{\sqrt{5}} \cdot \dfrac{\sqrt{5}}{\sqrt{5}}$

$\qquad = \dfrac{\sqrt{5}}{3} + \dfrac{\sqrt{5}}{5}$

$\qquad = \dfrac{1}{3}\sqrt{5} + \dfrac{1}{5}\sqrt{5}$

$\qquad = \left(\dfrac{1}{3} + \dfrac{1}{5}\right)\sqrt{5}$

$\qquad = \dfrac{8\sqrt{5}}{15}$

31. $\sqrt{x} - \dfrac{1}{\sqrt{x}} = \sqrt{x} - \dfrac{1}{\sqrt{x}} \cdot \dfrac{\sqrt{x}}{\sqrt{x}}$

$\qquad = \dfrac{x\sqrt{x}}{x} - \dfrac{\sqrt{x}}{x}$

$\qquad = \dfrac{(x-1)\sqrt{x}}{x}$

33. $\dfrac{\sqrt{18}}{6} + \sqrt{\dfrac{1}{2}} + \dfrac{\sqrt{2}}{2} = \dfrac{\sqrt{18}}{6} + \dfrac{1}{\sqrt{2}} \cdot \dfrac{\sqrt{2}}{\sqrt{2}} + \dfrac{\sqrt{2}}{2}$

$\qquad = \dfrac{3\sqrt{2}}{6} + \dfrac{\sqrt{2}}{2} + \dfrac{\sqrt{2}}{2}$

$\qquad = \dfrac{\sqrt{2}}{2} + \dfrac{\sqrt{2}}{2} + \dfrac{\sqrt{2}}{2}$

$\qquad = \dfrac{3\sqrt{2}}{2}$

35. $\sqrt{6} - \sqrt{\dfrac{2}{3}} + \sqrt{\dfrac{1}{6}} = \dfrac{6\sqrt{6}}{6} - \dfrac{\sqrt{2}}{\sqrt{3}} \cdot \dfrac{\sqrt{3}}{\sqrt{3}} + \dfrac{\sqrt{1}}{\sqrt{6}} \cdot \dfrac{\sqrt{6}}{\sqrt{6}}$

$\qquad = \dfrac{6\sqrt{6}}{6} - \dfrac{\sqrt{6}}{3} + \dfrac{\sqrt{6}}{6}$

$\qquad = \dfrac{6\sqrt{6}}{6} - \dfrac{2\sqrt{6}}{6} + \dfrac{\sqrt{6}}{6}$

$\qquad = \dfrac{5\sqrt{6}}{6}$

37. $\sqrt[3]{25} + \dfrac{3}{\sqrt[3]{5}} = \sqrt[3]{25} + \dfrac{3}{\sqrt[3]{5}} \cdot \dfrac{\sqrt[3]{25}}{\sqrt[3]{25}} = \sqrt[3]{25} + \dfrac{3\sqrt[3]{25}}{5} = \left(1 + \dfrac{3}{5}\right)\sqrt[3]{25} = \dfrac{8\sqrt[3]{25}}{5}$

39. $\sqrt{12} \approx 3.464$

$\quad 2\sqrt{3} \approx 2(1.732) = 3.464$

41. $\sqrt{8} + \sqrt{18} \approx 2.828 + 4.243 = 7.071$

$\qquad \sqrt{50} \approx 7.071$

$\qquad \sqrt{26} \approx 5.099$

$\quad \sqrt{8} + \sqrt{18} =$ the decimal approximation of $\sqrt{50}$

43. $3\sqrt{2x} + 5\sqrt{2x} = 8\sqrt{2x}$

45. $\sqrt{9+16} = \sqrt{25} = 5$ the right side corrected

47.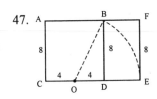

$OB = \sqrt{4^2 + 8^2} = \sqrt{16 + 64} = \sqrt{80} = 4\sqrt{5}$

Golden Ratio $= \dfrac{CE}{EF} = \dfrac{4 + 4\sqrt{5}}{8} = \dfrac{1 + \sqrt{5}}{2}$

49.

$OB = \sqrt{(1/2)^2 + (1)^2} = \sqrt{1/4 + 4/4} = \sqrt{5/2} = \dfrac{\sqrt{5}}{2}$

Golden Ratio $= \dfrac{CE}{EF} = \dfrac{1/2 + \sqrt{5}/2}{1} = \dfrac{1 + \sqrt{5}}{2}$

51.

$$OB = \sqrt{x^2 + (2x)^2} = \sqrt{5x^2} = x\sqrt{5}$$

$$\text{Golden Ratio} = \frac{CE}{EF} = \frac{x + x\sqrt{5}}{2x} = \frac{1 + \sqrt{5}}{2}$$

53.

$$h = \sqrt{l^2 + l^2} = \sqrt{2l^2} = \sqrt{2}l$$

$$\frac{h}{l} = \frac{\sqrt{2}l}{l} = \frac{\sqrt{2}}{1}$$

55. (a) $c^2 = a^2 + b^2$, $c = $ diagonal, $a = 5$, $b = 5$

$$c^2 = 5^2 + 5^2$$

$$c^2 = 25 + 25$$

$$c = \sqrt{50}$$

$$\frac{c}{a} = \frac{5\sqrt{2}}{5} = \sqrt{2}$$

(b) $\dfrac{a^2}{c} = \dfrac{5^2}{5\sqrt{2}} = \dfrac{5}{\sqrt{2}}$

(c) $\dfrac{a^2}{4a} = \dfrac{5^2}{4(5)} = \dfrac{5}{4}$

57. $2x + 3y < 6$

See the graph in the back of the textbook.

59. $y \geq -3x - 4$

See the graph in the back of the textbook.

61. $x \geq 3$

See the graph in the back of the textbook.

9.5 Multiplication and Division of Radical Expressions

1. $\sqrt{6}\,\sqrt{3} = \sqrt{6 \cdot 3} = \sqrt{18} = \sqrt{9 \cdot 2} = 3\sqrt{2}$

3. $(2\sqrt{3})(5\sqrt{7}) = 10\sqrt{21}$

5. $(4\sqrt{6})(2\sqrt{15})(3\sqrt{10}) = (4 \cdot 2 \cdot 3)(\sqrt{6} \cdot \sqrt{15} \cdot \sqrt{10}) = 24\sqrt{900} = 24 \cdot 30 = 720$

7. $\left(3\sqrt[3]{3}\right)\left(6\sqrt[3]{9}\right) = 18\sqrt[3]{27}$

$$= 54$$

9. $\sqrt{3}\,(\sqrt{2} - 3\sqrt{3}) = \sqrt{3} \cdot \sqrt{2} - \sqrt{3} \cdot 3\sqrt{3} = \sqrt{6} - 3\sqrt{9} = \sqrt{6} - 3 \cdot 3 = \sqrt{6} - 9$

11. $6\sqrt[3]{4}\left(2\sqrt[3]{2}+1\right)=12\sqrt[3]{8}+6\sqrt[3]{4}$

$\qquad = 24+6\sqrt[3]{4}$

13. $\left(\sqrt{3}+\sqrt{2}\right)\left(3\sqrt{3}-\sqrt{2}\right)=\sqrt{3}\cdot 3\sqrt{3}-\sqrt{3}\sqrt{2}+\sqrt{2}\cdot 3\sqrt{3}-\sqrt{2}\sqrt{2}$

$\qquad = 3\cdot 3-\sqrt{6}+3\sqrt{6}-2$

$\qquad = 9+2\sqrt{6}-2$

$\qquad = 7+2\sqrt{6}$

15. $\left(\sqrt{x}+5\right)\left(\sqrt{x}-3\right)=x-3\sqrt{x}+5\sqrt{x}-15$

$\qquad = x+2\sqrt{x}-15$

17. $\left(3\sqrt{6}+4\sqrt{2}\right)\left(\sqrt{6}+2\sqrt{2}\right)=3\sqrt{6}\cdot\sqrt{6}+3\sqrt{6}\cdot 2\sqrt{2}+4\sqrt{2}\cdot\sqrt{6}+4\sqrt{2}\cdot 2\sqrt{2}$

$\qquad = 3\cdot 6+6\sqrt{12}+4\sqrt{12}+8\cdot 2$

$\qquad = 18+10\sqrt{12}+16$

$\qquad = 10\sqrt{12}+34$

$\qquad = 10\sqrt{4\cdot 3}+34$

$\qquad = 34+20\sqrt{3}$

19. $\left(\sqrt{3}+4\right)^2=\left(\sqrt{3}\right)^2+2\left(\sqrt{3}\right)(4)+4^2$

$\qquad = 3+8\sqrt{3}+16$

$\qquad = 19+8\sqrt{3}$

21. $\left(\sqrt{x}-3\right)^2=\left(\sqrt{x}\right)^2-2\left(\sqrt{x}\right)(3)+(3)^2$

$\qquad = x-6\sqrt{x}+9$

23. $\left(2\sqrt{a}-3\sqrt{b}\right)^2$

$\qquad = \left(2\sqrt{a}\right)^2-2\left(2\sqrt{a}\right)\left(3\sqrt{b}\right)+\left(3\sqrt{b}\right)^2$

$\qquad = 4a-12\sqrt{ab}+9b$

25. $\left(\sqrt{x-4}+2\right)^2=\left(\sqrt{x-4}\right)^2+2\left(\sqrt{x-4}\right)(2)+2^2$

$\qquad = x-4+4\sqrt{x-4}+4$

$\qquad = x+4\sqrt{x-4}$

27. $\left(\sqrt{x-5}-3\right)^2$

$\qquad = \left(\sqrt{x-5}\right)^2-2\left(\sqrt{x-5}\right)(3)+3^2$

$\qquad = x-5-6\sqrt{x-5}+9$

$\qquad = x+4-6\sqrt{x-5}$

29. $\left(\sqrt{3}-\sqrt{2}\right)\left(\sqrt{3}+\sqrt{2}\right)=\left(\sqrt{3}\right)^2-\left(\sqrt{2}\right)^2$

$\qquad = 3-2=1$

31. $\left(\sqrt{a}+7\right)\left(\sqrt{a}-7\right)=\left(\sqrt{a}\right)^2-7^2$

$\qquad = a-49$

33. $\left(5-\sqrt{x}\right)\left(5+\sqrt{x}\right)=5^2-\left(\sqrt{x}\right)^2=25-x$

35. $\left(\sqrt{x-4}+2\right)\left(\sqrt{x-4}-2\right) = \left(\sqrt{x-4}\right)^2 - 2^2$
$$= x - 4 - 4$$
$$= x - 8$$

37. $\left(\sqrt{3}+1\right)^3 = \left(\sqrt{3}+1\right)\left(\sqrt{3}+1\right)^2$
$$= \left(\sqrt{3}+1\right)\left(3+2\sqrt{3}+1\right)$$
$$= \left(\sqrt{3}+1\right)\left(2\sqrt{3}+4\right)$$
$$= 2\cdot 3 + 4\sqrt{3} + 2\sqrt{3} + 4$$
$$= 10 + 6\sqrt{3}$$

39. $\dfrac{\sqrt{2}}{\sqrt{6}-\sqrt{2}} = \dfrac{\sqrt{2}}{\sqrt{6}-\sqrt{2}} \cdot \dfrac{\sqrt{6}+\sqrt{2}}{\sqrt{6}+\sqrt{2}}$
$$= \dfrac{\sqrt{12}+2}{6-2}$$
$$= \dfrac{2\sqrt{3}+2}{4}$$
$$= \dfrac{\sqrt{3}+1}{2}$$

41. $\dfrac{\sqrt{5}}{\sqrt{5}+1} = \dfrac{\sqrt{5}}{\sqrt{5}+1} \cdot \dfrac{\sqrt{5}-1}{\sqrt{5}-1}$
$$= \dfrac{\sqrt{5}\sqrt{5}-\sqrt{5}\cdot 1}{\left(\sqrt{5}\right)^2 - 1^2}$$
$$= \dfrac{5-\sqrt{5}}{5-1} = \dfrac{5-\sqrt{5}}{4}$$

43. $\dfrac{\sqrt{x}}{\sqrt{x}-3} = \dfrac{\sqrt{x}}{\sqrt{x}-3} \cdot \dfrac{\sqrt{x}+3}{\sqrt{x}+3}$
$$= \dfrac{x+3\sqrt{x}}{x-9}$$

45. $\dfrac{\sqrt{5}}{2\sqrt{5}-3} = \dfrac{\sqrt{5}}{2\sqrt{5}-3} \cdot \dfrac{2\sqrt{5}+3}{2\sqrt{5}+3}$
$$= \dfrac{\sqrt{5}\cdot 2\sqrt{5}+\sqrt{5}\cdot 3}{\left(2\sqrt{5}\right)^2 - 3^2}$$
$$= \dfrac{10+3\sqrt{5}}{20-9} = \dfrac{10+3\sqrt{5}}{11}$$

47. $\dfrac{3}{\sqrt{x}-\sqrt{y}} = \dfrac{3}{\sqrt{x}-\sqrt{y}} \cdot \dfrac{\sqrt{x}+\sqrt{y}}{\sqrt{x}+\sqrt{y}}$
$$= \dfrac{3\sqrt{x}+3\sqrt{y}}{x-y}$$

49. $\dfrac{\sqrt{6}+\sqrt{2}}{\sqrt{6}-\sqrt{2}} = \dfrac{\sqrt{6}+\sqrt{2}}{\sqrt{6}-\sqrt{2}} \cdot \dfrac{\sqrt{6}+\sqrt{2}}{\sqrt{6}+\sqrt{2}}$
$$= \dfrac{6+2\sqrt{12}+2}{6-2}$$
$$= \dfrac{2\sqrt{12}+8}{4}$$
$$= \dfrac{2\sqrt{4\cdot 3}+8}{4}$$
$$= \dfrac{4\sqrt{3}+8}{4}$$
$$= \dfrac{4\left(\sqrt{3}+2\right)}{4}$$
$$= 2 + \sqrt{3}$$

51. $\dfrac{\sqrt{7}-2}{\sqrt{7}+2} = \dfrac{\sqrt{7}-2}{\sqrt{7}+2} \cdot \dfrac{\sqrt{7}-2}{\sqrt{7}-2}$

$= \dfrac{7-4\sqrt{7}+4}{7-4}$

$= \dfrac{11-4\sqrt{7}}{3}$

53. $\dfrac{\sqrt{a}+\sqrt{b}}{\sqrt{a}-\sqrt{b}} = \dfrac{\sqrt{a}+\sqrt{b}}{\sqrt{a}-\sqrt{b}} \cdot \dfrac{\sqrt{a}+\sqrt{b}}{\sqrt{a}+\sqrt{b}}$

$= \dfrac{\left(\sqrt{a}\right)^2 + 2\sqrt{a}\sqrt{b} + \left(\sqrt{b}\right)^2}{\left(\sqrt{a}\right)^2 - \left(\sqrt{b}\right)^2}$

$= \dfrac{a + 2\sqrt{ab} + b}{a-b}$

55. $\dfrac{\sqrt{x}+2}{\sqrt{x}-2} = \dfrac{\sqrt{x}+2}{\sqrt{x}-2} \cdot \dfrac{\sqrt{x}+2}{\sqrt{x}+2}$

$= \dfrac{x + 4\sqrt{x} + 4}{x-4}$

57. $\dfrac{2\sqrt{3}-\sqrt{7}}{3\sqrt{3}+\sqrt{7}} = \dfrac{2\sqrt{3}-\sqrt{7}}{3\sqrt{3}+\sqrt{7}} \cdot \dfrac{3\sqrt{3}-\sqrt{7}}{3\sqrt{3}-\sqrt{7}}$

$= \dfrac{2\sqrt{3}\cdot 3\sqrt{3} - 2\sqrt{3}\cdot\sqrt{7} - \sqrt{7}\cdot 3\sqrt{3} + \left(\sqrt{7}\right)^2}{\left(3\sqrt{3}\right)^2 - \left(\sqrt{7}\right)^2}$

$= \dfrac{6\cdot 3 - 2\sqrt{21} - 3\sqrt{21} + 7}{9\cdot 3 - 7}$

$= \dfrac{18 - 5\sqrt{21} + 7}{27-7}$

$= \dfrac{25 - 5\sqrt{21}}{20}$

$= \dfrac{5\left(5 - \sqrt{21}\right)}{20}$

$= \dfrac{5 - \sqrt{21}}{4}$

59. $\dfrac{3\sqrt{x}+2}{1+\sqrt{x}} = \dfrac{3\sqrt{x}+2}{1+\sqrt{x}} \cdot \dfrac{1-\sqrt{x}}{1-\sqrt{x}}$

$= \dfrac{3\sqrt{x} - 3x + 2 - 2\sqrt{x}}{1-x}$

$= \dfrac{\sqrt{x} - 3x + 2}{1-x}$

61. $\sqrt[3]{4} - \sqrt[3]{6} + \sqrt[3]{9}$

$\underline{\sqrt[3]{2} + \sqrt[3]{3}}$

$\sqrt[3]{8} - \sqrt[3]{12} + \sqrt[3]{18}$

$\underline{\phantom{\sqrt[3]{8}} \sqrt[3]{12} - \sqrt[3]{18} + \sqrt[3]{27}}$

$\sqrt[3]{8} \qquad\qquad + \sqrt[3]{27} = 2 + 3 = 5$

The product is 5.

63. $5(2\sqrt{3}) = 10\sqrt{3}$

65. $\left(\sqrt{x}+3\right)^2 = \left(\sqrt{x}+3\right)\left(\sqrt{x}+3\right)$

   $\qquad = \left(\sqrt{x}\right)^2 + 2\sqrt{x}\cdot 3 + 3^2$

   $\qquad = x + 6\sqrt{x} + 9 \qquad$ The right side corrected

67. $(5\sqrt{3}\,)^2 = 25\cdot 3 = 75$

69. $t = \dfrac{\sqrt{100-h}}{4}, \quad h = 50$

   $t = \dfrac{\sqrt{100-50}}{4} = \dfrac{\sqrt{50}}{4} = \dfrac{\sqrt{25\cdot 2}}{4} = \dfrac{5\sqrt{2}}{4}$ seconds

   $t = \dfrac{\sqrt{100-h}}{4}, \quad h = 0$

   $t = \dfrac{\sqrt{100-0}}{4} = \dfrac{\sqrt{100}}{4} = \dfrac{10}{4} = \dfrac{5}{2}$ seconds

71. If $AC = 6$ and $\dfrac{CE}{AC} = \dfrac{1+\sqrt{5}}{2}$,

   then $OB = \sqrt{3^2 + 6^2} = \sqrt{45} = 3\sqrt{5}$

   and $DE = 3\sqrt{5} - 3$.

   Golden Ratio $= \dfrac{EF}{DE} = \dfrac{6}{3\sqrt{5}-3} = \dfrac{2}{\sqrt{5}-1}$

   $\qquad = \dfrac{2}{\sqrt{5}-1}\cdot\dfrac{\sqrt{5}+1}{\sqrt{5}+1} = \dfrac{1+\sqrt{5}}{2}$

73. If $BD = 2x$ and $OD = x$, then

   $OB^2 = \left(2x\right)^2 + x^2$

   $\qquad = 4x^2 + x^2$

   $\qquad = 5x^2$

   $OB = \sqrt{5}x$

   $OE = OB = \sqrt{5}x$

   Width $DE = \sqrt{5}x - x = \left(\sqrt{5}-1\right)x$

   Golden Ratio $= \dfrac{EF}{DE} = \dfrac{2x}{\left(\sqrt{5}-1\right)x} = \dfrac{2}{\sqrt{5}-1}$

   $\qquad = \dfrac{2}{\sqrt{5}-1}\cdot\dfrac{\sqrt{5}+1}{\sqrt{5}+1} = \dfrac{1+\sqrt{5}}{2}$

75. $y = Kx^2, \quad y = 75, \quad x = 5$

   $75 = K\left(5\right)^2 \Rightarrow K = 3$

   $y = 3x^2, \quad x = 7$

   $y = 3\left(7\right)^2$

   $y = 147$

77. $y = \dfrac{K}{x}, \quad y = 10, \quad x = 25$

   $10 = \dfrac{K}{25} \Rightarrow K = 250$

   $y = \dfrac{250}{x}, \quad y = 5$

   $5 = \dfrac{250}{x}$

   $x = 50$

79. $z = Kxy^2, \quad z = 40, \quad x = 5, \quad y = 2$

   $40 = K\left(5\right)\left(2\right)^2 \Rightarrow K = 2$

   $z = 2xy^2, \quad x = 2, \quad y = 5$

   $z = 2\left(2\right)\left(5\right)^2$

   $z = 100$

79.

9.6 Equations with Radicals

1.
$$\sqrt{2x+1} = 3$$
$$\left(\sqrt{2x+1}\right)^2 = 3^2$$
$$2x+1 = 9$$
$$2x = 8$$
$$x = 4$$

3.
$$\sqrt{4x+1} = -5$$
$$\left(\sqrt{4x+1}\right)^2 = (-5)^2$$
$$4x+1 = 25$$
$$4x = 24$$
$$x = 6$$
6 does not solve the original equation, $\varnothing$

5.
$$\sqrt{2y-1} = 3$$
$$\left(\sqrt{2y-1}\right)^2 = 3^2$$
$$2y-1 = 9$$
$$2y = 10$$
$$y = 5$$

7.
$$\sqrt{5x-7} = -1$$
$$\left(\sqrt{5x-7}\right)^2 = (-1)^2$$
$$5x-7 = 1$$
$$5x = 8$$
$$x = \frac{8}{5}$$
$\frac{8}{5}$ does not solve the original equation, $\varnothing$

9.
$$\sqrt{2x-3} - 2 = 4$$
$$\sqrt{2x-3} = 6$$
$$\left(\sqrt{2x-3}\right)^2 = 6^2$$
$$2x-3 = 36$$
$$2x = 39$$
$$x = \frac{39}{2}$$

11.
$$\sqrt{4a+1} + 3 = 2$$
$$\sqrt{4a+1} = -1$$
$$\left(\sqrt{4a+1}\right)^2 = (-1)^2$$
$$4a+1 = 1$$
$$4a = 0$$
$$a = 0$$
0 does not solve the original equation, $\varnothing$

13.
$$\sqrt[4]{3x+1} = 2$$
$$\left(\sqrt[4]{3x+1}\right)^4 = 2^4$$
$$3x+1 = 16$$
$$3x = 15$$
$$x = 5$$

15.
$$\sqrt[3]{2x-5} = 1$$
$$\left(\sqrt[3]{2x-5}\right)^3 = 1^3$$
$$2x-5 = 1$$
$$2x = 6$$
$$x = 3$$

17.$\quad \sqrt[3]{3a+5} = -3$

$$\left(\sqrt[3]{3a+5}\right)^3 = (-3)^3$$
$$3a+5 = -27$$
$$3a = -32$$
$$a = -\frac{32}{3}$$

19.$\quad \sqrt{y-3} = y-3$

$$\left(\sqrt{y-3}\right)^2 = (y-3)^2$$
$$y-3 = y^2 - 6y + 9$$
$$0 = y^2 - 7y + 12$$
$$0 = (y-3)(y-4)$$
$$y-3 = 0 \quad \text{or} \quad y-4 = 0$$
$$y = 3 \qquad\qquad y = 4$$

21.$\quad \sqrt{a+2} = a+2$

$$\left(\sqrt{a+2}\right)^2 = (a+2)^2$$
$$a+2 = a^2 + 4a + 4$$
$$0 = a^2 + 3a + 2$$
$$0 = (a+1)(a+2)$$
$$a+1 = 0 \quad \text{or} \quad a+2 = 0$$
$$a = -1 \qquad\qquad a = -2$$

23.$\quad \sqrt{2x+4} = \sqrt{1-x}$

$$\left(\sqrt{2x+4}\right)^2 = \left(\sqrt{1-x}\right)^2$$
$$2x+4 = 1-x$$
$$3x = -3$$
$$x = -1$$

25.$\quad \sqrt{4a+7} = -\sqrt{a+2}$

$$\left(\sqrt{4a+7}\right)^2 = \left(-\sqrt{a+2}\right)^2$$
$$4a+7 = a+2$$
$$3a+7 = 2$$
$$3a = -5$$
$$a = -\frac{5}{3}$$

Possible solution $-\dfrac{5}{3}$, which does

not check, $\varnothing$

27.$\quad \sqrt[4]{5x-8} = \sqrt[4]{4x-1}$

$$\left(\sqrt[4]{5x-8}\right)^4 = \left(\sqrt[4]{4x-1}\right)^4$$
$$5x-8 = 4x-1$$
$$x = 7$$

29.$\quad x+1 = \sqrt{5x+1}$

$$(x+1)^2 = \left(\sqrt{5x+1}\right)^2$$
$$x^2 + 2x + 1 = 5x+1$$
$$x^2 - 3x = 0$$
$$x(x-3) = 0$$
$$x = 0 \quad \text{or} \quad x-3 = 0$$
$$x = 3$$

31.$\quad t+5 = \sqrt{2t+9}$

$$(t+5)^2 = \left(\sqrt{2t+9}\right)^2$$
$$t^2 + 10t + 25 = 2t+9$$
$$t^2 + 8t + 16 = 0$$
$$(t+4)^2 = 0$$
$$t+4 = 0$$
$$t = -4$$

33. $\sqrt{y-8} = \sqrt{8-y}$

$\left(\sqrt{y-8}\right)^2 = \left(\sqrt{8-y}\right)^2$

$y - 8 = 8 - y$

$2y - 8 = 8$

$2y = 16$

$y = 8$

35. $\sqrt[3]{3x+5} = \sqrt[3]{5-2x}$

$\left(\sqrt[3]{3x+5}\right)^3 = \left(\sqrt[3]{5-2x}\right)^3$

$3x + 5 = 5 - 2x$

$5x = 0$

$x = 0$

37. $\sqrt{x-8} = \sqrt{x} - 2$

$\left(\sqrt{x-8}\right)^2 = \left(\sqrt{x}-2\right)^2$

$x - 8 = \left(\sqrt{x}-2\right)\left(\sqrt{x}-2\right)$

$x - 8 = x - 4\sqrt{x} + 4$

$-8 = -4\sqrt{x} + 4$

$-12 = -4\sqrt{x}$

$3 = \sqrt{x}$

$9 = x$

39. $\sqrt{x+1} = \sqrt{x} + 1$

$\left(\sqrt{x+1}\right)^2 = \left(\sqrt{x}+1\right)^2$

$x + 1 = x + 2\sqrt{x} + 1$

$0 = 2\sqrt{x}$

$0 = x$

41. $\sqrt{x+8} = \sqrt{x-4} + 2$

$\left(\sqrt{x+8}\right)^2 = \left(\sqrt{x-4}+2\right)^2$

$x + 8 = \left(\sqrt{x-4}+2\right)\left(\sqrt{x-4}+2\right)$

$x + 8 = x - 4 + 4\sqrt{x-4} + 4$

$x + 8 = x + 4\sqrt{x-4}$

$8 = 4\sqrt{x-4}$

$2 = \sqrt{x-4}$

$2^2 = \left(\sqrt{x-4}\right)^2$

$4 = x - 4$

$8 = x$

43. $\sqrt{x-5} - 3 = \sqrt{x-8}$

$\left(\sqrt{x-5}-3\right)^2 = \left(\sqrt{x-8}\right)^2$

$x - 5 - 6\sqrt{x-5} + 9 = x - 8$

$-6\sqrt{x-5} = -12$

$\sqrt{x-5} = 2$

$x - 5 = 4$

$x = 9$

Possible solution 9, which does not check, $\varnothing$

45. $\sqrt{x+4} = 2 - \sqrt{2x}$

$\left(\sqrt{x+4}\right)^2 = \left(2 - \sqrt{2x}\right)^2$

$x + 4 = 4 - 4\sqrt{2x} + 2x$

$-x + 4 = 4 - 4\sqrt{2x}$

$-x = -4\sqrt{2x}$

$\dfrac{x}{4} = \sqrt{2x}$

$\left(\dfrac{x}{4}\right)^2 = \left(\sqrt{2x}\right)^2$

$\dfrac{x^2}{16} = 2x$

$x^2 = 32x$

$x^2 - 32x = 0$

$x(x - 32) = 0$

$x = 0$ or $x - 32 = 0$

$x = 32$

Possible solution, 0 and 32; only 0 checks.

47. $\sqrt{2x+4} = \sqrt{x+3} + 1$

$\left(\sqrt{2x+4}\right)^2 = \left(\sqrt{x+3} + 1\right)^2$

$2x + 4 = x + 3 + 2\sqrt{x+3} + 1$

$x = 2\sqrt{x+3}$

$x^2 = 4x + 12$

$x^2 - 4x - 12 = 0$

$(x+2)(x-6) = 0$

$x + 2 = 0$ or $x - 6 = 0$

$x = -2$ $x = 6$

Possible solutions -2 and 6; only 6 checks.

49. $t = \dfrac{\sqrt{100-h}}{4}$

$4t = \sqrt{100-h}$

$\left(4t\right)^2 = \left(\sqrt{100-h}\right)^2$

$16t^2 = 100 - h$

$16t^2 - 100 = -h$

$-16t^2 + 100 = h$

$h = 100 - 16t^2$

51. $T = 2\pi\sqrt{\dfrac{L}{32}}$, $T = 2$

$2 = 2\left(\dfrac{22}{7}\right)\sqrt{\dfrac{L}{32}}$

$\dfrac{7}{22} = \sqrt{\dfrac{L}{32}}$

$\dfrac{49}{484} = \dfrac{L}{32}$

$\dfrac{49 \cdot 32}{484} = L$

$\dfrac{392}{121} = L$

$3.24 \text{ ft.} \approx L$

53. If $\dfrac{ED}{DC} = \dfrac{FG}{GB}$, then

$\dfrac{l}{10} = \dfrac{\sqrt{x+1}}{3}$

$l = \dfrac{10\sqrt{x+1}}{3}$

55. $y = \sqrt{x},\ x = 25$

$y = \sqrt{25} = 5$ meters

57. Halfway across the river is $x = \frac{1}{2}(100) = 50$,

$y = \sqrt{x}$

$50 = \sqrt{x}$

$x = 2500$ meters down river

59. $y = \sqrt{x},\ 0 \le y \le 100$

61. $y = 2\sqrt{x}$

See the graph in the back of the textbook.

63. $y = \sqrt{x} - 2$

See the graph in the back of the textbook.

65. $y = \sqrt{x-2}$

See the graph in the back of the textbook.

67. $y = 3\sqrt[3]{x}$

See the graph in the back of the textbook.

69. $y = \sqrt[3]{x} + 3$

See the graph in the back of the textbook.

71. $y = \sqrt[3]{x+3}$

See the graph in the back of the textbook.

73. $\sqrt{2}\left(\sqrt{3} - \sqrt{2}\right) = \sqrt{2}\left(\sqrt{3}\right) - \sqrt{2}\left(\sqrt{2}\right) = \sqrt{6} - 2$

75. $\left(\sqrt{x} + 5\right)^2 = \left(\sqrt{x}\right)^2 + 2\left(\sqrt{x}\right)(5) + 5^2$

$= x + 10\sqrt{x} + 25$

77. $\dfrac{\sqrt{x}}{\sqrt{x}+3} = \dfrac{\sqrt{x}}{\sqrt{x}+3} \cdot \dfrac{\sqrt{x}-3}{\sqrt{x}-3} = \dfrac{x - 3\sqrt{x}}{x-9}$

79. $\dfrac{x}{3\sqrt{2x-3}} - \dfrac{1}{\sqrt{2x-3}} = \dfrac{1}{3}$

$x - 3 = \sqrt{2x-3}$

$x^2 - 6x + 9 = 2x - 3$

$x^2 - 8x + 12 = 0$

$(x-2)(x-6) = 0$

$x - 2 = 0 \quad \text{or} \quad x - 6 = 0$

$x = 2 \qquad\qquad x = 6$

Possible solutions 2 and 6, only 6 checks.

81.
$$x+1 = \sqrt[3]{4x+4}$$
$$(x+1)^3 = \left(\sqrt[3]{4x+4}\right)^3$$
$$x^3 + 3x^2 + 3x + 1 = 4x + 4$$
$$x^3 + 3x^2 - x - 3 = 0$$
$$x^2(x+3) - 1(x+3) = 0$$
$$(x^2-1)(x+3) = 0$$
$$x+1 = 0 \quad \text{or} \quad x-1 = 0 \quad \text{or} \quad x+3 = 0$$
$$x = -1 \qquad x = 1 \qquad x = -3$$

83.
$$y+2 = \sqrt{x^2 + (y-2)^2}$$
$$(y+2)^2 = \left(\sqrt{x^2 + (y-2)^2}\right)^2$$
$$y^2 + 4y + 4 = x^2 + y^2 - 4y + 4$$
$$8y = x^2$$
$$y = \frac{1}{8}x^2$$

85. See the graph in the back of the textbook.

87. The constant, b, shifts the graph b units along the y-axis.

89. See the graph in the back of the textbook.

91. The constant, b, shifts the graph $-b$ units along the x-axis.

93. See the graph in the back of the textbook.

95. The smaller the absolute value of a, the more slowly the graph rises or falls. If a is negative, the graph lies below the x-axis.

9.7 Complex Numbers

1. $\sqrt{-36} = \sqrt{36(-1)} = \sqrt{36}\,\sqrt{-1} = 6i$

3. $-\sqrt{-25} = -\sqrt{25}\,\sqrt{-1} = -5i$

5. $\sqrt{-72} = \sqrt{72(-1)} = \sqrt{72}\,\sqrt{-1} = 6\sqrt{2}\,i = 6i\sqrt{2}$

7. $-\sqrt{-12} = -\sqrt{4\cdot3}\,\sqrt{-1} = -2i\sqrt{3}$

9. $i^{28} = (i^2)^{14} = (-1)^{14} = 1$

11. $i^{26} = (i^2)^{13} = (-1)^{13} = -1$

13. $i^{75} = (i^2)^{37}\cdot i = (-1)^{37}\cdot i = -1\cdot i = -i$

15. $2x + 3yi = 6 - 3i$
$$2x = 6 \quad 3y = -3$$
$$x = 3 \quad y = -1$$

17. $2 - 5i = -x + 10yi$
$$2 = -x \qquad -5i = 10yi$$
$$-2 = x \qquad -\frac{1}{2} = y$$

19. $2x + 10i = -16 - 2yi$
$$2x = -16 \quad 10 = -2y$$
$$x = -8 \quad -5 = y$$

21. $(2x-4)-3i=10-6yi$

$2x-4=10 \quad -3i=-6yi$

$2x=14 \quad \dfrac{1}{2}=y$

$x=7$

23. $(7x-1)+4i=2+(5y+2)i$

$7x-1=2 \quad 4=5y+2$

$7x=3 \quad 2=5y$

$x=\dfrac{3}{7} \quad \dfrac{2}{5}=y$

25. $(2+3i)+(3+6i)=(2+3)+(3+6)i=5+9i$

27. $(3-5i)+(2+4i)=5-i$

29. $(5+2i)-(3+6i)=(5-3)+(2-6)i=2-4i$

31. $(3-5i)-(2+i)=1-6i$

33. $[(3+2i)-(6+i)]+(5+i)=[(3-6)+(2-1)i]+(5+i)$

$=(-3+i)+(5+i)$

$=(-3+5)+(i+i)$

$=2+2i$

35. $[(7-i)-(2+4i)]-(6+2i)$

$=(5-5i)-(6+2i)$

$=-1-7i$

37. $(3+2i)-[(3-4i)-(6+2i)]=(3+2i)-[(3-6)+(-4-2)i]$

$=(3+2i)-(-3-6i)$

$=(3+3)+(2+6)i$

$=6+8i$

39. $(4-9i)+[(2-7i)-(4+8i)]$

$=(4-9i)+(-2-15i)$

$=2-24i$

41. $3i(4+5i)=3i\cdot4+3i(5i)=12i+15i^2=-15+12i$

43. $6i(4-3i)=24i-18i^2$

$=18+24i$

45. $(3+2i)(4+i)=3\cdot4+3\cdot i+2i\cdot4+2i\cdot i$

$=12+3i+8i+2i^2$

$=12+11i+2(-1)$

$=12+11i-2$

$=10+11i$

47. $(4+9i)(3-i)=12+23i-9i^2$

$=12+23i+9$

$=21+23i$

49. $(1+i)^3=(1+i)^2(1+i)$

$=(1+2i+i^2)(1+i)$

$=(1+2i-1)(1+i)$

$=(2i)(1+i)$

$=2i+2i^2$

$=2i+2(-1)$

$=-2+2i$

51. $(2-i)^3=(2-i)^2(2-i)$

$=(4-4i+i^2)(2-i)$

$=(4-4i-1)(2-i)$

$=(3-4i)(2-i)$

$=6-11i+4i^2$

$=6-11i-4$

$=2-11i$

53. $(2+5i)^2 = 2^2 + 2(2)(5i) + (5i)^2$
$= 4 + 20i + 25i^2$
$= 4 + 20i + 25(-1)$
$= 4 + 20i - 25$
$= -21 + 20i$

55. $(1-i)^2 = 1 - 2i + i^2$
$= 1 - 2i - 1$
$= -2i$

57. $(3-4i)^2 = 3^2 - 2(3)(4i) + (4i)^2$
$= 9 - 24i + 16i^2$
$= 9 - 24i + 16(-1)$
$= 9 - 24i - 16$
$= -7 - 24i$

59. $(2+i)(2-i) = 4 - i^2$
$= 4 + 1$
$= 5$

61. $(6-2i)(6+2i) = 6^2 - (2i)^2$
$= 36 - 4i^2$
$= 36 - 4(-1)$
$= 36 + 4$
$= 40$

63. $(2+3i)(2-3i) = 4 - 9i^2$
$= 4 + 9$
$= 13$

65. $(10+8i)(10-8i) = 10^2 - (8i)^2$
$= 100 - 64i^2$
$= 100 - 64(-1)$
$= 100 + 64$
$= 164$

67. $\dfrac{2-3i}{i} = \dfrac{2-3i}{i} \cdot \dfrac{i}{i}$
$= \dfrac{2i - 3i^2}{i^2}$
$= \dfrac{3+2i}{-1}$
$= -3 - 2i$

69. $\dfrac{5+2i}{-i} = \dfrac{5+2i}{-i} \cdot \dfrac{i}{i} = \dfrac{5i + 2i^2}{-i^2} = \dfrac{-2+5i}{1} = -2 + 5i$

71. $\dfrac{4}{2-3i} = \dfrac{4}{2-3i} \cdot \dfrac{2+3i}{2+3i}$
$= \dfrac{8+12i}{4-9i^2}$
$= \dfrac{8+12i}{13}$
$= \dfrac{8}{13} + \dfrac{12}{13}i$

73. $\dfrac{6}{-3+2i} = \dfrac{6}{-3+2i} \cdot \dfrac{(-3-2i)}{(-3-2i)}$

$= \dfrac{-18-12i}{(-3)^2 - (2i)^2}$

$= \dfrac{-18-12i}{9-4i^2}$

$= \dfrac{-18-12i}{9-4(-1)}$

$= \dfrac{-18-12i}{9+4}$

$= \dfrac{-18-12i}{13}$

$= -\dfrac{18}{13} - \dfrac{12}{13} i$

75. $\dfrac{2+3i}{2-3i} = \dfrac{2+3i}{2-3i} \cdot \dfrac{2+3i}{2+3i}$

$= \dfrac{4+12i+9i^2}{4-9i^2}$

$= \dfrac{-5+12i}{13}$

$= -\dfrac{5}{13} + \dfrac{12}{13} i$

77. $\dfrac{5+4i}{3+6i} = \dfrac{5+4i}{3+6i} \cdot \dfrac{3-6i}{3-6i}$

$= \dfrac{15-30i+12i-24i^2}{(3)^2 - (6i)^2}$

$= \dfrac{15-18i-24(-1)}{9-36i^2}$

$= \dfrac{15-18i+24}{9-36(-1)}$

$= \dfrac{39-18i}{9+36}$

$= \dfrac{39-18i}{45}$

$= \dfrac{39}{45} - \dfrac{18}{45} i$

$= \dfrac{13}{15} - \dfrac{2}{5} i$

79. $V = 80 + 20i$ and $I = -6 + 2i$

$V = RI \rightarrow R = \dfrac{V}{I}$

$R = \dfrac{80+20i}{-6+2i}$

$= \dfrac{80+20i}{-6+2i} \cdot \dfrac{-6-2i}{-6-2i}$

$= \dfrac{-480-280i-40i^2}{36-4i^2}$

$= \dfrac{-440-280i}{40}$

$= -11 - 7i$ ohms

81. Domain $= \{1, 3, 4\}$, Range $= \{2, 4\}$, Function

83. $\{(3,1), (2,3), (1,2)\}$

Domain $\{1, 2, 3\}$

Range $\{1, 2, 3\}$

Function

85. A function

87. $\dfrac{1}{i} = \dfrac{1}{i} \cdot \dfrac{i}{i} = \dfrac{i}{i^2} = \dfrac{i}{-1} = -i$

89. $\quad z = Kxy^2, \quad z = 40, \quad x = 5, \quad y = 2$

$\qquad 40 = K(5)(2)^2 \Rightarrow K = 2$

$\qquad z = 2xy^2, \quad x = 2, \quad y = 5$

$\qquad z = 2(2)(5)^2$

$\qquad z = 100$

91. $\qquad x^3 - 11x + 20 = 0, \quad x = 2 + i$

$\qquad (2+i)^3 - 11(2+i) + 20 = 0$

$\qquad (4+4i-1)(2+i) - 22 - 11i + 20 = 0$

$\qquad (4+4i-1)(2+i) - 2 - 11i = 0$

$\qquad (3+4i)(2+i) - 2 - 11i = 0$

$\qquad 6 + 8i + 3i + 4i^2 - 2 - 11i = 0$

$\qquad 11i + 2 - 2 - 11i = 0$

$\qquad 0 = 0$

Chapter 9 Review

1. $\sqrt{49} = 7$

3. $16^{1/4} = \sqrt[4]{16} = 2$

5. $\sqrt[5]{32x^{15}y^{10}} = \left(32x^{15}y^{10}\right)^{1/5}$

$\qquad\qquad\qquad = 32^{1/5}\left(x^{15}\right)^{1/5}\left(y^{10}\right)^{1/5}$

$\qquad\qquad\qquad = 2x^3 y^2$

7. $x^{2/3} \cdot x^{4/3} = x^{2/3 + 4/3}$

$\qquad\qquad\quad = x^{6/3}$

$\qquad\qquad\quad = x^2$

9. $\dfrac{a^{3/5}}{a^{1/4}} = a^{3/5 - 1/4}$

$\qquad\quad = a^{12/20 - 5/20}$

$\qquad\quad = a^{7/20}$

11. $\left(3x^{1/2} + 5y^{1/2}\right)\left(4x^{1/2} - 3y^{1/2}\right)$

$\qquad = 3x^{1/2}\left(4x^{1/2}\right) - 3x^{1/2}\left(3y^{1/2}\right) + 5y^{1/2}\left(4x^{1/2}\right) - 5y^{1/2}\left(3y^{1/2}\right)$

$\qquad = 12x^1 - 9x^{1/2}y^{1/2} + 20x^{1/2}y^{1/2} - 15y^1$

$\qquad = 12x + 11x^{1/2}y^{1/2} - 15y$

13. $\dfrac{28x^{5/6} + 14x^{7/6}}{7x^{1/3}} = \dfrac{28x^{5/6}}{7x^{1/3}} + \dfrac{14x^{7/6}}{7x^{1/3}}$

$\qquad\qquad\qquad = 4x^{5/6 - 1/3} + 2x^{7/6 - 1/3}$

$\qquad\qquad\qquad = 4x^{5/6 - 2/6} + 2x^{7/6 - 2/6}$

$\qquad\qquad\qquad = 4x^{3/6} + 2x^{5/6}$

$\qquad\qquad\qquad = 4x^{1/2} + 2x^{5/6}$

15. $x^{3/4} + \dfrac{5}{x^{1/4}} = \dfrac{x^{3/4}}{1} \cdot \dfrac{x^{1/4}}{x^{1/4}} + \dfrac{5}{x^{1/4}} = \dfrac{x}{x^{1/4}} + \dfrac{5}{x^{1/4}} = \dfrac{x+5}{x^{1/4}}$

17. $\sqrt{50} = \sqrt{25 \cdot 2}$

$\qquad\quad = \sqrt{25}\sqrt{2}$

$\qquad\quad = 5\sqrt{2}$

19. $\sqrt{18x^2} = \sqrt{9x^2}\sqrt{2} = 3x\sqrt{2}$

21. $\sqrt[4]{32a^4b^5c^6} = \sqrt[4]{16a^4b^4c^4}\,\sqrt[4]{2bc^2}$

                 $= 2abc\sqrt[4]{2bc^2}$

23. $\dfrac{6}{\sqrt[3]{2}} = \dfrac{6}{\sqrt[3]{2}} \cdot \dfrac{\sqrt[3]{2^2}}{\sqrt[3]{2^2}}$

          $= \dfrac{6\sqrt[3]{2^2}}{\sqrt[3]{2^3}}$

          $= \dfrac{6\sqrt[3]{4}}{2}$

          $= 3\sqrt[3]{4}$

25. $\sqrt[3]{\dfrac{40x^2y^3}{3z}} = \dfrac{\sqrt[3]{8y^3}\ \sqrt[3]{5x^2}}{\sqrt[3]{3z}}$

              $= \dfrac{2y\sqrt[3]{5x^2}}{\sqrt[3]{3z}} \cdot \dfrac{\sqrt[3]{3^2z^2}}{\sqrt[3]{3^2z^2}}$

              $= \dfrac{2y\sqrt[3]{45x^2z^2}}{3z}$

27. $\sqrt{12} + \sqrt{3} = \sqrt{4}\sqrt{3} + \sqrt{3}$

              $= 2\sqrt{3} + 1\sqrt{3}$

              $= (2+1)\sqrt{3}$

              $= 3\sqrt{3}$

29. $3\sqrt{8} - 4\sqrt{72} + 5\sqrt{50} = 3\sqrt{4}\sqrt{2} - 4\sqrt{36}\sqrt{2} + 5\sqrt{25}\sqrt{2}$

                         $= 6\sqrt{2} - 24\sqrt{2} + 25\sqrt{2}$

                         $= (6 - 24 + 25)\sqrt{2}$

                         $= 7\sqrt{2}$

31. $2x\sqrt[3]{xy^3z^2} - 6y\sqrt[3]{x^4z^2} = 2x\sqrt[3]{y^3}\,\sqrt[3]{xz^2} - 6y\sqrt[3]{x^3}\,\sqrt[3]{xz^2}$

                        $= 2xy\sqrt[3]{xz^2} - 6xy\sqrt[3]{xz^2}$

                        $= -4xy\sqrt[3]{xz^2}$

33. $(\sqrt{x} - 2)(\sqrt{x} - 3) = (\sqrt{x})^2 + \sqrt{x}(-3) + (-2)\sqrt{x} - 2(-3)$

                       $= x - 3\sqrt{x} - 2\sqrt{x} + 6$

                       $= x - 5\sqrt{x} + 6$

35. $\dfrac{\sqrt{7} + \sqrt{5}}{\sqrt{7} - \sqrt{5}} = \dfrac{\sqrt{7} + \sqrt{5}}{\sqrt{7} - \sqrt{5}} \cdot \dfrac{\sqrt{7} + \sqrt{5}}{\sqrt{7} + \sqrt{5}}$

           $= \dfrac{7 + \sqrt{7}\sqrt{5} + \sqrt{5}\sqrt{7} + 5}{(\sqrt{7})^2 - (\sqrt{5})^2}$

           $= \dfrac{12 + 2\sqrt{35}}{2}$

           $= \dfrac{2(6 + \sqrt{35})}{2}$

           $= 6 + \sqrt{35}$

37. $\sqrt{4a+1} = 1$

       $\left(\sqrt{4a+1}\right)^2 = 1^2$

            $4a + 1 = 1$

                $4a = 0$

                  $a = 0$

39. $$\sqrt{3x+1}-3=1$$
$$\sqrt{3x+1}=4$$
$$\left(\sqrt{3x+1}\right)^2=4^2$$
$$3x+1=16$$
$$3x=15$$
$$x=5$$
The solution is $x=5$

41. $y=3\sqrt{x}$

See the graph in the back of the textbook.

43. $i^{24}=\left(i^2\right)^{12}=(-1)^{12}=1$

45. $3-4i=-2x+8yi$

$$3=-2x \qquad -4i=8yi$$
$$-\frac{3}{2}=x \qquad\qquad -\frac{4i}{8i}=y$$
$$\qquad\qquad\qquad -\frac{1}{2}=y$$

47. $(3+5i)+(6-2i)=(3+6)+(5-2)i=9+3i$

49. $3i(4+2i)=3i\cdot4+(3i)(2i)=12i+6i^2=-6+12i$

51. $(4+2i)^2=4^2+2(4)(2i)+(2i)^2=16+16i+4i^2=16+16i-4=12+16i$

53. $\dfrac{3+i}{i}=\dfrac{3+i}{i}\cdot\dfrac{-i}{-i}=\dfrac{-3i-i^2}{-i^2}=\dfrac{-3i-(-1)}{-(-1)}=1-3i$

55. Let s be the length of one side of the roof.
$$s^2=13.5^2+18^2$$
$$s=\sqrt{182.25+324}=22.5 \text{ feet}$$

Cumulative Review: Chapters 1-9

1. $33 - 22 - (-11) + 1 = 33 + (-22) + 11 + 1$ Add the opposites

$$= 23$$

3. $-6 + 5[3 - 2(-4 - 1)] = -6 + 5[3 - 2(-5)]$ Parentheses first

$$= -6 + 5[3 + 10]$$
$$= -6 + 5[13]$$
$$= -6 + 65$$
$$= 59$$

5. $\left(2y^{-3}\right)^{-1}\left(4y^{-3}\right)^2 = 2^{-1}y^3 \cdot 4^2 y^{-6} = \dfrac{16y^{-3}}{2} = \dfrac{8}{y^3}$

7. $\sqrt{72y^5} = \sqrt{36 \cdot 2 \cdot y^4 y} = 6y^2\sqrt{2y}$

9.   Opposite of 12 is -12

     Reciprocal of 12 is $\dfrac{1}{12}$

11. $41,500 = 4.15 \times 10^4$

13. $625a^4 - 16b^4 = \left(25a^2 + 4b^2\right)\left(25a^2 - 4b^2\right)$

$$= \left(25a^2 + 4b^2\right)(5a + 2b)(5a - 2b)$$

15. $\dfrac{246}{861} = \dfrac{2 \cdot 3 \cdot 41}{3 \cdot 7 \cdot 41} = \dfrac{2}{7}$

17. $\dfrac{x^2 - 9x + 20}{x^2 - 7x + 12} = \dfrac{(x-5)(x-4)}{(x-3)(x-4)} = \dfrac{x-5}{x-3}$

19. $(2x - 5y)(3x - 2y) = 6x^2 - 4xy - 15xy + 10y^2$

$$= 6x^2 - 19xy + 10y^2$$

21. $\left(\sqrt{x} - 2\right)^2 = \left(\sqrt{x}\right)^2 - 2 \cdot 2\sqrt{x} + (2)^2$

$$= x - 4\sqrt{x} + 4$$

23. $\dfrac{27x^3 y^2}{13x^2 y^4} \div \dfrac{9xy}{26y} = \dfrac{27x^3 y^2}{13x^2 y^4} \cdot \dfrac{26y}{9xy}$

$$= \dfrac{3 \cdot 3 \cdot 3 \cdot 2 \cdot 13 x^3 y^2 y}{13 \cdot 3 \cdot 3 x^2 y^4 xy}$$
$$= \dfrac{6x^3 y^3}{x^3 y^5}$$
$$= 6x^0 y^{-2}$$
$$= \dfrac{6}{y^2}$$

25. $\left(\dfrac{2}{3}x^3 + \dfrac{1}{6}x^2 + \dfrac{1}{2}\right) - \left(\dfrac{1}{4}x^2 - \dfrac{1}{3}x + \dfrac{1}{12}\right) = \dfrac{2}{3}x^3 + \left(\dfrac{1}{6} - \dfrac{1}{4}\right)x^2 + \dfrac{1}{3}x + \dfrac{1}{2} - \dfrac{1}{12}$

$$= \dfrac{2}{3}x^3 + \left(\dfrac{2}{12} - \dfrac{3}{12}\right)x^2 + \dfrac{1}{3}x + \dfrac{6}{12} - \dfrac{1}{12}$$

$$= \dfrac{2}{3}x^3 - \dfrac{1}{12}x^2 + \dfrac{1}{3}x + \dfrac{5}{12}$$

27. $3y - 8 = -4y + 6$

$\quad 7y - 8 = 6$

$\quad\quad 7y = 14$

$\quad\quad\, y = 2$

29. $|3x - 1| - 2 = 6$

$\quad |3x - 1| = 8$

$\quad 3x - 1 = 8 \quad \text{or} \quad 3x - 1 = -8$

$\quad\quad 3x = 9 \quad\quad\quad\quad 3x = -7$

$\quad\quad\, x = 3 \quad\quad\quad\quad\, x = -\dfrac{7}{3}$

31. $\quad\dfrac{x+2}{x+1} - 2 = \dfrac{1}{x+1}$

$\quad\quad\quad \text{LCD} = x + 1$

$\quad x + 2 - 2(x+1) = 1$

$\quad\quad x + 2 - 2x - 2 = 1$

$\quad\quad\quad\quad\quad -x = 1$

$\quad\quad\quad\quad\quad\;\; x = -1$

Possible solution -1, which does not check; $\varnothing$

33. $(x+3)^2 - 3(x+3) - 70 = 0$

$\quad$ Let $y = x + 3$

$\quad\quad y^2 - 3y - 70 = 0$

$\quad\quad (y+7)(y-10) = 0$

$\quad\quad y + 7 = 0 \quad \text{or} \quad y - 10 = 0$

$\quad\quad\quad y = -7 \quad\quad\quad\; y = 10$

$\quad$ Substitute $x + 3$ for y

$\quad\quad x + 3 = -7 \quad \text{or} \quad x + 3 = 10$

$\quad\quad\quad x = -10 \quad\quad\quad x = 7$

35. $P = 2L + 2W, \; W = 3, \; P = 18$

$\quad 18 = 2L + 2(3)$

$\quad 18 = 2L + 6$

$\quad 12 = 2L$

$\quad\; 6 = L$

37. $|5x - 4| \geq 6$

$\quad 5x - 4 \geq 6 \quad \text{or} \quad 5x - 4 \leq -6$

$\quad\quad 5x \geq 10 \quad \text{or} \quad\quad 5x \leq -2$

$\quad\quad\; x \geq 2 \quad \text{or} \quad\quad\;\; x \leq -\dfrac{2}{5}$

See the graph in the back of the textbook.

39. $4x + 9y = 2$

$\qquad y = 2x - 12$

Substitute $y = 2x - 12$ into the first equation

$4x + 9(2x - 12) = 2$

$4x + 18x - 108 = 2$

$\qquad 22x = 110$

$\qquad\quad x = 5$

Substitute $x = 5$ into the second equation

$\quad y = 2(5) - 12$

$\quad y = -2$

The solution to the system is $(5, -2)$.

41. $x + y \qquad = -2 \qquad (1)$

$\quad\; y + 10z = -1 \qquad (2)$

$2x \quad - 13z = 5 \qquad (3)$

$-x - y \qquad = 2 \qquad -1 \text{ times } (1)$

$\quad\;\; y + 10z = -1 \qquad (2)$

$\overline{}$

$-x \quad + 10z = 1 \qquad (4)$

$2x - 13z = 5 \qquad (3)$

$\underline{-2x + 20z = 2} \qquad 2 \text{ times } (4)$

$\quad 7z = 7$

$\quad\; z = 1$

$y + 10(1) = -1 \qquad (2),\; z = 1$

$\qquad y = -11$

$x - 11 = -2 \qquad (1),\; y = -1$

$\qquad x = 9$

The solution is $(9, -11, 1)$.

43. $y = \dfrac{1}{3}x - 3$

See the graph in the back of the textbook.

45. $y = \dfrac{4}{x}$

See the graph in the back of the textbook.

47. $4x - 5y = 15$

$\quad -5y = -4x + 15$

$\qquad y = \dfrac{4}{5} - 3$

The slope is $\dfrac{4}{5}$ and the y-intercept is -3.

49. $m = \dfrac{2}{3}$ and $(9, 3)$

$y - y_1 = m(x - x_1)$

$y - 3 = \dfrac{2}{3}(x - 9)$

$y - 3 = \dfrac{2}{3}x - 6$

$\qquad y = \dfrac{2}{3}x - 3$

Chapter 9 Test

1. $27^{-2/3} = \dfrac{1}{27^{2/3}}$

$\qquad = \dfrac{1}{\left(27^{1/3}\right)^2}$

$\qquad = \dfrac{1}{3^2}$

$\qquad = \dfrac{1}{9}$

2. $\left(\dfrac{25}{49}\right)^{-1/2} = \left(\dfrac{49}{25}\right)^{1/2} = \dfrac{7}{5}$

3. $a^{3/4} \cdot a^{-1/3} = a^{3/4-1/3}$

$\qquad\qquad = a^{9/12-4/12}$

$\qquad\qquad = a^{5/12}$

4. $\dfrac{\left(x^{2/3}y^{-3}\right)^{1/2}}{\left(x^{3/4}y^{1/2}\right)^{-1}} = \dfrac{x^{1/3}y^{-3/2}}{x^{-3/4}y^{-1/2}}$

$\qquad\qquad = x^{1/3+3/4}\,y^{-3/2+1/2}$

$\qquad\qquad = x^{4/12+9/12}\,y^{-1}$

$\qquad\qquad = \dfrac{x^{13/12}}{y}$

5. $\sqrt{49x^8y^{10}} = 7x^4y^5$

6. $\sqrt[5]{32x^{10}y^{20}} = 2x^2y^4$

7. $\dfrac{\left(36a^8b^4\right)^{1/2}}{\left(27a^9b^6\right)^{1/3}} = \dfrac{6a^4b^2}{3a^3b^2} = 2a$

8. $\dfrac{\left(x^n y^{1/n}\right)^n}{\left(x^{1/n}y^n\right)^{n^2}} = \dfrac{x^{n^2}y}{x^n y^{n^3}} = x^{n^2-n}y^{1-n^3}$

9. $2a^{1/2}\left(3a^{3/2} - 5a^{1/2}\right) = 6a^{1/2+3/2} - 10a^{1/2+1/2}$

$\qquad\qquad\qquad\qquad = 6a^2 - 10a$

10. $\left(4a^{3/2} - 5\right)^2 = \left(4a^{3/2}\right)^2 + 2\left(4a^{3/2}\right)(-5) + (-5)^2$

$\qquad\qquad\qquad = 16a^3 - 40a^{3/2} + 25$

11. $3x^{2/3} + 5x^{1/3} - 2 = \left(3x^{1/3} - 1\right)\left(x^{1/3} + 2\right)$

12. $9x^{2/3} - 49 = \left(3x^{1/3} - 7\right)\left(3x^{1/3} + 7\right)$

13. $\dfrac{4}{x^{1/2}} + x^{1/2} = \dfrac{4}{x^{1/2}} + \dfrac{x^{1/2}}{1} \cdot \dfrac{x^{1/2}}{x^{1/2}} = \dfrac{4}{x^{1/2}} + \dfrac{x}{x^{1/2}} = \dfrac{x+4}{x^{1/2}}$

14. $\dfrac{x^2}{\left(x^2-3\right)^{1/2}} - \left(x^2-3\right)^{1/2} = \dfrac{x^2}{\left(x^2-3\right)^{1/2}} - \dfrac{\left(x^2-3\right)^{1/2}}{1} \cdot \dfrac{\left(x^2-3\right)^{1/2}}{\left(x^2-3\right)^{1/2}} = \dfrac{x^2 - x^2 + 3}{\left(x^2-3\right)^{1/2}} = \dfrac{3}{\left(x^2-3\right)^{1/2}}$

15. $\sqrt{125x^3y^5} = \sqrt{25x^2y^4}\sqrt{5xy} = 5xy^2\sqrt{5xy}$

16. $\sqrt[3]{40x^7y^8} = \sqrt[3]{8x^6y^6}\sqrt[3]{5xy^2} = 2x^2y^2\sqrt[3]{5xy^2}$

17. $\sqrt{\dfrac{2}{3}} = \dfrac{\sqrt{2}}{\sqrt{3}} \cdot \dfrac{\sqrt{3}}{\sqrt{3}} = \dfrac{\sqrt{6}}{3}$

18. $\sqrt{\dfrac{12a^4b^3}{5c}} = \dfrac{\sqrt{12a^4b^3}}{\sqrt{5c}} \cdot \dfrac{\sqrt{5c}}{\sqrt{5c}} = \dfrac{\sqrt{4a^4b^2}\sqrt{3b}\sqrt{5c}}{5c} = \dfrac{2a^2b\sqrt{15bc}}{5c}$

19. $3\sqrt{12} - 4\sqrt{27} = 3\sqrt{4}\sqrt{3} - 4\sqrt{9}\sqrt{3} = 6\sqrt{3} - 12\sqrt{3} = -6\sqrt{3}$

20. $2\sqrt[3]{24a^3b^3} - 5a\sqrt[3]{3b^3} = 2\sqrt[3]{8a^3b^3}\sqrt[3]{3} - 5a\sqrt[3]{b^3}\sqrt[3]{3} = 4ab\sqrt[3]{3} - 5ab\sqrt[3]{3} = -ab\sqrt[3]{3}$

21. $\left(\sqrt{x}+7\right)\left(\sqrt{x}-4\right) = x - 4\sqrt{x} + 7\sqrt{x} - 28$
$$= x + 3\sqrt{x} - 28$$

22. $\left(3\sqrt{2} - \sqrt{3}\right)^2 = \left(3\sqrt{2}\right)^2 + 2\left(3\sqrt{2}\right)\left(-\sqrt{3}\right) + \left(-\sqrt{3}\right)^2 = 9 \cdot 2 - 6\sqrt{6} + 3 = 21 - 6\sqrt{6}$

23. $\dfrac{5}{\sqrt{3}-1} = \dfrac{5}{\sqrt{3}-1} \cdot \dfrac{\sqrt{3}+1}{\sqrt{3}+1}$

$= \dfrac{5\sqrt{3}+5}{\left(\sqrt{3}\right)^2 - 1^2}$

$= \dfrac{5+5\sqrt{3}}{2}$

24. $\dfrac{\sqrt{x}-\sqrt{2}}{\sqrt{x}+\sqrt{2}} = \dfrac{\sqrt{x}-\sqrt{2}}{\sqrt{x}+\sqrt{2}} \cdot \dfrac{\sqrt{x}-\sqrt{2}}{\sqrt{x}-\sqrt{2}}$

$= \dfrac{x - 2\sqrt{2x} + 2}{x-2}$

25. $\sqrt{3x+1} = x-3$

$\left(\sqrt{3x+1}\right)^2 = \left(x-3\right)^2$

$3x+1 = x^2 - 6x + 9$

$0 = x^2 - 9x + 8$

$0 = \left(x-1\right)\left(x-8\right)$

$x-1 = 0 \quad\text{or}\quad x-8 = 0$

$x = 1 \qquad\qquad x = 8$

Possible solutions 1 and 8, only 8 checks; 8.

26. $\sqrt[3]{2x+7} = -1$

$\left(\sqrt[3]{2x+7}\right)^3 = \left(-1\right)^3$

$2x+7 = -1$

$2x = -8$

$x = -4$

27. $\sqrt{x+3} = \sqrt{x+4} - 1$

$\left(\sqrt{x+3}\right)^2 = \left(\sqrt{x+4} - 1\right)^2$

$x + 3 = x + 4 - 2\sqrt{x+4} + 1$

$-2 = -2\sqrt{x+4}$

$1 = \sqrt{x+4}$

$1^2 = \left(\sqrt{x+4}\right)^2$

$1 = x + 4$

$-3 = x$

28. $y = \sqrt{x} - 2$

See the graph in the back of the textbook.

29. $y = \sqrt{x} + 3$

See the graph in the back of the textbook.

30. $(2x+5) - 4i = 6 - (y-3)i$

$2x + 5 = 6 \qquad -4i = -(y-3)i$

$2x = 1 \qquad -4i = -yi + 3i$

$x = \dfrac{1}{2} \qquad -7i = -yi$

$\qquad\qquad\quad 7 = y$

31. $(3+2i) - [(7-i)-(4+3i)] = (3+2i) - (3-4i) = 6i$

32. $(2-3i)(4+3i) = 2(4) + 2(3i) + (-3i)(4) + (-3i)(3i) = 8 - 6i - 9i^2 = 17 - 6i$

33. $(5-4i)^2 = 5^2 + 2(5)(-4i) + (-4i)^2 = 25 - 40i + 16i^2 = 9 - 40i$

34. $\dfrac{2-3i}{2+3i} = \dfrac{2-3i}{2+3i} \cdot \dfrac{2-3i}{2-3i}$

$= \dfrac{4 - 12i + 9i^2}{4 - 9i^2}$

$= \dfrac{-5 - 12i}{13}$

$= -\dfrac{5}{13} - \dfrac{12}{13}i$

35. $i^{38} = \left(i^2\right)^{19} = (-1)^{19} = -1$

CHAPTER 10

Quadratic Functions

10.1 Completing the Square

1. $x^2 = 25$

   $x = \pm\sqrt{25}$

   $x = \pm 5$

3. $a^2 = -9$

   $a = \pm\sqrt{-9}$

   $a = \pm 3i$

5. $y^2 = \dfrac{3}{4}$

   $y = \pm\sqrt{\dfrac{3}{4}} = \dfrac{\pm\sqrt{3}}{\sqrt{4}} = \dfrac{\pm\sqrt{3}}{2}$

7. $x^2 + 12 = 0$

   $x^2 = -12$

   $x = \pm\sqrt{-12}$

   $x = \pm 2i\sqrt{3}$

9. $4a^2 - 45 = 0$

   $4a^2 = 45$

   $a^2 = \dfrac{45}{4}$

   $a = \pm\sqrt{\dfrac{45}{4}}$

   $a = \pm\dfrac{3\sqrt{5}}{2}$

11. $(2y-1)^2 = 25$

   $2y - 1 = \pm\sqrt{25}$

   $2y - 1 = \pm 5$

   $2y = 1 \pm 5$

   $y = \dfrac{1 \pm 5}{2}$

   $y = \dfrac{1+5}{2}$ or $y = \dfrac{1-5}{2}$

   $y = 3$ $\qquad\quad y = -2$

13. $(2a+3)^2 = -9$

   $2a + 3 = \pm\sqrt{-9}$

   $2a + 3 = \pm 3i$

   $2a = -3 \pm 3i$

   $a = \dfrac{-3 \pm 3i}{2}$

15. $(5x+2)^2 = -8$

   $5x + 2 = \pm\sqrt{-8}$

   $5x + 2 = \pm 2i\sqrt{2}$

   $5x = -2 \pm 2i\sqrt{2}$

   $x = \dfrac{-2 \pm 2i\sqrt{2}}{5}$

17. $x^2 + 8x + 16 = -27$

$(x+4)^2 = -27$

$x + 4 = \pm\sqrt{-27}$

$x + 4 = \pm 3i\sqrt{3}$

$x = -4 \pm 3i\sqrt{3}$

19. $4a^2 - 12a + 9 = -4$

$(2a-3)^2 = -4$

$2a - 3 = \pm\sqrt{-4}$

$2a - 3 = \pm 2i$

$2a = 3 \pm 2i$

$a = \dfrac{3 \pm 2i}{2}$

21. $(x+5)^2 + (x-5)^2 = 52$

$x^2 + 10x + 25 + x^2 - 10x + 25 = 52$

$2x^2 + 50 = 52$

$2x^2 = 2$

$x^2 = 1$

$x = \pm\sqrt{1}$

$x = \pm 1$

23. $(2x+3)^2 + (2x-3)^2 = 26$

$4x^2 + 12x + 9 + 4x^2 - 12x + 9 = 26$

$8x^2 + 18 = 26$

$8x^2 = 8$

$x^2 = 1$

$x = \pm\sqrt{1}$

$x = \pm 1$

25. $(3x+4)(3x-4) - (x+2)(x-2) = -4$

$9x^2 - 16 - (x^2 - 4) = -4$

$9x^2 - 16 - x^2 + 4 = -4$

$8x^2 - 12 = -4$

$8x^2 = 8$

$x^2 = 1$

$x = \pm\sqrt{1}$

$x = \pm 1$

27. $x^2 + 12x + \underline{36} = (x + \underline{6})^2$

29. $x^2 - 4x + \underline{4} = (x - \underline{2})^2$

31. $a^2 - 10a + \underline{25} = (a - \underline{5})^2$

33. $x^2 + 5x + \dfrac{25}{\underline{4}} = \left(x + \dfrac{5}{\underline{2}}\right)^2$

35. $y^2 - 7y + \dfrac{49}{\underline{4}} = \left(y - \dfrac{7}{\underline{2}}\right)^2$

37. $\quad x^2 + 4x = 12$

$\quad x^2 + 4x + 4 = 12 + 4$

$\quad (x+2)^2 = 16$

$\quad x + 2 = \pm\sqrt{16}$

$\quad x + 2 = \pm 4$

$\quad x = -2 \pm 4$

$\quad x = -2 + 4 \quad$ or $\quad x = -2 - 4$

$\quad x = 2 \qquad\qquad x = -6$

39. $\quad x^2 + 12x = -27$

$\quad x^2 + 12x + 36 = -27 + 36$

$\quad (x+6)^2 = 9$

$\quad x + 6 = \pm\sqrt{9}$

$\quad x + 6 = \pm 3$

$\quad x = -6 \pm 3$

$\quad x = -6 - 3 \quad$ or $\quad x = -6 + 3$

$\quad x = -9 \qquad\qquad x = -3$

41.

$\quad a^2 - 2a + 5 = 0$

$\quad a^2 - 2a = -5$

$\quad a^2 - 2a + 1 = -5 + 1$

$\quad (a-1)^2 = -4$

$\quad a - 1 = \pm\sqrt{-4}$

$\quad a - 1 = \pm 2i$

$\quad a = 1 \pm 2i$

43. $\quad y^2 - 8y + 1 = 0$

$\quad y^2 - 8y + 16 = -1 + 16$

$\quad (y-4)^2 = 15$

$\quad y - 4 = \pm\sqrt{15}$

$\quad y = 4 \pm \sqrt{15}$

45. $\quad x^2 - 5x - 3 = 0$

$\quad x^2 - 5x = 3$

$\quad x^2 - 5x + \dfrac{25}{4} = 3 + \dfrac{25}{4}$

$\quad \left(x - \dfrac{5}{2}\right)^2 = \dfrac{37}{4}$

$\quad x - \dfrac{5}{2} = \pm\sqrt{\dfrac{37}{4}}$

$\quad x - \dfrac{5}{2} = \pm\dfrac{\sqrt{37}}{2}$

$\quad x = \dfrac{5}{2} \pm \dfrac{\sqrt{37}}{2}$

$\quad x = \dfrac{5 \pm \sqrt{37}}{2}$

47. $\quad 2x^2 - 4x - 8 = 0$

$\quad x^2 - 2x - 4 = 0$

$\quad x^2 - 2x + 1 = 4 + 1$

$\quad (x-1)^2 = 5$

$\quad x - 1 = \pm\sqrt{5}$

$\quad x = 1 \pm \sqrt{5}$

49. $3t^2 - 8t + 1 = 0$

$$3t^2 - 8t = -1$$

$$t^2 - \frac{8}{3}t = -\frac{1}{3}$$

$$t^2 - \frac{8}{3}t + \frac{16}{9} = -\frac{1}{3} + \frac{16}{9}$$

$$\left(t - \frac{4}{3}\right)^2 = \frac{13}{9}$$

$$t - \frac{4}{3} = \pm\sqrt{\frac{13}{9}}$$

$$t - \frac{4}{3} = \pm\frac{\sqrt{13}}{3}$$

$$t = \frac{4}{3} \pm \frac{\sqrt{13}}{3}$$

$$t = \frac{4 \pm \sqrt{13}}{3}$$

51. $4x^2 - 3x + 5 = 0$

$$x^2 - \frac{3}{4}x = -\frac{5}{4}$$

$$x^2 - \frac{3}{4}x + \frac{9}{64} = -\frac{5}{4} + \frac{9}{64}$$

$$\left(x - \frac{3}{8}\right)^2 = \frac{-71}{64}$$

$$x - \frac{3}{8} = \pm\sqrt{\frac{-71}{64}}$$

$$x = \frac{3}{8} \pm \frac{i\sqrt{71}}{8}$$

$$x = \frac{3 \pm i\sqrt{71}}{8}$$

53. Let the shortest side $= \frac{1}{2}$ inch long, then the hypotenuse $= 1$ inch long.

$$a^2 + b^2 = c^2$$

$$x^2 + \left(\frac{1}{2}\right)^2 = 1^2$$

$$x^2 + \frac{1}{4} = 1$$

$$x^2 = \frac{3}{4}$$

$$x = \frac{\sqrt{3}}{2}$$

The other two sides are $\frac{\sqrt{3}}{2}$ inch and 1 inch.

55. $x = $ length of shortest side

$2x = $ length of hypotenuse

$A = $ length of the other side

$$(2x)^2 = x^2 + A^2$$

$$4x^2 = x^2 + A^2$$

$$4x^2 - x^2 = A^2$$

$$x\sqrt{3} = A$$

The length of the other two sides are

$2x$ and $x\sqrt{3}$.

57. The lengths of the shorter sides are 1 inch.

$$a^2 + b^2 = c^2$$

$$1^2 + 1^2 = x^2$$

$$2 = x^2$$

$$\sqrt{2} = x$$

The length of the hypotenuse is $\sqrt{2}$ inches.

59. Let $x = $ length of shorter sides

$$x^2 + x^2 = 1^2$$

$$2x^2 = 1$$

$$x^2 = \frac{1}{2}$$

$$x = \frac{\sqrt{2}}{2} \text{ inch}$$

61. $a^2 + b^2 = c^2$

$x^2 + x^2 = c^2$

$2x^2 = c^2$

$x\sqrt{2} = c$

63. Let x = horizontal distance

$790^2 = 120^2 + x^2$

$624,100 = 14,400 + x^2$

$609,700 = x^2$

$x = 781$ feet to the nearest foot

65. Let x = horizontal distance

$5750^2 = 120^2 + x^2$

$x^2 = 31,693,600$

$x = 56630$

$\dfrac{1170}{5630} = \dfrac{\text{(vertical rise)}}{\text{(horizontal distance)}}$

$= 0.21$ to the nearest hundredth

67. Let x = length of escalator, base = height = 20

$x^2 = 20^2 + 20^2$

$x^2 = 800$

$x = \sqrt{800} = 20\sqrt{2} = 28.3$ feet

69. $3456 = 3000(1+r)^2$

$\dfrac{3456}{3000} = (1+r)^2$

$\sqrt{\dfrac{3456}{3000}} = 1+r$

$1.073 = 1 + r$

$0.073 = r$

The annual interest rate is 7.3%

71. $AC^2 = 12^2 + 12^2$

$AC = 12\sqrt{2}$

$AD^2 = \left(12\sqrt{2}\right)^2 + \left(\dfrac{AD}{2}\right)^2$

$AD^2 = 288 + \dfrac{AD^2}{4}$

$\dfrac{3AD^2}{4} = 288$

$AD^2 = 384$

$AD = 8\sqrt{6} = 19.6$ inches

73. $\sqrt{45} = \sqrt{9 \cdot 5} = 3\sqrt{5}$

75. $\sqrt{27y^5} = \sqrt{9y^4 \cdot 3y} = 3y^2\sqrt{3y}$

77. $\sqrt[3]{54x^6 y^5} = \sqrt[3]{27x^6 y^3 \cdot 2y^2}$

$= 3x^2 y \sqrt[3]{2y^2}$

79. $a = 6,\ b = 7,\ c = -5$

$\sqrt{b^2 - 4ac} = \sqrt{7^2 - 4(6)(-5)}$

$= \sqrt{49 + 120}$

$= \sqrt{169}$

$= 13$

81. $\dfrac{3}{\sqrt{2}} = \dfrac{3}{\sqrt{2}} \cdot \dfrac{\sqrt{2}}{\sqrt{2}} = \dfrac{3\sqrt{2}}{2}$

83. $\dfrac{2}{\sqrt[3]{4}} = \dfrac{2}{\sqrt[3]{4}} \cdot \dfrac{\sqrt[3]{2}}{\sqrt[3]{2}} = \dfrac{2\sqrt[3]{2}}{\sqrt[3]{8}} = \dfrac{2\sqrt[3]{2}}{2} = \sqrt[3]{2}$

85.
$$(x+a)^2 + (x-a)^2 = 10a^2$$
$$x^2 + 2ax + a^2 + x^2 - 2ax + a^2 = 10a^2$$
$$2x^2 + 2a^2 = 10a^2$$
$$2x^2 - 8a^2 = 0$$
$$2(x^2 - 4a^2) = 0$$
$$x^2 - 4a^2 = 0$$
$$x^2 = 4a^2$$
$$x = \pm\sqrt{4a^2}$$
$$x = \pm 2a$$

87.
$$x^2 + 2ax = -a^2$$
$$x^2 + 2ax + \left(\frac{2a}{2}\right)^2 = -a^2 + \left(\frac{2a}{2}\right)^2$$
$$(x+a)^2 = -a^2 + a^2$$
$$(x+a)^2 = 0$$
$$x + a = 0$$
$$x = -a$$

89.
$$x^2 + 2ax = 0$$
$$x^2 + 2ax + \left(\frac{2a}{2}\right)^2 = \left(\frac{2a}{2}\right)^2$$
$$(x+a)^2 = a^2$$
$$x + a = \pm a$$
$$x = -a \pm a$$
$$x = -a + a \quad \text{or} \quad x = -a - a$$
$$x = 0 \qquad\qquad x = -2a$$

91.
$$x^2 + px + q = 0$$
$$x^2 + px = -q$$
$$x^2 + px + \left(\frac{p}{2}\right)^2 = -q + \left(\frac{p}{2}\right)^2$$
$$\left(x + \frac{p}{2}\right)^2 = -q + \frac{p^2}{4}$$
$$x + \frac{p}{2} = \pm\sqrt{\frac{-4q + p^2}{4}}$$
$$x = -\frac{p}{2} \pm \frac{\sqrt{p^2 - 4q}}{2}$$
$$x = \frac{-p \pm \sqrt{p^2 - 4q}}{2}$$

10.2 The Quadratic Formula

1.
$$x^2 + 5x + 6 = 0$$
$$(x+2)(x+3) = 0$$
$$x + 2 = 0 \quad \text{or} \quad x + 3 = 0$$
$$x = -2 \qquad\qquad x = -3$$

3. $a^2 - 4a + 1 = 0$
$$a = \frac{-(-4) \pm \sqrt{(-4)^2 - 4(1)(1)}}{2(1)}$$
$$= \frac{4 \pm \sqrt{16 - 4}}{2}$$
$$= \frac{4 \pm \sqrt{12}}{2}$$
$$= \frac{4 \pm 2\sqrt{3}}{2}$$
$$= 2 \pm \sqrt{3}$$

5.
$$\frac{1}{6}x^2 - \frac{1}{2}x + \frac{1}{3} = 0$$

$$6\left(\frac{1}{6}x^2\right) - 6\left(\frac{1}{2}x\right) + 6\left(\frac{1}{3}\right) = 6(0)$$

$$x^2 - 3x + 2 = 0$$

$$(x-1)(x-2) = 0$$

$$x - 1 = 0 \quad \text{or} \quad x - 2 = 0$$

$$x = 1 \qquad\qquad x = 2$$

7.
$$\frac{x^2}{2} + 1 = \frac{2x}{3}$$

$$3x^2 - 4x + 6 = 0$$

$$x = \frac{-(-4) \pm \sqrt{(-4)^2 - 4(3)(6)}}{2(3)}$$

$$= \frac{4 \pm \sqrt{16 - 72}}{6}$$

$$= \frac{4 \pm \sqrt{-56}}{6}$$

$$= \frac{4 \pm 2i\sqrt{14}}{6}$$

$$= \frac{2 \pm i\sqrt{14}}{3}$$

9.
$$y^2 - 5y = 0$$

$$y(y - 5) = 0$$

$$y = 0 \quad \text{or} \quad y - 5 = 0$$

$$y = 5$$

11.
$$30x^2 + 40x = 0$$

$$10x(3x + 4) = 0$$

$$10x = 0 \quad \text{or} \quad 3x + 4 = 0$$

$$x = 0 \qquad\qquad 3x = -4$$

$$x = -\frac{4}{3}$$

13.
$$\frac{2t^2}{3} - t = -\frac{1}{6}$$

$$6 \cdot \frac{2t^2}{3} - 6t = -\frac{1}{6} \cdot 6$$

$$4t^2 - 6t = -1$$

$$4t^2 - 6t + 1 = 0$$

$$t = \frac{-b \pm \sqrt{b^2 - 4ac}}{2a}$$

$$t = \frac{-(-6) \pm \sqrt{(-6)^2 - 4(4)(1)}}{2(4)}$$

$$= \frac{6 \pm \sqrt{36 - 16}}{8}$$

$$= \frac{6 \pm \sqrt{20}}{8}$$

$$= \frac{6 \pm 2\sqrt{5}}{8}$$

$$= \frac{2(3 \pm \sqrt{5})}{8}$$

$$= \frac{3 \pm \sqrt{5}}{4}$$

15.
$$0.01x^2 + 0.06x - 0.08 = 0$$

$$x^2 + 6x - 8 = 0$$

$$x = \frac{-6 \pm \sqrt{6^2 - 4(1)(-8)}}{2(1)}$$

$$= \frac{-6 \pm \sqrt{36 + 32}}{2}$$

$$= \frac{-6 \pm \sqrt{68}}{2}$$

$$= \frac{-6 \pm 2\sqrt{17}}{2}$$

$$= -3 \pm \sqrt{17}$$

17. $\quad 2x + 3 = -2x^2$

$2x^2 + 2x + 3 = 0$

$$x = \frac{-b \pm \sqrt{b^2 - 4ac}}{2a}$$

$$= \frac{-2 \pm \sqrt{4 - 4(2)(3)}}{2(2)}$$

$$= \frac{-2 \pm \sqrt{4 - 24}}{4}$$

$$= \frac{-2 \pm \sqrt{-20}}{4}$$

$$= \frac{-2 \pm 2i\sqrt{5}}{4}$$

$$= \frac{2\left(-1 \pm i\sqrt{5}\right)}{4}$$

$$= \frac{-1 \pm i\sqrt{5}}{2}$$

19. $100x^2 - 200x + 100 = 0$

$x^2 - 2x + 1 = 0$

$(x - 1)(x - 1) = 0$

$x - 1 = 0$

$x = 1$

21. $\quad \dfrac{1}{2}r^2 = \dfrac{1}{6}r - \dfrac{2}{3}$

$6\left(\dfrac{1}{2}r^2\right) = 6\left(\dfrac{1}{6}r\right) - 6\left(\dfrac{2}{3}\right)$

$3r^2 = r - 4$

$3r^2 - r + 4 = 0$

$$r = \frac{-b \pm \sqrt{b^2 - 4ac}}{2a}$$

$$= \frac{-(-1) \pm \sqrt{(-1)^2 - 4(3)(4)}}{2(3)}$$

$$= \frac{1 \pm \sqrt{1 - 48}}{6}$$

$$= \frac{1 \pm \sqrt{-47}}{6}$$

$$= \frac{1 \pm i\sqrt{47}}{6}$$

23. $\quad (x - 3)(x - 5) = 1$

$x^2 - 8x + 15 - 1 = 0$

$x^2 - 8x + 14 = 0$

$$x = \frac{-(-8) \pm \sqrt{(-8)^2 - 4(1)(14)}}{2(1)}$$

$$= \frac{8 \pm \sqrt{64 - 56}}{2}$$

$$= \frac{8 \pm \sqrt{8}}{2}$$

$$= \frac{8 \pm 2\sqrt{2}}{2}$$

$$= 4 \pm \sqrt{2}$$

25. $(x+3)^2 + (x-8)(x-1) = 16$

$x^2 + 6x + 9 + x^2 - 9x + 8 = 16$

$2x^2 - 3x + 1 = 0$

$(2x-1)(x-1) = 0$

$2x - 1 = 0 \quad \text{or} \quad x - 1 = 0$

$2x = 1 \qquad\qquad x = 1$

$x = \dfrac{1}{2}$

27. $\dfrac{x^2}{3} - \dfrac{5x}{6} = \dfrac{1}{2}$

$2x^2 - 5x - 3 = 0$

$(2x+1)(x-3) = 0$

$2x + 1 = 0 \quad \text{or} \quad x - 3 = 0$

$x = -\dfrac{1}{2} \qquad\qquad x = 3$

29.

$\dfrac{1}{x+1} - \dfrac{1}{x} = \dfrac{1}{2}$

$2x(x+1) \cdot \dfrac{1}{x+1} - 2x(x+1)\dfrac{1}{x} = \dfrac{1}{2} \cdot 2x(x+1)$

$2x - 2(x+1) = x(x+1)$

$2x - 2x - 2 = x^2 + x$

$-2 = x^2 + x$

$0 = x^2 + x + 2$

$x = \dfrac{-1 \pm \sqrt{1 - 4(1)(2)}}{2(1)}$

$= \dfrac{-1 \pm \sqrt{-7}}{2}$

$= \dfrac{-1 \pm i\sqrt{7}}{2}$

31. $\dfrac{1}{y-1} + \dfrac{1}{y+1} = 1$

$y + 1 + y - 1 = (y-1)(y+1)$

$2y = y^2 - 1$

$0 = y^2 - 2y - 1$

$y = \dfrac{-(-2) \pm \sqrt{(-2)^2 - 4(1)(-1)}}{2(1)}$

$= \dfrac{2 \pm \sqrt{4+4}}{2}$

$= \dfrac{2 \pm 2\sqrt{2}}{2}$

$= 1 \pm \sqrt{2}$

33.

$\dfrac{1}{x+2} + \dfrac{1}{x+3} = 1$

$(x+2)(x+3)\left(\dfrac{1}{x+2} + \dfrac{1}{x+3}\right) = 1(x+2)(x+3)$

$(x+2)(x+3)\dfrac{1}{x+2} + (x+2)(x+3)\dfrac{1}{x+3} = (x+2)(x+3)$

$x + 3 + x + 2 = x^2 + 5x + 6$

$2x + 5 = x^2 + 5x + 6$

$0 = x^2 + 3x + 1$

$x = \dfrac{-3 \pm \sqrt{(3)^2 - 4(1)(1)}}{2(1)} = \dfrac{-3 \pm \sqrt{5}}{2}$

35. $\dfrac{6}{r^2-1} - \dfrac{1}{2} = \dfrac{1}{r+1}$

$12 - (r^2 - 1) = 2(r-1)$

$12 - r^2 + 1 = 2r - 2$

$0 = r^2 + 2r - 15$

$0 = (r+5)(r-3)$

$r + 5 = 0 \quad \text{or} \quad r - 3 = 0$

$r = -5 \qquad\qquad r = 3$

37.
$$x^3 - 8 = 0$$
$$(x-2)(x^2 + 2x + 4) = 0$$
$$x - 2 = 0$$
$$x = 2$$
or
$$x^2 + 2x + 4 = 0$$
$$x = \frac{-2 \pm \sqrt{4 - 4(1)(4)}}{2(1)}$$
$$x = \frac{-2 \pm \sqrt{4 - 16}}{2}$$
$$= \frac{-2 \pm \sqrt{-12}}{2}$$
$$= \frac{-2 \pm 2i\sqrt{3}}{2}$$
$$= \frac{2(-1 \pm i\sqrt{3})}{2}$$
$$= -1 \pm i\sqrt{3}$$

The solutions are $2, \; -1 \pm i\sqrt{3}$

39.
$$8a^3 + 27 = 0$$
$$(2a + 3)(4a^2 - 6a + 9) = 0$$
$$2a + 3 = 0$$
$$a = -\frac{3}{2}$$
or
$$4a^2 - 6a + 9 = 0$$
$$a = \frac{-(-6) \pm \sqrt{(-6)^2 - 4(4)(9)}}{2(4)}$$
$$= \frac{6 \pm \sqrt{36 - 144}}{8}$$
$$= \frac{6 \pm \sqrt{-108}}{8}$$
$$= \frac{6 \pm 6i\sqrt{3}}{8}$$
$$= \frac{3 \pm 3i\sqrt{3}}{4}$$

The solutions are $-\frac{3}{2}, \; \frac{3 \pm 3i\sqrt{3}}{4}$

41.
$$125t^3 - 1 = 0$$
$$(5t - 1)(25t^2 + 5t + 1) = 0$$
$$5t - 1 = 0 \quad \text{or} \quad 25t^2 + 5t + 1 = 0$$

The first equation leads to a solution of $t = \frac{1}{5}$.

$$t = \frac{-5 \pm \sqrt{25 - 4(25)(1)}}{2(25)}$$
$$= \frac{-5 \pm \sqrt{25 - 100}}{50}$$
$$= \frac{-5 \pm \sqrt{-75}}{50}$$
$$= \frac{-5 \pm 5i\sqrt{3}}{50}$$
$$= \frac{5(-1 \pm i\sqrt{3})}{50}$$
$$= \frac{-1 \pm i\sqrt{3}}{10}$$

The solutions are $\frac{1}{5}, \; \frac{-1 \pm i\sqrt{3}}{10}$

43.
$$2x^3 + 2x^2 + 3x = 0$$
$$x(2x^2 + 2x + 3) = 0$$
$$x = 0 \quad \text{or} \quad 2x^2 + 2x + 3 = 0$$
$$x = \frac{-2 \pm \sqrt{2^2 - 4(2)(3)}}{2(2)}$$
$$= \frac{-2 \pm \sqrt{4 - 24}}{4}$$
$$= \frac{-2 \pm \sqrt{-20}}{4}$$
$$= \frac{-2 \pm 2i\sqrt{5}}{4}$$
$$= \frac{-1 \pm i\sqrt{5}}{2}$$

The solutions are $0, \; \frac{-1 \pm i\sqrt{5}}{2}$

45.
$$3y^4 = 6y^3 - 6y^2$$
$$3y^4 - 6y^3 + 6y^2 = 0$$
$$3y^2\left(y^2 - 2y + 2\right) = 0$$
$$3y^2 = 0 \quad \text{or} \quad y^2 - 2y + 2 = 0$$

The first equation leads to a solution of $y = 0$.

$$y = \frac{-(-2) \pm \sqrt{4 - 4(1)(2)}}{2(1)}$$

$$= \frac{2 \pm \sqrt{4 - 8}}{2}$$

$$= \frac{2 \pm \sqrt{-4}}{2}$$

$$= \frac{2 \pm 2i}{2}$$

$$= \frac{2(1 \pm i)}{2}$$

$$= 1 \pm i$$

The solutions are $0, \ 1 \pm i$

47.
$$6t^5 + 4t^4 = -2t^3$$
$$6t^5 + 4t^4 + 2t^3 = 0$$
$$2t^3\left(3t^2 + 2t + 1\right) = 0$$
$$2t^3 = 0$$
$$t^3 = 0$$
$$t = 0$$

or

$$3t^2 + 2t + 1 = 0$$

$$t = \frac{-2 \pm \sqrt{2^2 - 4(3)(1)}}{2(3)}$$

$$= \frac{-2 \pm \sqrt{4 - 12}}{6}$$

$$= \frac{-2 \pm \sqrt{-8}}{6}$$

$$= \frac{-2 \pm 2i\sqrt{2}}{6}$$

$$= \frac{-1 \pm i\sqrt{2}}{3}$$

The solutions are $0, \ \dfrac{-1 \pm i\sqrt{2}}{3}$

49. $\dfrac{-3 - 2i}{5}$

51. $s(t) = 5t + 16t^2$
$$74 = 5t + 16t^2$$
$$0 = 16t^2 + 5t - 74$$
$$0 = (t - 2)(16t + 37)$$
$$t - 2 = 0 \quad \text{or} \quad 16t + 37 = 0$$
$$t = 2 \text{ sec} \qquad t = -\frac{37}{16}\left(\text{not a solution}\right)$$

53.
$$h(t) = 20t - 16t^2, \quad h(t) = 4$$
$$4 = 20t - 16t^2$$
$$16t^2 - 20t + 4 = 0$$
$$4t^2 - 5t + 1 = 0$$
$$(4t - 1)(t - 1) = 0$$
$$4t - 1 = 0 \quad \text{or} \quad t - 1 = 0$$
$$4t = 1 \qquad\qquad t = 1$$
$$t = \frac{1}{4}$$

This occurs at $\frac{1}{4}$ second and 1 second

55.
$$P(x) = R(x) - C(x)$$
$$P(x) = \$300, \quad R(x) = 100x - 0.5x^2, \quad C(x) = 60x + 300$$
$$300 = 100x - 0.5x^2 - (60x + 300)$$
$$300 = 100x - 0.5x^2 - 60x - 300$$
$$0 = 0.5x^2 - 40x + 600$$
$$x = \frac{-(-40) \pm \sqrt{(-40)^2 - 4(0.5)(600)}}{2(0.5)}$$
$$= \frac{40 \pm \sqrt{1600 - 1200}}{1}$$
$$= 40 \pm \sqrt{400}$$
$$= 40 \pm 20$$
$$x = 20 \text{ items} \quad \text{or} \quad x = 60 \text{ items}$$

57.
$$P(x) = R(x) - C(x)$$
$$P(x) = 700, \quad R(x) = 10x - 0.002x^2 \text{ and } C(x) = 800 + 6.5x$$
$$700 = 10x - 0.002x^2 - (800 + 6.5x)$$
$$700 = 10x - 0.002x^2 - 800 - 6.5x$$
$$700 = 3.5x - 0.002x^2 - 800$$
$$0.002x^2 - 3.5x + 1500 = 0$$
$$x = \frac{-(-3.5) \pm \sqrt{12.25 - 4(0.002)(1500)}}{2(0.002)}$$
$$= \frac{3.5 \pm 0.5}{0.004}$$
$$x = 1000 \qquad \text{or} \qquad x = 750$$

The monthly profit will be $700 if they produce and sell 1,000 patterns or produce and sell 750 patterns.

59. Let x = width of the strip
$$A_{\text{new}} = 0.8 A_{\text{old}}$$
$$(10.5 - 2x)(8.2 - 2x) = 0.8(10.5)(8.2)$$
$$86.1 - 37.4x + 4x^2 = 68.88$$
$$4x^2 - 37.4x + 17.22 = 0$$
$$x = \frac{-(-37.4) \pm \sqrt{(-37.4)^2 - 4(4)(17.22)}}{2(4)}$$
$$= 8.86 \, (\text{impossible}) \quad \text{or} \quad 0.49 \text{ cm (width of strip)}$$

61. Perimeter=20 yards   Area=15 sq. yards

(a) $P = 2l + 2w$ $A = lw$

$2l + 2w = 20$ (1) $15 = lw$ (2)

(b) $2\left(\dfrac{15}{w}\right) + 2w = 20$ (1), $l = \dfrac{15}{w}$

$30 + 2w^2 = 20w$

$2w^2 - 20w + 30 = 0$

$w^2 - 10w + 15 = 0$

$w = \dfrac{-(-10) \pm \sqrt{(-10)^2 - 4(1)(15)}}{2(1)}$

$= \dfrac{10 \pm \sqrt{40}}{2}$

$w = 8.16$ or 1.84, $l = \dfrac{15}{8.16} = 1.84$ or $l = \dfrac{15}{1.84} = 8.16$

The dimensions are 1.84 yards by 8.16 yards

(c) There are two answers because the two variables are interchangeable in the two original equations.

63.

$$\begin{array}{r} 4y + 1 + \dfrac{-2}{2y-7} \\ 2y-7\overline{)\ 8y^2 - 26y - 9} \\ \underline{-(8y^2 - 28y)} \\ 2y - 9 \\ \underline{-(2y-7)} \\ -2 \end{array}$$

65.

$$\begin{array}{r} x^2 + 7x + 12 \\ x+2\overline{)\ x^3 + 9x^2 + 26x + 24} \\ \underline{x^3 + 2x^2} \\ 7x^2 + 26x \\ \underline{7x^2 + 14x} \\ 12x + 24 \\ \underline{12x + 24} \\ 0 \end{array}$$

67. $25^{1/2} = 5$

69. $\left(\dfrac{9}{25}\right)^{3/2} = \left[\left(\dfrac{9}{25}\right)^{1/2}\right]^3 = \left(\dfrac{3}{5}\right)^3 = \dfrac{27}{125}$

71. $8^{-2/3} = \dfrac{1}{8^{2/3}} = \dfrac{1}{\left[(8)^{1/3}\right]^2} = \dfrac{1}{2^2} = \dfrac{1}{4}$

73. $\dfrac{\left(49x^8 y^{-4}\right)^{1/2}}{\left(27x^{-3} y^9\right)^{-1/3}} = \dfrac{(49)^{1/2}\left(x^8\right)^{1/2}\left(y^{-4}\right)^{1/2}}{(27)^{-1/3}\left(x^{-3}\right)^{-1/3}\left(y^9\right)^{-1/3}}$

$= \dfrac{7x^4 y^{-2}}{(3)^{-1} xy^{-3}}$

$= 7 \cdot 3x^{4-1} y^{-2-(-3)}$

$= 21x^3 y$

75. $x^2 + \sqrt{3}x - 6 = 0$

$$x = \frac{-\sqrt{3} \pm \sqrt{\left(\sqrt{3}\right)^2 - 4(1)(-6)}}{2(1)}$$

$$= \frac{-\sqrt{3} \pm \sqrt{3 + 24}}{2}$$

$$= \frac{-\sqrt{3} \pm 3\sqrt{3}}{2}$$

$$x = \frac{2\sqrt{3}}{2} \quad \text{or} \quad x = \frac{-4\sqrt{3}}{2}$$

$$x = \sqrt{3} \qquad\qquad x = -2\sqrt{3}$$

77. $\sqrt{2}x^2 + 2x - \sqrt{2} = 0$

$$x = \frac{-2 \pm \sqrt{2^2 - 4\left(\sqrt{2}\right)\left(-\sqrt{2}\right)}}{2\left(\sqrt{2}\right)}$$

$$= \frac{-2 \pm \sqrt{12}}{2\sqrt{2}}$$

$$= \frac{-2 \pm 2\sqrt{3}}{2\sqrt{2}}$$

$$= \frac{2\left(-1 \pm \sqrt{3}\right)}{2\sqrt{2}}$$

$$= \frac{-1 \pm \sqrt{3}}{\sqrt{2}} \cdot \frac{\sqrt{2}}{\sqrt{2}}$$

$$= \frac{-\sqrt{2} \pm \sqrt{6}}{2}$$

79. $x^2 + ix + 2 = 0$

$$x = \frac{-i \pm \sqrt{i^2 - 4(1)(2)}}{2(1)}$$

$$= \frac{-i \pm \sqrt{-1 - 8}}{2}$$

$$= \frac{-i \pm 3i}{2}$$

$$x = -2i \quad \text{or} \quad x = i$$

81. $ix^2 + 3x + 4i = 0$

$$x = \frac{-3 \pm \sqrt{3^2 - 4(i)(4i)}}{2i}$$

$$= \frac{-3 \pm \sqrt{9 - 16i^2}}{2i}$$

$$= \frac{-3 \pm \sqrt{9 + 16}}{2i}$$

$$= \frac{-3 \pm \sqrt{25}}{2i}$$

$$= \frac{-3 \pm 5}{2i}$$

$$x = \frac{2}{2i} \quad \text{or} \quad x = \frac{-8}{2i}$$

$$x = \frac{1}{i} \cdot \frac{i}{i} \qquad\qquad x = \frac{-4}{i} \cdot \frac{i}{i}$$

$$x = -i \qquad\qquad\qquad x = 4i$$

10.3 Additional Items Involving Solutions to Equations

1. $x^2 - 6x + 5 = 0$
$a = 1,\ b = -6,\ c = 5$
$(-6)^2 - 4(1)(5) = 36 - 20 = 16$
The discriminant is a perfect square.
Therefore, the equation has two rational solutions.

3. $4x^2 - 4x = -1$
$4x^2 - 4x + 1 = 0$
$a = 4,\ b = -4,\ c = 1$
$(-4)^2 - 4(4)(1) = 16 - 16 = 0$
The discriminant is zero. Therefore, the
equation has one rational solution.

5. $x^2 + x - 1 = 0$
$a = 1,\ b = 1,\ c = -1$
$1^2 - 4(1)(-1) = 1 + 4 = 5$
The discriminant is a positive number, but not a perfect square.
Therefore, the equation will have two irrational solutions.

7. $2y^2 = 3y + 1$
$2y^2 - 3y - 1 = 0$
$a = 2,\ b = -3,\ c = -1$
$(-3)^2 - 4(2)(-1) = 9 + 8 = 17$
The discriminant is a positive number, but not a
perfect square. Therefore, the equation will have
two irrational solutions.

9. $x^2 - 9 = 0$
$a = 1,\ b = 0,\ c = -9$
$0^2 - 4(1)(-9) = 0 + 36 = 36$
The discriminant is a perfect square. Therefore,
the equation has two rational solutions.

11. $5a^2 - 4a = 5$
$5a^2 - 4a - 5 = 0$
$a = 5,\ b = -4,\ c = -5$
$(-4)^2 - 4(5)(-5) = 116$
The discriminant is a positive number that is
not a perfect square. Therefore, the equation
will have two irrational solutions.

13. $x^2 - kx + 25 = 0$
$a = 1,\ b = -k,\ c = 25$
$(-k)^2 - 4(1)(25) = k^2 - 100 \ \text{ must} = 0$
$\quad k^2 - 100 = 0$
$\quad\quad k^2 = 100$
$\quad\quad\ k = \pm 10$

15. $x^2 = kx - 36$
$x^2 - kx + 36 = 0$
$a = 1,\ b = -k,\ c = 36$
$(-k)^2 - 4(1)(36) = k^2 - 144 \ \text{ must} = 0$
$\quad k^2 - 144 = 0$
$\quad\quad k^2 = 144$
$\quad\quad\ k = \pm 12$

17. $4x^2 - 12x + k = 0$

$a = 4,\ b = -12,\ c = k$

$(-12)^2 - 4(4)(k) = 144 - 16k$ must $= 0$

$144 - 16k = 0$

$-16k = -144$

$k = 9$

19. $kx^2 - 40x = 25$

$kx^2 - 40x - 25 = 0$

$a = k,\ b = -40,\ c = -25$

$(-40)^2 - 4(k)(-25) = 1600 + 100k$ must $= 0$

$1600 + 100k = 0$

$1600 = -100k$

$-16 = k$

21. $3x^2 - kx + 2 = 0$

$a = 3,\ b = -k,\ c = 2$

$(-k)^2 - 4(3)(2) = k^2 - 24$ must $= 0$

$k^2 - 24 = 0$

$k^2 = 24$

$k = \pm\sqrt{24}$

$k = \pm 2\sqrt{6}$

23. $x = 5 \qquad x = 2$

$x - 5 = 0 \qquad x - 2 = 0$

$(x-5)(x-2) = 0$

$x^2 - 7x + 10 = 0$

25. $t = -3 \qquad t = 6$

$t + 3 = 0 \qquad t - 6 = 0$

$(t+3)(t-6) = 0$

$t^2 - 3t - 18 = 0$

27. $y = 2 \qquad y = -2 \qquad y = 4$

$y - 2 = 0 \qquad y + 2 = 0 \qquad y - 4 = 0$

$(y-2)(y+2)(y-4) = 0$

$y^3 - 4y^2 - 4y + 16 = 0$

29. $x = \dfrac{1}{2} \qquad x = 3$

$2x - 1 = 0 \qquad x - 3 = 0$

$(2x-1)(x-3) = 0$

$2x^2 - 7x + 3 = 0$

31. $t = -\dfrac{3}{4} \qquad t = 3$

$4t + 3 = 0 \qquad t - 3 = 0$

$(4t+3)(t-3) = 0$

$4t^2 - 9t - 9 = 0$

33. $x = 3 \qquad x = -3 \qquad x = \dfrac{5}{6}$

$x - 3 = 0 \qquad x + 3 = 0 \qquad 6x - 5 = 0$

$(x-3)(x+3)(6x-5) = 0$

$(x^2 - 9)(6x-5) = 0$

$6x^3 - 5x^2 - 54x + 45 = 0$

35. $a = -\dfrac{1}{2} \qquad a = \dfrac{3}{5}$

$2a + 1 = 0 \qquad 5a - 3 = 0$

$(2a+1)(5a-3) = 0$

$10a^2 - a - 3 = 0$

37. $\quad x = -\dfrac{2}{3} \qquad x = \dfrac{2}{3} \qquad x = 1$

$\quad 3x + 2 = 0 \qquad 3x - 2 = 0 \qquad x - 1 = 0$

$\qquad\qquad (3x+2)(3x-2)(x-1) = 0$

$\qquad\qquad\qquad (9x^2 - 4)(x-1) = 0$

$\qquad\qquad\qquad 9x^3 - 9x^2 - 4x + 4 = 0$

39. $\quad x = 2 \qquad x = -2 \qquad x = 3 \qquad x = -3$

$\quad x - 2 = 0 \quad x + 2 = 0 \quad x - 3 = 0 \quad x + 3 = 0$

$\qquad\qquad (x-2)(x+2)(x-3)(x+3) = 0$

$\qquad\qquad\qquad x^4 - 13x^2 + 36 = 0$

41. $\quad x = 3 \qquad\qquad x = -5$

$\quad x - 3 = 0 \qquad\qquad x + 5 = 0$

$\qquad\qquad (x-3)(x+5)^2 = 0$

$\qquad\qquad (x-3)(x^2 + 10x + 25) = 0$

$\qquad\qquad x^3 + 7x^2 - 5x - 75 = 0$

43. $\quad x = 3 \qquad x = 3 \qquad x = -3 \qquad x = -3$

$\quad x - 3 = 0 \quad x - 3 = 0 \quad x + 3 = 0 \quad x + 3 = 0$

$\qquad\qquad (x-3)(x-3)(x+3)(x+3) = 0$

$\qquad\qquad\qquad x^4 - 18x^2 + 81 = 0$

45.

$$\begin{array}{r}
x^2 + 3x + 2 \\
x+3\overline{)\,x^3 + 6x^2 + 11x + 6} \\
\underline{x^3 + 3x^2} \\
3x^2 + 11x \\
\underline{3x^2 + 9x} \\
2x + 6 \\
\underline{2x + 6} \\
0
\end{array}$$

$x^3 + 6x^2 + 11x + 6 = 0$

$(x+3)(x^2 + 3x + 2) = 0$

$(x+3)(x+1)(x+2) = 0$

$x + 3 = 0 \quad \text{or} \quad x + 1 = 0 \quad \text{or} \quad x + 2 = 0$

$\quad x = -3 \qquad\qquad x = -1 \qquad\qquad x = -2$

Solutions are $x = -3,\ -2,\ -1$.

47.

$$\begin{array}{r}
y^2 + 2y - 8 \\
y+3\overline{)\,y^3 + 5y^2 - 2y - 24} \\
\underline{-\left(y^3 + 3y^2\right)} \\
2y^2 - 2y \\
\underline{-\left(2y^2 + 6y\right)} \\
-8y - 24 \\
\underline{-(-8y - 24)} \\
0
\end{array}$$

$y^2 + 2y - 8 = 0$

$(y+4)(y-2) = 0$

$y + 4 = 0 \quad \text{or} \quad y - 2 = 0$

$\quad y = -4 \qquad\qquad y = 2$

Solutions are $y = -4,\ -3,\ 2$

49.
$$\begin{array}{r} x^2 - 2x + 2 \\ x-3 \overline{\smash{)}\ x^3 - 5x^2 + 8x - 6} \end{array}$$
$$\underline{x^3 - 3x^2}$$
$$-2x^2 + 8x$$
$$\underline{-2x^2 + 6x}$$
$$2x - 6$$
$$\underline{2x - 6}$$
$$0$$

$$x = \frac{-(-2) \pm \sqrt{(-2)^2 - 4(1)(2)}}{2(1)}$$

$$= \frac{2 \pm \sqrt{-4}}{2}$$

$$= \frac{2 \pm 2i}{2}$$

$$= 1 \pm i$$

51. $t^3 = 13t^2 - 65t + 125$

$t^3 - 13t^2 + 65t - 125 = 0$

$$\begin{array}{r} t^2 - 8t + 25 \\ t-5 \overline{\smash{)}\ t^3 - 13t^2 + 65t - 125} \end{array}$$
$$\underline{-\left(t^3 - 5t^2\right)}$$
$$-8t^2 + 65t$$
$$\underline{-\left(-8t^2 + 40t\right)}$$
$$25t - 125$$
$$\underline{-\left(25t - 125\right)}$$
$$0$$

$t^2 - 8t + 25 = 0$

$$t = \frac{-(-8) \pm \sqrt{(-8)^2 - 4(1)(25)}}{2(1)}$$

$$= \frac{8 \pm \sqrt{-36}}{2}$$

$$= \frac{8 \pm 6i}{2}$$

$$= 4 \pm 3i$$

Solutions are $t = 5,\ 4 \pm 3i$

53. $a^4\left(a^{3/2} - a^{1/2}\right) = a^4 a^{3/2} - a^4 a^{1/2}$

$$= a^{4+3/2} - a^{4+1/2}$$

$$= a^{11/2} - a^{9/2}$$

55. $\left(x^{3/2} - 3\right)^2 = \left(x^{3/2}\right)^2 - 2\left(x^{3/2}\right)(3) + (-3)^2$

$$= x^3 - 6x^{3/2} + 9$$

57. $\dfrac{30x^{3/4} - 25x^{5/4}}{5x^{1/4}} = \dfrac{30x^{3/4}}{5x^{1/4}} - \dfrac{25x^{5/4}}{5x^{1/4}}$

$$= 6x^{2/4} - 5x^{4/4}$$

$$= 6x^{1/2} - 5x$$

59. $\dfrac{10(x-3)^{3/2} - 15(x-3)^{1/2}}{5(x-3)^{1/2}}$

$$= \frac{10(x-3)^{3/2}}{5(x-3)^{1/2}} - \frac{15(x-3)^{1/2}}{5(x-3)^{1/2}}$$

$$= 2(x-3) - 3$$

$$= 2x - 9$$

Factorization $= 5(x-3)^{1/2}(2x-9)$

61. $2x^{2/3} - 11x^{1/3} + 12 = \left(2x^{1/3} - 3\right)\left(x^{1/3} - 4\right)$

63.
$$x+2 \overline{)\begin{array}{r} x^3 - x^2 + x - 1 \\ x^4 + x^3 - x^2 + x - 2 \end{array}}$$

$$\underline{-\left(x^4 + 2x^3\right)}$$
$$-x^3 - x^2$$
$$\underline{-\left(-x^3 - 2x^2\right)}$$
$$x^2 + x$$
$$\underline{-\left(x^2 + 2x\right)}$$
$$-x - 2$$
$$\underline{-\left(-x - 2\right)}$$
$$0$$

$$x^3 - x^2 + x - 1 = 0$$
$$x^2\left(x - 1\right) + x - 1 = 0$$
$$\left(x - 1\right)\left(x^2 + 1\right) = 0$$
$$x - 1 = 0 \quad \text{or} \quad x^2 + 1 = 0$$
$$x = 1 \qquad\qquad x^2 = -1$$
$$x = \pm i$$

Solutions are $x = -2, 1, \pm i$

65.
$$x + a \overline{)\begin{array}{r} x^2 + 2ax + a^2 \\ x^3 + 3ax^2 + 3a^2 x + a^3 \end{array}}$$
$$\underline{x^3 + ax^2}$$
$$2ax^2 + 3a^2 x$$
$$\underline{2ax^2 + 2a^2 x}$$
$$a^2 x + a^3$$
$$\underline{a^2 x + a^3}$$
$$0$$

$$x^3 + 3ax^2 + 3a^2 x + a^3 = 0$$
$$\left(x + a\right)\left(x^2 + 2ax + a^2\right) = 0$$
$$\left(x + a\right)\left(x + a\right)\left(x + a\right) = 0$$
$$x + a = 0$$
$$x = -a$$

$x = -a$ is a solution of multiplicity 3.

67. $x^2 = 4x + 5$

$Y_1 = X^2, \quad Y_2 = 4X + 5$

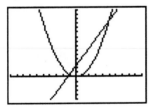

$$x^2 - 4x - 5 = 0$$
$$\left(x - 5\right)\left(x + 1\right) = 0$$
$$x - 5 = 0 \quad \text{or} \quad x + 1 = 0$$
$$x = 5 \qquad\qquad x = -1$$

Solutions are $x = -1, 5$

69. $x^2 - 1 = 2x$

$Y_1 = X^2 - 1, \ Y_2 = 2X$

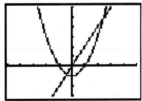

$$x^2 - 2x - 1 = 0$$
$$x = \frac{-(-2) \pm \sqrt{(-2)^2 - 4(1)(-1)}}{2(1)}$$
$$= \frac{2 \pm \sqrt{8}}{2}$$
$$= \frac{2 \pm 2\sqrt{2}}{2}$$
$$= 1 \pm \sqrt{2}$$

Solutions are

$x = 1 + \sqrt{2} \approx 2.41, \quad x = 1 - \sqrt{2} \approx -0.41$

71. $2x^3 - x^2 - 2x + 1 = 0$

$x^2(2x-1) - (2x-1) = 0$

$(2x-1)(x^2-1) = 0$

$2x - 1 = 0$ or $x^2 - 1 = 0$

$x = \dfrac{1}{2}$ $x = \pm 1$

Solutions are $x = -1, \ 0.5, \ 1$

73. $2x^3 + 2 = x^2 + 4x$

$2x^3 - x^2 - 4x + 2 = 0$

$x^2(2x-1) - 2(2x-1) = 0$

$(x^2-2)(2x-1) = 0$

$x^2 - 2 = 0$ or $2x - 1 = 0$

$x^2 = 2$ $2x = 1$

$x = \pm\sqrt{2}$ $x = \dfrac{1}{2}$

$x = \sqrt{2} \approx 1.41, \ x = -\sqrt{2} \approx -1.41, \ x = \dfrac{1}{2}$

75. $3x^3 - 8x^2 + 10x - 4 = 0$

Real solution is $x = \dfrac{2}{3}$

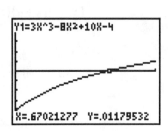

$3x-2 \overline{\big)\ 3x^3 - 8x^2 + 10x - 4}$, quotient $x^2 - 2x + 2$

$\underline{-(3x^3 - 2x^2)}$

$-6x^2 + 10x$

$\underline{-(-6x^2 + 4x)}$

$6x - 4$

$\underline{-(6x - 4)}$

0

$x^2 - 2x + 2 = 0$

$x = \dfrac{-(-2) \pm \sqrt{(-2)^2 - 4(1)(2)}}{2(1)}$

$= \dfrac{2 \pm \sqrt{-4}}{2}$

$= \dfrac{2 \pm 2i}{2}$

$= 1 \pm i$

Solutions are $x = 2/3, \ 1 \pm i$

10.4 Equations Quadratic in Form

1. $(x-3)^2 + 3(x-3) + 2 = 0, \quad y = x - 3$

$y^2 + 3y + 2 = 0$

$(y+1)(y+2) = 0$

$y + 1 = 0$ or $y + 2 = 0$

$y = -1$ $y = -2$

$x - 3 = -1$ or $x - 3 = -2$

$x = 2$ $x = 1$

3. $2(x+4)^2 + 5(x+4) - 12 = 0$

$2x^2 + 16x + 32 + 5x + 20 - 12 = 0$

$2x^2 + 21x + 40 = 0$

$(2x+5)(x+8) = 0$

$2x + 5 = 0$ or $x + 8 = 0$

$x = -\dfrac{5}{2}$ $x = -8$

5. $x^4 - 6x^2 - 27 = 0, \quad y = x^2$

$\quad y^2 - 6y - 27 = 0$

$\quad (y+3)(y-9) = 0$

$\quad y+3 = 0 \quad \text{or} \quad y-9 = 0$

$\qquad y = -3 \qquad\qquad y = 9$

$\quad x^2 = -3 \quad \text{or} \quad x^2 = 9$

$\quad x = \pm\sqrt{-3} \qquad x = \pm\sqrt{9}$

$\quad x = \pm i\sqrt{3} \qquad x = \pm 3$

7. $\qquad x^4 + 9x^2 = -20$

$\qquad x^4 + 9x^2 + 20 = 0$

$\qquad (x^2 + 5)(x^2 + 4) = 0$

$\qquad x^2 + 5 = 0 \quad \text{or} \quad x^2 + 4 = 0$

$\qquad\qquad x^2 = -5 \qquad\qquad x^2 = -4$

$\qquad\qquad x = \pm i\sqrt{5} \qquad\qquad x = \pm 2i$

9. $(2a-3)^2 - 9(2a-3) = -20, \quad x = 2a-3$

$\qquad x^2 - 9x = -20$

$\quad x^2 - 9x + 20 = 0$

$\quad (x-4)(x-5) = 0$

$\qquad x-4 = 0 \quad \text{or} \quad x-5 = 0$

$\qquad\qquad x = 4 \qquad\qquad x = 5$

$\qquad 2a-3 = 4 \quad \text{or} \quad 2a-3 = 5$

$\qquad\qquad 2a = 7 \qquad\qquad 2a = 8$

$\qquad\qquad a = \dfrac{7}{2} \qquad\qquad a = 4$

11. $\qquad 2(4a+2)^2 = 3(4a+2) + 20$

$\qquad 32a^2 + 32a + 8 = 12a + 6 + 20$

$\qquad 32a^2 + 20a - 18 = 0$

$\qquad 2(16a^2 + 10a - 9) = 0$

$\qquad (8a+9)(2a-1) = 0$

$\qquad\qquad 8a+9 = 0 \quad \text{or} \quad 2a-1 = 0$

$\qquad\qquad a = -\dfrac{9}{8} \qquad\qquad a = \dfrac{1}{2}$

13. $\qquad 6t^4 = -t^2 + 5, \quad x = t^2$

$\qquad 6x^2 = -x + 5$

$\quad 6x^2 + x - 5 = 0$

$\quad (x+1)(6x-5) = 0$

$\qquad 6x-5 = 0 \quad \text{or} \quad x+1 = 0$

$\qquad\qquad 6x = 5 \qquad\qquad x = -1$

$\qquad\qquad x = \dfrac{5}{6}$

$\qquad t^2 = \dfrac{5}{6} \quad \text{or} \quad t^2 = -1$

$\qquad t = \pm\sqrt{\dfrac{5}{6}} \qquad t = \pm\sqrt{-1}$

$\qquad = \pm\dfrac{\sqrt{30}}{6} \qquad = \pm i$

15. $\qquad 9x^4 - 49 = 0, \quad t = x^2$

$\qquad 9t^2 - 49 = 0$

$\quad (3t+7)(3t-7) = 0$

$\qquad 3t+7 = 0 \quad \text{or} \quad 3t-7 = 0$

$\qquad\qquad t = -\dfrac{7}{3} \qquad\qquad t = \dfrac{7}{3}$

$\qquad\qquad x^2 = -\dfrac{7}{3} \quad \text{or} \quad x^2 = \dfrac{7}{3}$

$\qquad\qquad x = \pm i\sqrt{\dfrac{7}{3}} \qquad x = \pm\sqrt{\dfrac{7}{3}}$

$\qquad\qquad x = \pm\dfrac{i\sqrt{21}}{3} \qquad x = \pm\dfrac{\sqrt{21}}{3}$

17. $x - 7\sqrt{x} + 10 = 0, \quad y = \sqrt{x}$

$y^2 - 7y + 10 = 0$

$(y - 2)(y - 5) = 0$

$y - 2 = 0 \quad \text{or} \quad y - 5 = 0$

$y = 2 \qquad\qquad y = 5$

$\sqrt{x} = 2 \quad \text{or} \quad \sqrt{x} = 5$

$x = 4 \qquad\qquad x = 25$

19. $t - 2\sqrt{t} - 15 = 0, \quad x = \sqrt{t}$

$x^2 - 2x - 15 = 0$

$(x + 3)(x - 5) = 0$

$x + 3 = 0 \quad \text{or} \quad x - 5 = 0$

$x = -3 \qquad\qquad x = 5$

$\sqrt{t} = -3 \quad \text{or} \quad \sqrt{t} = 5$

$t = 9 \qquad\qquad t = 25$

Possible solutions 9 and 25, only 25 checks.

21. $\qquad 6x + 11\sqrt{x} = 35$

$6x + 11\sqrt{x} - 35 = 0, \quad y = \sqrt{x}$

$6y^2 + 11y - 35 = 0$

$(2y + 7)(3y - 5) = 0$

$2y + 7 = 0 \quad \text{or} \quad 3y - 5 = 0$

$2y = -7 \qquad\qquad 3y = 5$

$y = -\dfrac{7}{2} \qquad\qquad y = \dfrac{5}{3}$

$\sqrt{x} = -\dfrac{7}{2} \quad \text{or} \quad \sqrt{x} = \dfrac{5}{3}$

$x = \dfrac{49}{4} \qquad\qquad x = \dfrac{25}{9}$

Possible solutions $\dfrac{49}{4}, \dfrac{25}{9}$, only $\dfrac{25}{9}$ checks.

23. $(a - 2) - 11\sqrt{a - 2} + 30 = 0, \quad x = \sqrt{a - 2}$

$x^2 - 11x + 30 = 0$

$(x - 5)(x - 6) = 0$

$x - 5 = 0 \quad \text{or} \quad x - 6 = 0$

$x = 5 \qquad\qquad x = 6$

$\sqrt{a - 2} = 5 \quad \text{or} \quad \sqrt{a - 2} = 6$

$a - 2 = 25 \qquad\qquad a - 2 = 36$

$a = 27 \qquad\qquad a = 38$

25. $(2x + 1) - 8\sqrt{2x + 1} + 15 = 0, \quad y = \sqrt{2x + 1}$

$y^2 - 8y + 15 = 0$

$(y - 3)(y - 5) = 0$

$y - 3 = 0 \quad \text{or} \quad y - 5 = 0$

$y = 3 \qquad\qquad y = 5$

$\sqrt{2x + 1} = 3 \quad \text{or} \quad \sqrt{2x + 1} = 5$

$2x + 1 = 9 \qquad\qquad 2x + 1 = 25$

$2x = 8 \qquad\qquad 2x = 24$

$x = 4 \qquad\qquad x = 12$

27. $16t^2 - vt - h = 0$

$t = \dfrac{-(-v) \pm \sqrt{(-v)^2 - 4(16)(-h)}}{2(16)}$

$= \dfrac{v \pm \sqrt{v^2 + 64h}}{32}$

29. $kx^2 + 8x + 4 = 0$

$$x = \frac{-8 \pm \sqrt{8^2 - 4(k)(4)}}{2(k)}$$

$$= \frac{-8 \pm \sqrt{64 - 16k}}{2k}$$

$$= \frac{-8 \pm \sqrt{16(4-k)}}{2k}$$

$$= \frac{-8 \pm 4\sqrt{4-k}}{2k}$$

$$= \frac{-4 \pm 2\sqrt{4-k}}{k}$$

31. $x^2 + 2xy + y^2 = 0$

$$x = \frac{-2y \pm \sqrt{(2y)^2 - 4(1)y^2}}{2(1)}$$

$$= \frac{-2y \pm \sqrt{0}}{2}$$

$$= \frac{-2y}{2}$$

$$= -y$$

33. $16t^2 - 8t - h = 0$

$$t = \frac{-(-8) \pm \sqrt{(-8)^2 - 4(16)(-h)}}{2(16)}$$

$$= \frac{8 \pm \sqrt{64 + 64h}}{32}$$

$$= \frac{8 \pm 8\sqrt{1+h}}{32}$$

$$= \frac{1 \pm \sqrt{1+h}}{4}$$

35. $16t^2 - vt - 20 = 0$

$a = 16$, $b = -v$ and $c = -20$.

$$t = \frac{-(-v) \pm \sqrt{(-v)^2 - 4(16)(-20)}}{2(16)}$$

$$= \frac{v \pm \sqrt{v^2 + 1{,}280}}{32}$$

37. $$\frac{AB}{BC} = \frac{BC}{AC}, \quad BC = x$$

$$\frac{4}{x} = \frac{x}{x+4}$$

$$(x+4)4 = x^2$$

$$0 = x^2 - 4x - 16$$

$$x = \frac{4 \pm \sqrt{(-4)^2 - 4(1)(-16)}}{2(1)}$$

$$= \frac{4 \pm 4\sqrt{5}}{2} = 2 + 2\sqrt{5}$$

$$4\,(\text{golden ratio}) = 4\left(\frac{1+\sqrt{5}}{2}\right)$$

$$= 2 + 2\sqrt{5}$$

39. Let $BC = x$, then

$$\frac{2}{x} = \frac{x}{x+2}$$

$$2(x+2) = x^2$$

$$x^2 - 2x - 4 = 0$$

Using the quadratic formula we get

$x = 1 \pm \sqrt{5}$ but negative length isn't allowed, so

$$\frac{BC}{AB} = \frac{1+\sqrt{5}}{2}, \text{ the golden ratio}$$

41. (a) See the graph in the back of the textbook.

(b) $-\dfrac{1}{150}x^2 + \dfrac{21}{5}x = 0$

$x\left(-\dfrac{1}{150}x + \dfrac{21}{5}\right) = 0$

$x = 0 \ \text{ or } \ -\dfrac{1}{150}x + \dfrac{21}{5} = 0$

$-x + 630 = 0$

$x = 630$

You have to walk 630 feet.

43. (a) $l + 2w = 160$

(b) $A = lw, \ \ l = 160 - 2w$

$A = (160 - 2w)w \ \text{ or } \ A = -2w^2 + 160w$

(c) See the table in the back of the textbook.

(d) The largest area is 3200 square yards.

45. $5\sqrt{7} - 2\sqrt{7} = (5 - 2)\sqrt{7} = 3\sqrt{7}$

47. $\sqrt{18} - \sqrt{8} + \sqrt{32} = 3\sqrt{2} - 2\sqrt{2} + 4\sqrt{2} = 5\sqrt{2}$

49. $9x\sqrt{20x^3 y^2} + 7y\sqrt{45x^5}$

$= 9x\sqrt{4x^2 y^2 \cdot 5x} + 7y\sqrt{9x^4 \cdot 5x}$

$= 9x \cdot 2xy\sqrt{5x} + 7y \cdot 3x^2 \sqrt{5x}$

$= 18x^2 y\sqrt{5x} + 21x^2 y\sqrt{5x}$

$= 39x^2 y\sqrt{5x}$

51. $\left(\sqrt{5} - 2\right)\left(\sqrt{5} + 8\right) = 5 + 6\sqrt{5} - 16 = -11 + 6\sqrt{5}$

53. $\left(\sqrt{x} + 2\right)^2 = \left(\sqrt{x}\right)^2 + 2\left(\sqrt{x}\right)2 + 2^2 = x + 4\sqrt{x} + 4$

55. $\dfrac{\sqrt{7}}{\sqrt{7} - 2} = \dfrac{\sqrt{7}}{\sqrt{7} - 2} \cdot \dfrac{\sqrt{7} + 2}{\sqrt{7} + 2}$

$= \dfrac{7 + 2\sqrt{7}}{7 - 4}$

$= \dfrac{7 + 2\sqrt{7}}{3}$

57. $y = x^3 - 4x$

y-intercept: $y = 0^3 - 4(0) = 0$

x-intercepts:

$0 = x^3 - 4x$

$0 = x\left(x^2 - 4\right)$

$0 = x(x + 2)(x - 2)$

$x = 0 \quad \text{ or } \quad x + 2 = 0 \quad \text{ or } \quad x - 2 = 0$

$x = 0 \qquad\qquad x = -2 \qquad\qquad x = 2$

59. $y = 3x^3 + x^2 - 27x - 9$

y-intercept: $y = 3(0)^3 + (0)^2 - 27(0) - 9 = -9$

x-intercept:

$3x^3 + x^2 - 27x - 9 = 0$

$x^2(3x + 1) - 9(3x + 1) = 0$

$\left(x^2 - 9\right)(3x + 1) = 0$

$x^2 - 9 = 0 \quad \text{ or } \quad 3x + 1 = 0$

$x^2 = 9 \qquad\qquad 3x = -1$

$x = \pm 3 \qquad\qquad x = -\dfrac{1}{3}$

61.

$$\begin{array}{r} 2x^2 + x - 1 \\ x-4\overline{\smash)2x^3 - 7x^2 - 5x + 4} \\ \underline{2x^3 - 8x^2} \\ x^2 - 5x \\ \underline{x^2 - 4x} \\ -x + 4 \\ \underline{-x + 4} \\ 0 \end{array}$$

$$2x^3 - 7x^2 - 5x + 4 = 0$$

$$(x-4)(2x^2 + x - 1) = 0$$

$$(x-4)(x+1)(2x-1) = 0$$

$x - 4 = 0$ or $x + 1 = 0$ or $2x - 1 = 0$

$x = 4$ $x = -1$ $2x = 1$

                                 $x = 1/2$

It also crosses the x-axis at -1 and $1/2$.

10.5 Graphing Parabolas

1. $y = x^2 + 2x - 3$

$0 = x^2 + 2x - 3$ $y = 0$

$0 = (x+3)(x-1)$

$x = -3$ or $x = 1$

The x-intercepts are -3 and 1

$x = \dfrac{-b}{2a} = \dfrac{-2}{2(1)} = -1$

$y = (-1)^2 + 2(-1) - 3 = 1 - 2 - 3 = -4$

The vertex is $(-1, -4)$

See the graph in the back of the textbook.

3. $y = -x^2 - 4x + 5, \quad y = 0$

$0 = -x^2 - 4x + 5$

$0 = x^2 + 4x - 5$

$0 = (x+5)(x-1)$

$x = -5$ or $x = 1$

The x-intercepts are -5 and 1

$x = \dfrac{-b}{2a} = \dfrac{-(-4)}{2(-1)} = -2$

$y = -(-2)^2 - 4(-2) + 5 = 9$

The vertex is $(-2, 9)$.

See the graph in the back of the textbook.

5. $y = x^2 - 1, \quad y = 0$

$0 = x^2 - 1$

$0 = (x+1)(x-1)$

$x = -1$ or $x = 1$

The x-intercepts are -1 and 1

$x = \dfrac{-b}{2a} = \dfrac{-0}{2(1)} = 0$

$y = (0)^2 - 1 = 0 - 1 = -1$

The vertex is $(0, -1)$

See the graph in the back of the textbook.

7. $y = -x^2 + 9, \quad y = 0$

$0 = -x^2 + 9$

$0 = x^2 - 9$

$0 = (x+3)(x-3)$

$x = -3$ or $x = 3$

The x-intercepts are -3 and 3

$x = \dfrac{-b}{2a} = \dfrac{-0}{2(-1)} = 0$

$y = -0^2 + 9 = 9$

The vertex is $(0, 9)$.

See the graph in the back of the textbook.

9. $y = 2x^2 - 4x - 6, \quad y = 0$

$0 = 2x^2 - 4x - 6$

$0 = x^2 - 2x - 3$

$0 = (x-3)(x+1)$

$x = 3 \quad \text{or} \quad x = -1$

The x-intercepts are 3 and -1

$x = \dfrac{-b}{2a} = \dfrac{-(-4)}{2(2)} = 1$

$y = 2(1)^2 - 4(1) - 6 = 2 - 4 - 6 = -8$

The vertex is $(1, -8)$

See the graph in the back of the textbook.

11. $y = x^2 - 2x - 4, \quad y = 0$

$0 = x^2 - 2x - 4$

$x = \dfrac{-(-2) \pm \sqrt{(-2)^2 - 4(1)(-4)}}{2(1)}$

$= \dfrac{2 \pm 2\sqrt{5}}{2}$

$= 1 \pm \sqrt{5}$

The x-intercepts are $1 + \sqrt{5}$ and $1 - \sqrt{5}$

$x = \dfrac{-b}{2a} = \dfrac{-(-2)}{2(1)} = 1$

$y = 1^2 - 2(1) - 4 = -5$

The vertex is $(1, -5)$.

See the graph in the back of the textbook

13. $y = x^2 - 4x - 4$

$x = \dfrac{-b}{2a} = \dfrac{-(-4)}{2(1)} = 2$

$y = 2^2 - 4 \cdot 2 - 4 = -8$

The vertex is $(2, -8)$.

$x = 0: \quad y = 0^2 - 4 \cdot 0 - 4 = -4$

$x = 4: \quad y = 4^2 - 4 \cdot 4 - 4 = -4$

See the graph in the back of the textbook.

15. $y = -x^2 + 2x - 5$

$x = \dfrac{-2}{2(-1)} = 1$

$y = -(1)^2 + 2(1) - 5 = -4$

The vertex is $(1, -4)$.

$x = 0, \quad y = -0^2 + 2(0) - 5 = -5$

$x = 2, \quad y = -2^2 + 2(2) - 5 = -5$

See the graph in the back of the textbook.

17. $y = x^2 + 1$

$x = \dfrac{-b}{2a} = \dfrac{-0}{2(1)} = 0$

$y = 0^2 + 1 = 1$

The vertex is $(0, 1)$.

$x = -2: y = (-2)^2 + 1 = 5$

$x = 2: \quad y = 2^2 + 1 = 5$

See the graph in the back of the textbook.

19. $y = -x^2 - 3$

$x = \dfrac{-0}{2(-1)} = 0$

$y = -0^2 - 3 = -3$

The vertex is $(0, -3)$.

$x = -1, \quad y = -(-1)^2 - 3 = -4$

$x = 1, \quad y = -(1)^2 - 3 = -4$

See the graph in the back of the textbook.

21. $y = 3x^2 + 4x + 1$

$x = \dfrac{-b}{2a} = \dfrac{-4}{2(3)} = -\dfrac{4}{6} = -\dfrac{2}{3}$

$y = 3\left(-\dfrac{2}{3}\right)^2 + 4\left(-\dfrac{2}{3}\right) + 1 = \dfrac{4}{3} + \dfrac{-8}{3} + 1 = -\dfrac{1}{3}$

The vertex is $\left(-\dfrac{2}{3}, -\dfrac{1}{3}\right)$.

$x = -2: \quad y = 3(-2)^2 + 4(-2) + 1 = 5$

$x = 1: \quad y = 3 \cdot 1^2 + 4 \cdot 1 + 1 = 8$

See the graph in the back of the textbook.

23. $y = x^2 - 6x + 5$

$y = (x^2 - 6x + 9) + 5 - 9$

$y = (x - 3)^2 - 4$

The vertex is $(3, -4)$.

The graph will be concave up because $a > 0$. The vertex is the lowest point.

25. $y = -x^2 + 2x + 8$

$y = -1(x^2 - 2x + 1) + 1 + 8$

$y = -1(x - 1)^2 + 9$

The vertex is $(1, 9)$.

The graph will be concave-down because $a < 0$.

This means the vertex is the highest point.

27. $y = 12 + 4x - x^2$

$y = -x^2 + 4x + 12$

$y = -\left(x^2 - 4x + 4\right) + 4 + 12$

$y = -(x - 2)^2 + 16$

The vertex is $(2, 16)$.

The graph will be concave down because $a < 0$. The vertex is the highest point.

29. $y = -x^2 - 8x$

$y = -1\left(x^2 + 8x + 16\right) + 0 + 16$

$y = -1(x + 4)^2 + 16$

The vertex is $(-4, 16)$.

The graph will be concave-down because $a < 0$.

This means the vertex is the highest point.

31. $P(x) = -0.5x^2 + 40x - 300$

$x = \dfrac{-(40)}{2(-0.5)} = 40$ items to sell each week

$P(40) = -0.5(40)^2 + 40(40) - 300$

$= -800 + 1600 - 300$

$= \$500$ maximum weekly profit

33. $P(x) = -0.002x^2 + 3.5x - 800$

$x = \dfrac{-b}{2a} = \dfrac{-3.5}{2(-0.002)} = 875$ patterns to sell

$P(875) = -0.002(875)^2 + 3.5(875) - 800$

$= -0.002(765,625) + 3,062.5 - 800$

$= -1,531.25 + 3,062.5 - 800$

$= \$731.25$ maximum monthly profit

35. $h(t) = 32t - 16t^2$ for $0 \le t \le 2$

In her hand at $h = 0$.

$0 = 32t - 16t^2$

$0 = 16t(2 - t)$

$16t = 0 \quad$ or $\quad 2 - t = 0$

$t = 0$ seconds $\qquad t = 2$ seconds

Vertex: $t = -\dfrac{32}{2(-16)} = 1$

$h(1) = 32(1) - 16(1)^2$

$= 16$ feet maximum height

37. $h(t) = 128t - 16t^2$

Vertex: $t = \dfrac{-b}{2a} = \dfrac{-128}{2(-16)} = 4$

$h(4) = 128(4) - 16(4)^2 = 512 - 256 = 256$

The maximum height is 256 feet.

39. Let $x =$ width, $80 - 2x =$ length

$A(x) = x(80 - 2x) = 80x - 2x^2$

Maximum area is at the vertex.

Vertex: $x = \dfrac{-80}{2(-2)} = 20$

$80 - 2x = 80 - 2(20) = 40$

Dimensions: 40 feet by 20 feet

41. $R = xp$ when $x = 1200 - 100p$

$R = (1200 - 100p)p$

$R = 1200p - 100p^2$

See the graph in the back of the textbook.

A price of $6.00 per calculator brings a
maximum revenue of $3,600.

43. $R = xp = (1,700 - 100p)p$

$\quad = 1,700p - 100p^2$

See the graph in the back of the textbook.

Maximum $R = \$7225$ when $p = \$8.50$

45. $(3 - 5i) - (2 - 4i) = 3 - 5i - 2 + 4i = 3 - 2 - 5i + 4i = 1 - i$

47. $(3 + 2i)(7 - 3i) = 21 + 14i - 9i - 6i^2$

$\quad = 21 + 5i - 6(-1)$

$\quad = 27 + 5i$

49. $\dfrac{i}{3+i} = \dfrac{i}{3+i} \cdot \dfrac{3-i}{3-i} = \dfrac{3i - i^2}{3^2 - i^2} = \dfrac{3i - (-1)}{9 - (-1)} = \dfrac{1}{10} + \dfrac{1}{3}i$

51. $y = (x - 2)^2 - 4$

53. See the graph in the back of the textbook.

$y = a(x - h)^2 + k$

Vertex is at $(90, 60)$ so $h = 90$ and $k = 60$

$y = a(x - 90)^2 + 60$

The landing point is $(180, 0)$, which gives

$0 = a(180 - 90)^2 + 60$

$-60 = 8100a$

$a = -\dfrac{1}{135}$

The equation is $y = -\dfrac{1}{135}(x - 90)^2 + 60$

55. 1. $x^2 - 4 = (x + 2)(x - 2)$

2. $x^2 - 4 = 0$

$(x + 2)(x - 2) = 0$

$x = -2 \qquad x = 2$

4. $f(x) = 0$ at $x = -2$, and 2

The solutions to the equation in Problem 2 are
the values of x where the graph crosses the
x-axis and the function is zero.

10.6 Quadratic Inequalities

1. $x^2 + x - 6 > 0$

    $(x-2)(x+3) > 0$ same signs

    $x < -3$ or $x > 2$

    See the graph in the back of the textbook.

    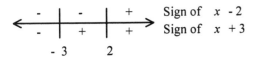

3. $x^2 - x - 12 \le 0$

    $(x+3)(x-4) \le 0$ negative

    $-3 \le x \le 4$

    See the graph in the back of the textbook.

    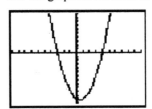

5. $x^2 + 5x \ge -6$

    $x^2 + 5x + 6 \ge 0$

    $(x+2)(x+3) \ge 0$ same signs

    $x \le -3$ or $x \ge -2$

    See the graph in the back of the textbook.

    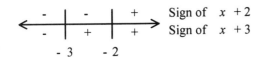

7. $6x^2 < -5x - 1$

    $6x^2 - 5x + 1 < 0$

    $(2x-1)(3x-1) \le 0$ negative

    $\dfrac{1}{3} < x < \dfrac{1}{2}$

    See the graph in the back of the textbook.

    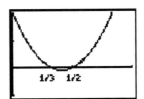

9. $x^2 - 9 < 0$

    $(x-3)(x+3) < 0$ opposite signs

    $-3 < x < 3$

    See the graph in the back of the textbook.

    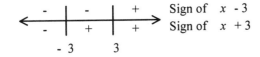

11. $4x^2 - 9 \ge 0$

    $(2x+3)(2x-3) \ge 0$ positive

    $x \le -\dfrac{3}{2}$ or $x \ge \dfrac{3}{2}$

    See the graph in the back of the textbook.

    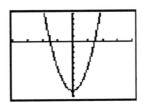

13. $2x^2 - x - 3 < 0$

$(2x-3)(x+1) < 0$ opposite signs

$$-1 < x < \frac{3}{2}$$

See the graph in the back of the textbook.

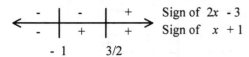

15. $x^2 - 4x + 4 \geq 0$ positive

$$(x-2)^2 \geq 0$$

All real numbers

17. $x^2 - 10x + 25 < 0$

$$(x-5)^2 < 0$$

No real solution, $\varnothing$

19. $(x-2)(x-3)(x-4) > 0$ positive

$2 < x < 3$ or $x > 4$

See the graph in the back of the textbook.

21. $(x+1)(x+2)(x+3) \leq 0$ either 3 or 1 negative signs

See the graph in the back of the textbook.

$x \leq -3$ or $-2 \leq x \leq -1$

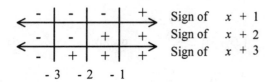

23. $\dfrac{x-1}{x+4} \leq 0$ negative

$-4 < x \leq 1$

See the graph in the back of the textbook.

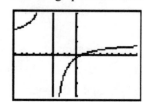

25. $\dfrac{3x}{x+6} - \dfrac{8}{x+6} < 0$

$$\dfrac{3x-8}{x+6} < 0$$ opposite signs

$$-6 < x < \frac{8}{3}$$

See the graph in the back of the textbook.

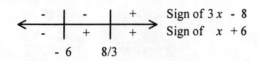

27. $\dfrac{4}{x-6} + 1 > 0$

$$\dfrac{4+(x-6)}{x-6} > 0$$

$$\dfrac{x-2}{x-6} > 0$$ positive

$x < 2$ or $x > 6$

See the graph in the back of the textbook.

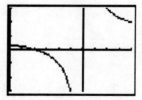

29. $\dfrac{x-2}{(x+3)(x-4)} < 0$ either 3 or 1 negative signs

$x < -3$ or $2 < x < 4$

See the graph in the back of the textbook.

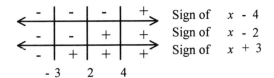

- 3   2   4

31. $\dfrac{2}{x-4} - \dfrac{1}{x-3} > 0$

$\dfrac{4(x+2)-3(x+3)}{(x-4)(x-3)} > 0$

$\dfrac{x-2}{(x-4)(x-3)} > 0$ positive

$2 < x < 3$ or $x > 4$

See the graph in the back of the textbook.

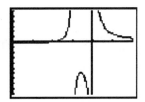

33. $\dfrac{x+7}{2x+12} + \dfrac{6}{x^2-36} \le 0$

$\dfrac{x+7}{2(x+6)} + \dfrac{6}{(x+6)(x-6)} \le 0$

$\dfrac{x+7}{2(x+6)} \cdot \dfrac{x-6}{x-6} + \dfrac{6}{(x+6)(x-6)} \cdot \dfrac{2}{2} \le 0$

$\dfrac{x^2+x-30}{2(x+6)(x-6)} \le 0$

$\dfrac{(x+6)(x-5)}{2(x+6)(x-6)} \le 0$

$\dfrac{x-5}{2(x-6)} \le 0$ opposite signs

$5 \le x < 6$

See the graph in the back of the textbook.

- | - | +   Sign of   x - 6
- | + | +   Sign of   x - 5

5   6

35. (a) $x^2 - 4 < 0 \Rightarrow -2 < x < 2$

(b) $x^2 - 4 > 0 \Rightarrow x < -2$ or $x > 2$

(c) $x^2 - 4 = 0 \Rightarrow x = -2$ or $x = 2$

37. (a) $x^2 - 3x - 10 < 0 \Rightarrow -2 < x < 5$

(b) $x^2 - 3x - 10 > 0 \Rightarrow x < -2$ or $x > 5$

(c) $x^2 - 3x - 10 = 0 \Rightarrow x = -2$ or $x = 5$

39. (a) $x^3 - 3x^2 - x + 3 < 0 \Rightarrow x < -1$ or $1 < x < 3$

(b) $x^3 - 3x^2 - x + 3 > 0 \Rightarrow -1 < x < 1$ or $x > 3$

(c) $x^3 - 3x^2 - x + 3 = 0 \Rightarrow x = -1$ or $x = 1$ or $x = 3$

41.　Let $x =$ width and $2x+3 =$ length

$$A \leq l \times w$$
$$44 \leq (2x+3)(x)$$
$$44 \leq 2x^2 + 3x$$
$$0 \leq 2x^2 + 3x - 44$$
$$0 \leq (2x+11)(x-4) \quad \text{same signs}$$
$$x \leq -\frac{11}{2} \quad \text{or} \quad x \geq 4$$

Width cannot be negative.

Width is greater than or equal to 4 inches

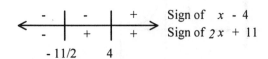

43.　$R = 1300p - 100p^2 \quad R \geq \$4000$

$$4000 \leq 1300p - 100p^2$$
$$0 \geq 100p^2 - 1300p + 4000$$
$$0 \geq p^2 - 13p + 40$$
$$0 \geq (p-5)(p-8) \quad \text{negative}$$
$$5 \leq p \leq 8$$

She should charge at least $5,

but no more than $8 for each radio.

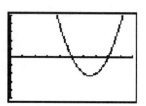

45.　Let $x =$ number of \$10-increases in dues.

$$I = dn = (10,000 - 200x)(100 + 10x)$$
$$= 1,000,000 + 80,000x - 2,000x^2$$

Vertex: $x = -\dfrac{b}{2a} = -\dfrac{80,000}{2(-2,000)} = 20$

$$d(20) = 100 + 10(20) = \$300.$$
$$I(20) = \big[10,000 - 200(20)\big]\big[100 + 10(20)\big]$$
$$= \$1,800,000$$

Maximum income of $1,800,000 at $300 dues.

47.　Let $x =$ number of \$2 increases in price

$$I = pn = (20 + 2x)(40 - 2x) = 80 + 40x - 4x^2$$

Vertex: $x = \dfrac{-40}{2(-4)} = 5$

$$I(5) = 80 + 40(5) - 4(5)^2 = 900$$
$$p = 20 + 2(5) = 30$$

Maximum income of $900 at $30 per oil change.

49.　$\sqrt{3t-1} = 2$

$$3t - 1 = 4$$
$$3t = 5$$
$$t = \frac{5}{3}$$

51.　$\sqrt{x+3} = x - 3$

$$x + 3 = (x-3)^2$$
$$x + 3 = x^2 - 6x + 9$$
$$0 = x^2 - 7x + 6$$
$$0 = (x-1)(x-6)$$
$$x - 1 = 0 \quad \text{or} \quad x - 6 = 0$$
$$x = 1 \qquad\qquad x = 6$$

Possible solutions 1 and 6; only 6 checks.

53.　$y = \sqrt[3]{x-1}$

See the graph in the back of the textbook.

55. $x^2 - 2x - 1 < 0$ negative

$$x = \frac{-(-2) \pm \sqrt{(-2)^2 - 4(1)(-1)}}{2(1)}$$

$$= \frac{2 \pm \sqrt{4 + 4}}{2}$$

$$= \frac{2 \pm 2\sqrt{2}}{2}$$

$$= 1 \pm \sqrt{2}$$

$$1 - \sqrt{2} < x < 1 + \sqrt{2}$$

See the graph in the back of the textbook.

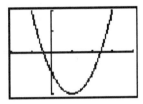

57. $x^2 - 8x + 13 > 0$ same signs

$$x = \frac{-(-8) \pm \sqrt{(-8)^2 - 4(1)(13)}}{2(1)}$$

$$= \frac{8 \pm \sqrt{12}}{2}$$

$$= \frac{8 \pm 2\sqrt{3}}{2}$$

$$= \frac{2(4 \pm \sqrt{3})}{2}$$

$$= 4 \pm \sqrt{3}$$

$$x < 4 - \sqrt{3} \quad \text{or} \quad x > 4 + \sqrt{3}$$

See the graph in the back of the textbook.

$$\text{Sign of } x - \left(4 + \sqrt{3}\right)$$
$$\text{Sign of } x - \left(4 - \sqrt{3}\right)$$

$$4 - \sqrt{3} \qquad 4 + \sqrt{3}$$

Chapter 10 Review

1. $(2t-5)^2 = 25$

$$2t-5 = \pm\sqrt{25}$$
$$2t-5 = \pm 5$$
$$2t = 5 \pm 5$$
$$t = \frac{5 \pm 5}{2}$$
$$t = \frac{5+5}{2} \quad \text{or} \quad t = \frac{5-5}{2}$$
$$t = \frac{10}{2} \qquad\qquad t = \frac{0}{2}$$
$$t = 5 \qquad\qquad t = 0$$

The solutions are 0 and 5.

3. $(3y-4)^2 = -49$

$$3y-4 = \pm\sqrt{-49} \quad \text{Theorem 7.1}$$
$$3y-4 = \pm 7i$$
$$3y = 4 \pm 7i$$
$$y = \frac{4 \pm 7i}{3}$$

The solutions are $\dfrac{4+7i}{3}$ and $\dfrac{4-7i}{3}$.

5. $2x^2 + 6x - 20 = 0$

$$x^2 + 3x = 10$$
$$x^2 + 3x + \frac{9}{4} = 10 + \frac{9}{4}$$
$$\left(x + \frac{3}{2}\right)^2 = \frac{49}{4}$$
$$x + \frac{3}{2} = \pm\sqrt{\frac{49}{4}}$$
$$x + \frac{3}{2} = \pm\frac{7}{2}$$
$$x = -\frac{3}{2} \pm \frac{7}{2}$$
$$x = -\frac{3}{2} + \frac{7}{2} \quad \text{or} \quad x = -\frac{3}{2} - \frac{7}{2}$$
$$x = 2 \qquad\qquad x = -5$$

7. $a^2 + 9 = 6a$

$$a^2 - 6a = -9$$
$$a^2 - 6a + 9 = -9 + 9$$
$$(a-3)^2 = 0$$
$$a-3 = \pm\sqrt{0}$$
$$a-3 = 0$$
$$a = 3$$

9. $2y^2 + 6y = -3$

$$y^2 + 3y = -\frac{3}{2}$$
$$y^2 + 3y + \frac{9}{4} = -\frac{3}{2} + \frac{9}{4}$$
$$\left(y + \frac{3}{2}\right)^2 = \frac{3}{4}$$
$$y + \frac{3}{2} = \pm\frac{\sqrt{3}}{2}$$
$$y = \frac{-3 \pm \sqrt{3}}{2}$$

11.

$$\frac{1}{6}x^2 + \frac{1}{2}x - \frac{5}{3} = 0$$
$$6\left(\frac{1}{6}x^2\right) + 6\left(\frac{1}{2}x\right) - 6\left(\frac{5}{3}\right) = 6(0)$$
$$x^2 + 3x - 10 = 0$$
$$(x+5)(x-2) = 0$$
$$x+5 = 0 \quad \text{or} \quad x-2 = 0$$
$$x = -5 \qquad\qquad x = 2$$

13. $4t^2 - 8t + 19 = 0$

$$t = \frac{-(-8) \pm \sqrt{(-8)^2 - 4(4)(19)}}{2(4)}$$

$$= \frac{8 \pm \sqrt{64 - 304}}{8}$$

$$= \frac{8 \pm \sqrt{-240}}{8}$$

$$= \frac{8 \pm 4i\sqrt{15}}{8}$$

$$= \frac{4(2 \pm i\sqrt{15})}{8}$$

$$= \frac{2 \pm i\sqrt{15}}{2}$$

15.
$$0.06a^2 + 0.05a = 0.04$$
$$0.06a^2 + 0.05a - 0.04 = 0$$
$$6a^2 + 5a - 4 = 0$$
$$(3a + 4)(2a - 1) = 0$$
$$3a + 4 = 0 \quad \text{or} \quad 2a - 1 = 0$$
$$3a = -4 \qquad\qquad 2a = 1$$
$$a = -\frac{4}{3} \qquad\qquad a = \frac{1}{2}$$

17. $(2x+1)(x-5) - (x+3)(x-2) = -17$

$$2x^2 - 9x - 5 - (x^2 + x - 6) = -17$$
$$2x^2 - 9x - 5 - x^2 - x + 6 = -17$$
$$x^2 - 10x + 1 = -17$$
$$x^2 - 10x + 18 = 0$$

$$x = \frac{-(-10) \pm \sqrt{(-10)^2 - 4(1)(18)}}{2(1)}$$

$$= \frac{10 \pm \sqrt{100 - 72}}{2}$$

$$= \frac{10 \pm \sqrt{28}}{2}$$

$$= \frac{10 \pm 2\sqrt{7}}{2}$$

$$= \frac{2(5 \pm \sqrt{7})}{2}$$

$$= 5 \pm \sqrt{7}$$

19.
$$5x^2 = -2x + 3$$
$$5x^2 + 2x - 3 = 0$$

$$x = \frac{-2 \pm \sqrt{2^2 - 4(5)(-3)}}{2(5)}$$

$$= \frac{-2 \pm \sqrt{4 + 60}}{10}$$

$$= \frac{-2 \pm \sqrt{64}}{10}$$

$$= \frac{-2 \pm 8}{10}$$

$$= \frac{-1 \pm 4}{5}$$

$$x = \frac{-1 + 4}{5} \quad \text{or} \quad x = \frac{-1 - 4}{5}$$

$$x = \frac{3}{5} \qquad\qquad x = -1$$

21.
$$3 - \frac{2}{x} + \frac{1}{x^2} = 0$$

Multiply by the LCD, x^2

$$x^2(3) - x^2\left(\frac{2}{x}\right) + x^2\left(\frac{1}{x^2}\right) = x^2(0)$$

$$3x^2 - 2x + 1 = 0$$

Let $a = 3$, $b = -2$ and $c = 1$

$$x = \frac{-(-2) \pm \sqrt{(-2)^2 - 4(3)(1)}}{2(3)}$$

$$= \frac{2 \pm \sqrt{4 - 12}}{6}$$

$$= \frac{2 \pm \sqrt{-8}}{6}$$

$$= \frac{2 \pm i\sqrt{4}\sqrt{2}}{6}$$

$$= \frac{2 \pm 2i\sqrt{2}}{6}$$

$$= \frac{1 \pm i\sqrt{2}}{3}$$

23.
$$C(x) = 7x + 400, \quad R(x) = 34x - 0.1x^2$$
$$P = R - C$$
$$1300 = (34x - 0.1x^2) - (7x + 400)$$
$$1300 = 34x - 0.1x^2 - 7x - 400$$
$$1300 = -0.1x^2 + 27x - 400$$
$$0.1x^2 - 27x + 1700 = 0$$

Let $a = 0.1$, $b = -27$ and $c = 1700$

$$x = \frac{-(-27) \pm \sqrt{(-27)^2 - 4(0.1)(1700)}}{2(0.1)}$$

$$= \frac{27 \pm \sqrt{729 - 680}}{0.2}$$

$$= \frac{27 \pm \sqrt{49}}{0.2}$$

$$= \frac{27 \pm 7}{0.2}$$

$$x = \frac{27 + 7}{0.2} \quad \text{or} \quad x = \frac{27 - 7}{0.2}$$
$$= 170 \qquad\qquad = 100$$

They must produce and sell 100 or 170 items.

25. $2x^2 - 8x = -8$

$2x^2 - 8x + 8 = 0$ Standard form

Using $a = 2$, $b = -8$, and $c = 8$, we have:

$b^2 - 4ac = (-8)^2 - 4(2)(8) = 64 - 64 = 0$

The discriminant is zero, implying the equation will have one rational solution.

27. $2x^2 + x - 3 = 0$

Using $a = 2$, $b = 1$, and $c = -3$, we have:

$b^2 - 4ac = 1^2 - 4(2)(-3) = 1 + 24 = 25$

The discriminant is a perfect square. Therefore, the equation has two rational solutions.

29. $x^2 - x = 1$

$x^2 - x - 1 = 0$ Standard form

Using $a = 1$, $b = -1$, and $c = -1$, we have:

$b^2 - 4ac = (-1)^2 - 4(1)(-1) = 5$

The discriminant is a positive number that is not a perfect square, implying the equation will have two irrational solutions.

31. $3x^2 + 5x = -4$

$3x^2 + 5x + 4 = 0$ Standard form

Using $a = 3$, $b = 5$, and $c = 4$, we have:

$b^2 - 4ac = 5^2 - 4(3)(4) = 25 - 48 = -23$

The discriminant is negative, implying the equation will have two complex solutions that contain i.

33. $25x^2 - kx + 4 = 0$

Using $a = 25$, $b = -k$ and $c = 4$,

$b^2 - 4ac = (-k)^2 - 4(25)(4) = k^2 - 400$

An equation has exactly one real solution when the discriminant is 0.

$k^2 - 400 = 0$

$\qquad k^2 = 400$

$\qquad k = \pm 20$

35. $kx^2 + 12x + 9 = 0$

Using $a = k$, $b = 12$, and $c = 9$, we have:

$b^2 - 4ac = 12^2 - 4(k)(9) = 144 - 36k$

An equation has exactly one real solution when the discriminant is 0. We set the discriminant equal to 0 and solve:

$144 - 36k = 0$

$\qquad 144 = 36k$

$\qquad \dfrac{144}{36} = k$

$\qquad k = 4$

37. $9x^2 + 30x + k = 0$

Using $a = 9$, $b = 30$, and $c = k$, we have:

$b^2 - 4ac = (30)^2 - 4(9)(k) = 900 - 36k$

An equation has exactly one real solution when the discriminant is 0. We set the discriminant equal to 0 and solve:

$900 - 36k = 0$

$\qquad 900 = 36k$

$\qquad k = 25$

39.   If $x = 3$ $x = 5$

then $x - 3 = 0$ $x - 5 = 0$

Since $x - 3$ and $x - 5$ are 0, their product is also 0 by the zero-factor property.

$(x - 3)(x - 5) = 0$ Zero-factor property

$x^2 - 8x + 15 = 0$ Multiply the binomials

41.   If $y = \dfrac{1}{2}$ $y = -4$

$\qquad 2y = 1$

then $2y - 1 = 0$ $y + 4 = 0$

Since $2y - 1$ and $y + 4$ are 0, their product is also 0 by the zero-factor property.

$(2y - 1)(y + 4) = 0$ Zero-factor property

$2y^2 + 7y - 4 = 0$ Multiply the binomials

43. $(x - 2)^2 - 4(x - 2) - 60 = 0$

Replacing $x - 2$ with y, we have

$\qquad y^2 - 4y - 60 = 0$

$(y + 6)(y - 10) = 0$ Factor

$y + 6 = 0$ or $y - 10 = 0$ Set factors to 0

$\qquad y = -6$ $y = 10$

Replacing y with $x - 2$, and then solving for x,

we have $x - 2 = -6$ or $x - 2 = 10$

$\qquad\qquad x = -4$ $x = 12$

The solutions are -4 and 12.

45. $x^4 - x^2 = 12$

$x^4 - x^2 - 12 = 0$ Standard form

Replacing x^2 with y, we have

$y^2 - y - 12 = 0$

$(y+3)(y-4) = 0$ Factor

$y + 3 = 0$ or $y - 4 = 0$

  $y = -3$ $y = 4$

Replacing y with x^2, we have

$x^2 = -3$ or $x^2 = 4$

  $x = \pm i\sqrt{3}$ $x = \pm 2$

The solution set is $\{i\sqrt{3},\ -i\sqrt{3},\ 2 \text{ and } -2\}$.

47. $2x - 11\sqrt{x} = -12$

$2x - 11\sqrt{x} + 12 = 0$

Replacing $\sqrt{x}$ with y, we have

$2y^2 - 11y + 12 = 0$

$(2y - 3)(y - 4) = 0$

$2y - 3 = 0$ or $y - 4 = 0$

  $2y = 3$ $y = 4$

  $y = \dfrac{3}{2}$

Replacing y with $\sqrt{x}$, we have

$\sqrt{x} = \dfrac{3}{2}$ or $\sqrt{x} = 4$

$x = \dfrac{9}{4}$ $x = 16$

The solution set is $\left\{\dfrac{9}{4},\ 16\right\}$.

Both solutions check.

49. $\sqrt{y+21} + \sqrt{y} = 7$

$\left(\sqrt{y+21}\right)^2 = \left(7 - \sqrt{y}\right)^2$

$y + 21 = 49 - 14\sqrt{y} + y$

$-28 = -14\sqrt{y}$

$2 = \sqrt{y}$

$y = 4$

51. $16t^2 - 10t - h = 0$

$t = \dfrac{-(-10) \pm \sqrt{(-10)^2 - 4(16)(-h)}}{2(16)}$

$t = \dfrac{10 \pm \sqrt{100 + 64h}}{32}$

$t = \dfrac{10 \pm \sqrt{4(25 + 16h)}}{32}$

$t = \dfrac{10 \pm 2\sqrt{25 + 16h}}{32}$

$t = \dfrac{5 \pm \sqrt{25 + 16h}}{16}$

53. $x^2 - x - 2 < 0$ opposite signs

$(x+1)(x-2) < 0$

$-1 < x < 2$

See the graph in the back of the textbook

55. $2x^2 + 5x - 12 \geq 0$

$(x+4)(2x-3) \geq 0$ same signs

$x \leq -4$ or $x \geq \dfrac{3}{2}$

See the graph in the back of the textbook

57. $y = x^2 - 6x + 8$

x-intercepts

$0 = x^2 - 6x + 8$

$0 = (x-2)(x-4)$

$x - 2 = 0$ or $x - 4 = 0$

$x = 2$ $x = 4$

Vertex: $x = \dfrac{-b}{2a} = \dfrac{-(-6)}{2(1)} = 3$

$y = 3^2 - 6(3) + 8$

$y = -1$

See the graph in the back of the textbook

Cumulative Review: Chapters 1-10

1. $11 + 20 \div 5 - 3 \cdot 5 = 11 + 4 - 15 = 0$

3. $4(15-19)^2 - 3(17-19)^3 = 4(-4)^2 - 3(-2)^3$

$= 4(16) - 3(-8)$

$= 64 + 24$

$= 88$

5. $3 - 5\big[2x - 4(x-2)\big] = 3 - 5\big[2x - 4x + 8\big]$

$= 3 - 5\big[-2x + 8\big]$

$= 3 + 10x - 40$

$= 10x - 37$

7. $\sqrt[3]{32} = \sqrt[3]{8 \cdot 4} = \sqrt[3]{8} \cdot \sqrt[3]{4} = 2\sqrt[3]{4}$

9. $\dfrac{1 - \dfrac{3}{4}}{1 + \dfrac{3}{4}} = \dfrac{4\left(1 - \dfrac{3}{4}\right)}{4\left(1 + \dfrac{3}{4}\right)} = \dfrac{4 - 3}{4 + 3} = \dfrac{1}{7}$

11. $\dfrac{5x^2 - 26xy - 24y^2}{5x + 4y} = \dfrac{(5x + 4y)(x - 6y)}{5x + 4y} = x - 6y$

13. $(3x-2)(x^2-3x-2) = 3x^3 - 9x^2 - 6x - 2x^2 + 6x + 4 = 3x^3 - 11x^2 + 4$

15. $\dfrac{7-i}{3-2i} = \dfrac{7-i}{3-2i} \cdot \dfrac{3+2i}{3+2i} = \dfrac{21+11i-2i^2}{9-4i^2} = \dfrac{23+11i}{13} = \dfrac{23}{13} + \dfrac{11}{13}i$

17. $\dfrac{7}{5}a - 6 = 15$

$\quad\quad \dfrac{7}{5}a = 21$

$\quad\quad 7a = 105$

$\quad\quad a = 15$

19. $\quad \dfrac{a}{2} + \dfrac{3}{a-3} = \dfrac{a}{a-3}$

$\quad\quad\quad \text{LCD} = 2(a-3)$

$\quad a(a-3) + 3(2) = 2a$

$\quad\quad a^2 - 3a + 6 = 2a$

$\quad\quad a^2 - 5a + 6 = 0$

$\quad\quad (a-2)(a-3) = 0$

$\quad\quad a-2 = 0 \quad \text{or} \quad a-3 = 0$

$\quad\quad\quad a = 2 \quad\quad\quad\quad a = 3$

$\quad\quad \text{Only 2 checks; 2}$

21. $(3x-4)^2 = 18$

$\quad 3x-4 = \pm\sqrt{18}$

$\quad 3x-4 = \pm 3\sqrt{2}$

$\quad\quad 3x = 4 \pm 3\sqrt{2}$

$\quad\quad x = \dfrac{4 \pm 3\sqrt{2}}{3}$

23. $\quad 3y^3 - y = 5y^2$

$\quad 3y^3 - 5y^2 - y = 0$

$\quad y(3y^2 - 5y - 1) = 0$

$\quad\quad y = 0 \quad \text{or} \quad 3y^2 - 5y - 1 = 0$

Let $a = 3$, $b = -5$ and $c = -1$

$\quad y = \dfrac{-(-5) \pm \sqrt{(-5)^2 - 4(3)(-1)}}{2(3)}$

$\quad\quad = \dfrac{5 \pm \sqrt{25+12}}{6}$

$\quad\quad = \dfrac{5 \pm \sqrt{37}}{6}$

25. $\quad \sqrt{x-2} = 2 - \sqrt{x}$

$\quad (\sqrt{x-2})^2 = (2-\sqrt{x})^2$

$\quad\quad x - 2 = 4 - 4\sqrt{x} + x$

$\quad\quad -6 = -4\sqrt{x}$

$\quad\quad 3/2 = \sqrt{x}$

$\quad (3/2)^2 = (\sqrt{x})^2$

$\quad\quad 9/4 = x$

27. $5 \leq \dfrac{1}{4}x + 3 \leq 8$

$\quad 2 \leq \dfrac{1}{4}x \leq 5$

$\quad 8 \leq x \leq 20$

See the graph in the back of the textbook.

29. $3x - y = 2$ Multiply by 2 $\rightarrow$ $6x - 2y = 4$

    $-6x + 2y = -4$ No change $\rightarrow$ $\underline{-6x + 2y = -4}$

                                               $0 = 0$

A true statement means the lines coincide.

31. $3x - 2y = 5$

    $y = 3x - 7$

Substitute $y = 3x - 7$ into the first equation

    $3x - 2(3x - 7) = 5$

      $3x - 6x + 14 = 5$

            $-3x = -9$

               $x = 3$

Substitute $x = 3$ into the second equation

    $y = 3(3) - 7$

    $y = 2$

The solution to the system is $(3, 2)$.

33.   When $x = 0$, $2x - 3y = 12$ gives

    $2(0) - 3y = 12$

          $y = -4$

    When $y = 0$, $2x - 3y = 12$ gives

    $2x - 3(0) = 12$

          $x = 6$

See the graph in the back of the textbook.

35. $\left(\dfrac{3}{2}, \dfrac{4}{3}\right), \left(\dfrac{1}{4}, -\dfrac{1}{3}\right)$

$$m = \frac{y_2 - y_1}{x_2 - x_1} = \frac{-\dfrac{1}{3} - \left(\dfrac{4}{3}\right)}{\dfrac{1}{4} - \left(\dfrac{3}{2}\right)} = \frac{-\dfrac{5}{3}}{-\dfrac{5}{4}} = \frac{4}{3}$$

Using $(x_1, y_1) = \left(\dfrac{1}{4}, -\dfrac{1}{3}\right)$ and $m = \dfrac{4}{3}$

in $y - y_1 = m(x - x_1)$ yields

$$y - \left(-\frac{1}{3}\right) = \frac{4}{3}\left(x - \frac{1}{4}\right)$$

$$y + \frac{1}{3} = \frac{4}{3}x - \frac{1}{3}$$

$$3y + 1 = 4x - 1$$

$$4x - 3y = 2$$

37. $\dfrac{7}{\sqrt[3]{9}} = \dfrac{7}{\sqrt[3]{9}} \cdot \dfrac{\sqrt[3]{3}}{\sqrt[3]{3}} = \dfrac{7\sqrt[3]{3}}{\sqrt[3]{27}} = \dfrac{7\sqrt[3]{3}}{3}$

39.   Let x = largest angle

    $\dfrac{1}{4}x$ = smallest angle

    $\dfrac{1}{4}x + 30$ = third angle

    $\dfrac{1}{4}x + \dfrac{1}{4}x + 30 + x = 180$

    $x + x + 120 + 4x = 720$

           $6x + 120 = 720$

                 $6x = 600$

                   $x = 100$

             $\dfrac{1}{4}x = 25$

          $\dfrac{1}{4}x + 30 = 55$

The angles are $25°, 55°$, and $100°$.

Chapter 10 Test

1. $(2x+4)^2 = 25$

$$2x+4 = \pm\sqrt{25}$$
$$2x+4 = \pm 5$$
$$2x = -4 \pm 5$$
$$x = \frac{-4 \pm 5}{2}$$
$$x = \frac{-4+5}{2} \quad \text{or} \quad x = \frac{-4-5}{2}$$
$$x = \frac{1}{2} \qquad\qquad x = -\frac{9}{2}$$

2. $(2x-6)^2 = -8$

$$2x-6 = \pm\sqrt{-8}$$
$$2x-6 = \pm 2i\sqrt{2}$$
$$2x = 6 \pm 2i\sqrt{2}$$
$$x = \frac{6 \pm 2i\sqrt{2}}{2}$$
$$x = 3 \pm i\sqrt{2}$$

3. $y^2 - 10y + 25 = -4$

$$(y-5)^2 = -4$$
$$y-5 = \pm\sqrt{-4}$$
$$y-5 = \pm 2i$$
$$y = 5 \pm 2i$$

4. $(y+1)(y-3) = -6$

$$y^2 - 2y - 3 + 6 = 0$$
$$y^2 - 2y + 3 = 0$$

$$y = \frac{-(-2) \pm \sqrt{(-2)^2 - 4(1)(3)}}{2(1)}$$
$$= \frac{2 \pm \sqrt{-8}}{2}$$
$$= \frac{2 \pm 2i\sqrt{2}}{2}$$
$$= 1 \pm i\sqrt{2}$$

5.
$$8t^3 - 125 = 0$$
$$(2t-5)(4t^2 + 10t + 25) = 0$$
$$2t - 5 = 0$$
$$t = \frac{5}{2}$$

or

$$4t^2 + 10t + 25 = 0$$
$$t = \frac{-10 \pm \sqrt{(10)^2 - 4(4)(25)}}{2(4)}$$
$$= \frac{-10 \pm \sqrt{-300}}{8}$$
$$= \frac{-10 \pm 10i\sqrt{3}}{8}$$
$$t = \frac{-5 \pm 5i\sqrt{3}}{4}, \quad t = \frac{5}{2}$$

6.
$$\frac{1}{a+2} - \frac{1}{3} = \frac{1}{a}$$
$$3a - a(a+2) = 3(a+2)$$
$$3a - a^2 - 2a = 3a + 6$$
$$a^2 + 2a + 6 = 0$$

$$a = \frac{-(2) \pm \sqrt{(2)^2 - 4(1)(6)}}{2(1)}$$
$$a = \frac{-2 \pm \sqrt{-20}}{2}$$
$$a = \frac{-2 \pm 2i\sqrt{5}}{2}$$
$$a = -1 \pm i\sqrt{5}$$

7. $64(1+r)^2 = A$
$$(1+r)^2 = \frac{A}{64}$$
$$1 + r = \pm \frac{\sqrt{A}}{8}$$
$$r = \pm \frac{\sqrt{A}}{8} - 1$$

8.
$$x^2 - 4x = -2$$
$$x^2 - 4x + 4 = -2 + 4$$
$$(x-2)^2 = 2$$
$$x - 2 = \pm\sqrt{2}$$
$$x = 2 \pm \sqrt{2}$$

9.
$$s(t) = 32t - 16t^2, s(t) = 12$$
$$12 = 32t - 16t^2$$
$$16t^2 - 32t + 12 = 0$$
$$4(4t^2 - 8t + 3) = 0$$
$$(2t - 3)(2t - 1) = 0$$
$$2t - 3 = 0 \quad \text{or} \quad 2t - 1 = 0$$
$$2t = 3 \qquad\qquad 2t = 1$$
$$t = \frac{3}{2} \text{ seconds} \qquad t = \frac{1}{2} \text{ seconds}$$

10. $C(x) = 2x + 100, \quad R(x) = 25x - 0.2x^2$
$$P(x) = R(x) - C(x)$$
$$200 = 25x - 0.2x^2 - (2x + 100)$$
$$200 = -0.2x^2 + 23x - 100$$

$$0.2x^2 - 23x + 300 = 0$$
$$x = \frac{-(-23) \pm \sqrt{(-23)^2 - 4(0.2)(300)}}{2(0.2)}$$
$$= \frac{23 \pm \sqrt{289}}{0.4}$$
$$= \frac{23 \pm 17}{0.4}$$
$$x = \frac{23 + 17}{0.4} \quad \text{or} \quad x = \frac{23 - 17}{0.4}$$
$$x = 100 \text{ cups} \qquad x = 15 \text{ cups}$$

They must sell 15 or 100 cups a week.

11. $kx^2 = 12x - 4$

$kx^2 - 12x + 4 = 0$

Using $a = k$, $b = -12$ and $c = 4$, we have:

$b^2 - 4ac = (-12)^2 - 4(k)(4) = 144 - 16k$

An equation has exactly one rational solution when the discriminant is 0. We set the discriminant equal to 0 and solve:

$144 - 16k = 0$

$144 = 16k$

$9 = k$

12. $2x^2 - 5x = 7$

$2x^2 - 5x - 7 = 0$

$b^2 - 4ac = (-5)^2 - 4(2)(-7) = 81$

The discriminant is a positive number that is a perfect square, implying the equation will have two rational solutions.

13. $x = 5$ $x = -\dfrac{3}{2}$

$x - 5 = 0$ $3x = -2$

$3x + 2 = 0$

$(x - 5)(3x + 2) = 0$

$3x^2 - 13x - 10 = 0$

14. $x = 2$ $x = -2$ $x = 7$

$x - 2 = 0$ $x + 2 = 0$ $x - 7 = 0$

$(x - 2)(x + 2)(x - 7) = 0$

$(x^2 - 4)(x - 7) = 0$

$x^3 - 7x^2 - 4x + 28 = 0$

15. $4x^4 - 7x^2 - 2 = 0$

Replacing x^2 with y, we have:

$4y^2 - 7y - 2 = 0$

$(4y + 1)(y - 2) = 0$

$4y + 1 = 0$ or $y - 2 = 0$

$y = -\dfrac{1}{4}$ $y = 2$

Replacing y with x^2, we have:

$x^2 = -\dfrac{1}{4}$ or $x^2 = 2$

$x = \pm\sqrt{-\dfrac{1}{4}}$ $x = \pm\sqrt{2}$

$x = \pm\dfrac{1}{2}i$

16. $(2t + 1)^2 - 5(2t + 1) + 6 = 0$

Replacing $2t + 1$ with y, we have:

$y^2 - 5y + 6 = 0$

$(y - 2)(y - 3) = 0$

$y - 2 = 0$ or $y - 3 = 0$

$y = 2$ $y = 3$

Replacing y with $2t + 1$, we have:

$2t + 1 = 2$ or $2t + 1 = 3$

$t = \dfrac{1}{2}$ $t = 1$

17. $2t - 7\sqrt{t} + 3 = 0$

Replacing $\sqrt{t}$ with x, we have:

$2x^2 - 7x + 3 = 0$

$(2x - 1)(x - 3) = 0$

$2x - 1 = 0$ or $x - 3 = 0$

$x = \dfrac{1}{2}$ $x = 3$

Replacing x with $\sqrt{t}$, we have:

$\sqrt{t} = \dfrac{1}{2}$ or $\sqrt{t} = 3$

$t = \dfrac{1}{4}$ $t = 9$

Both solutions check.

The solution set is $\left\{\dfrac{1}{4}, 9\right\}$.

18. $16t^2 - 14t - h = 0$

Letting $a = 16$, $b = -14$ and $c = -h$, we have:

$$t = \frac{-(-14) \pm \sqrt{(-14)^2 - 4(16)(-h)}}{2(16)}$$

$$= \frac{14 \pm \sqrt{196 + 64h}}{32}$$

$$= \frac{14 \pm 2\sqrt{49 + 16h}}{32}$$

$$= \frac{7 \pm \sqrt{49 + 16h}}{16}$$

19. $y = x^2 - 2x - 3$

x-intercepts: when $y = 0$

$$0 = x^2 - 2x - 3$$

$$0 = (x + 1)(x - 3)$$

$x + 1 = 0$ or $x - 3 = 0$

$x = -1$ $x = 3$

Vertex: $x = \dfrac{-b}{2a} = \dfrac{-(-2)}{2(1)} = 1$

$$y = 1^2 - 2(1) - 3$$

$$y = -4$$

The vertex is $(1, -4)$.

See the graph in the back of the textbook.

20. $y = -x^2 + 2x + 8$

x-intercepts: when $y = 0$

$$0 = -x^2 + 2x + 8$$

$$0 = x^2 - 2x - 8$$

$$0 = (x + 2)(x - 4)$$

$x + 2 = 0$ or $x - 4 = 0$

$x = -2$ $x = 4$

Vertex: $x = \dfrac{-b}{2a} = \dfrac{-2}{2(-1)} = 1$

$$y = -1^2 + 2(1) + 8$$

$$y = 9$$

The vertex is $(1, 9)$.

See the graph in the back of the textbook.

21. $x^2 - x - 6 \le 0$

$(x + 2)(x - 3) \le 0$ opposite signs

$-2 \le x \le 3$

See the graph in the back of the textbook.

22. $2x^2 + 5x > 3$

$2x^2 + 5x - 3 > 0$

$(x + 3)(2x - 1) > 0$ same signs

$x < -3$ or $x > \dfrac{1}{2}$

See the graph in the back of the textbook

23. $P(x) = R(x) - C(x)$

    when $R = 25x - 0.1x^2$ and $C = 5x + 100$

    $P(x) = 25x - 0.1x^2 - (5x + 100)$

    $\quad = 25x - 0.1x^2 - 5x - 100$

    $\quad = -0.1x^2 + 20x - 100$

    The y-coordinate of the vertex represents the maximum profit.

    $x = \dfrac{-b}{2a} = \dfrac{-20}{2(-0.1)} = 100$

    $P_{max} = -0.1(100)^2 + 20(100) - 100$

    $\quad\quad = 900$

    Maximum profit $=$ \$900 by selling 100 items per week.

CHAPTER 11

Exponential and Logarithmic Functions

11.1 Exponential Functions

1. $g(x) = \left(\dfrac{1}{2}\right)^x$

   $g(0) = \left(\dfrac{1}{2}\right)^0 = 1$

3. $g(x) = \left(\dfrac{1}{2}\right)^x$

   $g(-1) = \left(\dfrac{1}{2}\right)^{-1} = 2$

5. $f(x) = 3^x$

   $f(-3) = 3^{-3} = \dfrac{1}{3^3} = \dfrac{1}{27}$

7. $f(x) = 3^x \qquad g(x) = \left(\dfrac{1}{2}\right)^x.$

   $f(2) = 3^2 = 9 \qquad g(-2) = \left(\dfrac{1}{2}\right)^{-2} = (2)^2 = 4$

   $f(2) + g(-2) = 9 + 4 = 13$

9. $y = 4^x$

   See the graph in the back of the textbook.

11. $y = 3^{-x}$

    See the graph in the back of the textbook.

13. $y = 2^{x+1}$

    See the graph in the back of the textbook.

15. $y = e^x$

    See the graph in the back of the textbook.

17. $y = 2x, \; y = x^2, \; y = 2^x$

    See the graph in the back of the textbook.

19. $y = b^x, \; b = 2, 4, 6, 8$

    See the graph in the back of the textbook.

21. $h(n) = A\left(\dfrac{2}{3}\right)^n$, $A = 6$ and $n = 5$,

    $h(5) = 6\left(\dfrac{2}{3}\right)^5 = \dfrac{6 \cdot 32}{243} \approx 0.79$

    It will bounce 0.79 ft.

    See the graph in the back of the textbook

23. $Q(t) = 0.85^t$ It will be 1/2 when $t = 4.3$

    See the graph in the back of the textbook.

25.. $A(t) = P\left(1 + \dfrac{r}{n}\right)^{nt}$

    (a) $A(t) = 1200\left(1 + \dfrac{.06}{4}\right)^{4t}$

    (b) $A(8) = 1200\left(1 + \dfrac{.06}{4}\right)^{4(8)} = \$1,932.39$

    (c) $2400 = 1200\left(1 + \dfrac{.06}{4}\right)^{4t}$, $t \approx 11.64$ years

    Let $Y_1 = 1200\left(1 + \dfrac{.06}{4}\right)^{4x}$ and $Y_2 = 2400$

    See the graph $\rightarrow$

    (d) $A(t) = Pe^{rt}$

       $A(8) = 1200e^{.06(8)} = \$1,939.29$

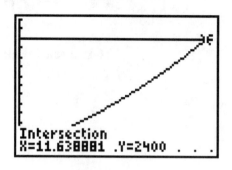

Intersection
X=11.638881 .Y=2400

27. $V(t) = 450,000(1 - 0.30)^t$

    (a) $V(3.5) = 450,000(1 - 0.30)^{3.5}$

         $= \$129,138.48$

    (b) Domain: $0 \le t \le 6$

    (c) See the graph in the back of the textbook.

    (d) Range: $52,942.05 \le V \le 450,000$

    (e) From the graph $t \approx 4.7$ years

29. $f(x) = 50 \cdot 4^x$

    $f(1) = 50 \cdot 4^1 = 200$

    $f(2) = 50 \cdot 4^2 = 800$

    $f(3) = 50 \cdot 4^3 = 3200$

31. $V(t) = 150 \cdot 2^{t/3}$ for $t \ge 0$

    See the graph in the back of the textbook.

33. $\{(-2, 6), (-2, 8), (2, 3)\}$

    $D = \{-2, 2\}$, $R = \{3, 6, 8\}$

    Not a function

35. $y = \dfrac{-4}{x^2 + 2x - 35}$

    $= \dfrac{-4}{(x+7)(x-5)}$

    Domain: $\{x | x \ne -7, x \ne 5\}$

37. $f(x) = 2x^2 - 18$

    $f(0) = 2(0)^2 - 18 = -18$

39. $g(x) = 2x - 6$

$\dfrac{g(x+h) - g(x)}{h} = \dfrac{2(x+h) - 6 - (2x - 6)}{h}$

$= \dfrac{2x + 2h - 6 - 2x + 6}{h}$

$= \dfrac{2h}{h}$

$= 2$

41. $f(x) = 2x^2 - 18$ and $g(x) = 2x - 6$

(a) $f(0) = 1$ (b) $f(-1) = \frac{1}{3}$ (c) $f(1) = 3$

(d) $g(0) = 1$ (e) $g(1) = \frac{1}{2}$ (f) $g(-1) = 2$

(g) $f[g(0)] = f(1) = 3$ (h) $g[f(0)] = g(1) = \frac{1}{2}$

43. (a) $y = 2^x$ and $y = x^2$

    See the graph in the back of the textbook.

    $y = x^2$ appears to be taking over.

(b) $2^2 = 4$; $2^4 = 16$ and $4^2 = 16$

(c) See the graph in the back of the textbook.

    $y = 2^x$ grows faster.

    $2^{100} = 1.3 \times 10^{30}$

    $100^2 = 1 \times 10^4$

11.2 The Inverse of a Function

1. $f(x) = 3x - 1$ means $y = 3x - 1$

    Inverse: $x = 3y - 1$.

        $x + 1 = 3y$

          $y = \dfrac{x+1}{3}$

    $f^{-1}(x) = \dfrac{x+1}{3}$

3. $f(x) = x^3$ means $y = x^3$

    Inverse: $x = y^3$

        $y = \sqrt[3]{x}$

    $f^{-1}(x) = \sqrt[3]{x}$

5. $f(x) = \dfrac{x-3}{x-1}$ means $y = \dfrac{x-3}{x-1}$

    Inverse: $x = \dfrac{y-3}{y-1}$

    $xy - x = y - 3$

    $y(x-1) = x - 3$

        $y = \dfrac{x-3}{x-1}$

    $f^{-1}(x) = \dfrac{x-3}{x-1}$

7. $f(x) = \dfrac{x-3}{4}$ means $y = \dfrac{x-3}{4}$

    Inverse: $x = \dfrac{y-3}{4}$

        $y = 4x + 3$

    $f^{-1}(x) = 4x + 3$

9. $f(x) = \dfrac{1}{2}x - 3$ means $y = \dfrac{1}{2}x - 3$

Inverse: $x = \dfrac{1}{2}y - 3.$

$$x + 3 = \dfrac{1}{2}y$$

$$y = 2(x + 3)$$

$$f^{-1}(x) = 2x + 6$$

11. $f(x) = \dfrac{2x+1}{3x+1}$ means $y = \dfrac{2x+1}{3x+1}$

Inverse: $x = \dfrac{2y+1}{3y+1}$

$$3xy + x = 2y + 1$$

$$3xy - 2y = 1 - x$$

$$y = \dfrac{1-x}{3x-2}$$

$$f^{-1}(x) = \dfrac{1-x}{3x-2}$$

13. $y = 2x - 1$

Inverse: $x = 2y - 1.$

$$x + 1 = 2y$$

$$y = \dfrac{x+1}{2}$$

$$f^{-1}x = \dfrac{x+1}{2}$$

See the graph in the back of the textbook.

15. $y = x^2 - 3$

Inverse: $x = y^2 - 3$

$$y^2 = x + 3$$

$$y = \pm\sqrt{x+3}$$

See the graph in the back of the textbook

17. $y = x^2 - 2x - 3$

Inverse: $x = y^2 - 2y - 3.$

See the graph in the back of the textbook.

19. $y = 3^x$

Inverse: $x = 3^y$

See the graph in the back of the textbook.

21. $y = 4$

Inverse: $x = 4$

See the graph in the back of the textbook.

23. $y = \dfrac{1}{2}x^3$

Inverse: $x = \dfrac{1}{2}y^3$

$$y^3 = 2x$$

$$y = \sqrt[3]{2x}$$

See the graph in the back of the textbook

25. $y = \dfrac{1}{2}x + 2$

Inverse: $x = \dfrac{1}{2}y + 2$

$$x - 2 = \dfrac{1}{2}y$$

$$y = 2(x - 2)$$

See the graph in the back of the textbook.

27. $y = \sqrt{x+2}$

Inverse: $x = \sqrt{y+2}$

See the graph in the back of the textbook

29. (a) One - to - one

(b) Not one - to - one.

Fails the horizontal line test.

(c) One - to - one

31. a. $f(x) = 3x - 2$

$f(2) = 3 \cdot 2 - 2 = 4$

b. $f^{-1}(x) = \dfrac{x+2}{3}$

$f^{-1}(2) = \dfrac{2+2}{3} = \dfrac{4}{3}$

c. $f^{-1}(2) = \dfrac{4}{3}$

$f(x) = 3x - 2$

$f\left(\dfrac{4}{3}\right) = 3\left(\dfrac{4}{3}\right) - 2 = 4 - 2 = 2$

$f\left[f^{-1}(2)\right] = f\left(\dfrac{4}{3}\right) = 2$

d. $f(2) = 4$

$f^{-1}(x) = \dfrac{x+2}{3}$

$f^{-1}(4) = \dfrac{4+2}{3} = 2$

$f^{-1}\left[f(2)\right] = f^{-1}(4) = 2$

33. $f(x) = \dfrac{1}{x}$ means $y = \dfrac{1}{x}$

Inverse: $x = \dfrac{1}{y}$.

$xy = 1$

$y = \dfrac{1}{x}$

$f^{-1}(x) = \dfrac{1}{x}$

35. $f(x) = \dfrac{x}{7} - 2$

Function; (a) Divide by 7

(b) Subtract 2

Inverse; (a) Add 2

(b) Multiply by 7

$f^{-1}(x) = 7(x+2)$

37. (a) $f\left(g(-3)\right) = f(2) = -3$

(b) $g\left(f(-6)\right) = g(3) = -6$

(c) $g\left(f(2)\right) = g(-3) = 2$

(d) $f\left(g(3)\right) = f(-6) = 3$

(e) $f\left(g(-2)\right) = f(3) = -2$

(f) $g\left(f(3)\right) = g(-2) = 3$

(g) f and g are inverse functions of each other.

39. $(2x-1)^2 = 25$

$2x - 1 = \pm 5$

$2x = 1 \pm 5$

$x = \dfrac{1 \pm 5}{2}$

$x = 3$ or $x = -2$

41. $x^2 - 10x + 25$ (add 25)

43.
$$x^2 - 10x + 8 = 0$$
$$x^2 - 10x = -8$$
$$x^2 - 10x + 25 = -8 + 25$$
$$(x-5)^2 = 17$$
$$x - 5 = \pm\sqrt{17}$$
$$x = 5 \pm \sqrt{17}$$

45.
$$3x^2 - 6x + 6 = 0$$
$$x^2 - 2x = -2$$
$$x^2 - 2x + 1 = -2 + 1$$
$$(x-1)^2 = -1$$
$$x - 1 = \pm i$$
$$x = 1 \pm i$$

47. $f(x) = 3x + 5$ means $y = 3x + 5$
Inverse: $x = 3y + 5$
$$x - 5 = 3y$$
$$\frac{x-5}{3} = y$$
$$f^{-1}(x) = \frac{x-5}{3}$$
$$f\left[f^{-1}(x)\right] = 3\left(\frac{x-5}{3}\right) + 5 = x - 5 + 5 = x$$

49. $f(x) = x^3 + 1$ means $y = x^3 + 1$
Inverse: $x = y^3 + 1$.
$$y^3 = x - 1$$
$$y = \sqrt[3]{x-1}$$
$$f^{-1}(x) = \sqrt[3]{x-1}$$
$$f\left(f^{-1}(x)\right) = \left(\sqrt[3]{x-1}\right)^3 + 1 = x - 1 + 1 = x$$

51. $f(x) = \dfrac{x-4}{x-2}$ means $y = \dfrac{x-4}{x-2}$

Inverse: $x = \dfrac{y-4}{y-2}$

$$(y-2)x = y - 4$$
$$xy - 2x = y - 4$$
$$xy - y = 2x - 4$$
$$y(x-1) = 2x - 4$$
$$y = \frac{2x-4}{x-1}$$

$$f^{-1}(x) = \frac{2x-4}{x-1}$$

$$f\left[f^{-1}(x)\right] = \frac{\left(\dfrac{2x-4}{x-1}\right) - 4}{\left(\dfrac{2x-4}{x-1}\right) - 2}$$

$$= \frac{2x - 4 - 4(x-1)}{2x - 4 - 2(x-1)}$$

$$= \frac{2x - 4 - 4x + 4}{2x - 4 - 2x + 2} = \frac{-2x}{-2} = x$$

53. (a) $f(0) = 1$ (b) $f(1) = 2$ (c) $f(2) = 5$

(d) $f^{-1}(1) = 0$ (e) $f^{-1}(2) = 1$ (f) $f^{-1}(5) = 2$

(g) $f^{-1}\left[f(2)\right] = f^{-1}(5) = 2$

(h) $f\left[f^{-1}(5)\right] = f(2) = 5$

11.3 Logarithms are Exponents

1. $2^4 = 16$
   $\log_2 16 = 4$

3. $125 = 5^3$
   $\log_5 125 = 3$

5. $0.01 = 10^{-2}$
   $\log_{10} 0.01 = -2$

7. $2^{-5} = \dfrac{1}{32}$
   $\log_2 \dfrac{1}{32} = -5$

9. $\left(\dfrac{1}{2}\right)^{-3} = 8$
   $\log_{1/2} 8 = -3$

11. $27 = 3^3$
    $\log_3 27 = 3$

13. $\log_{10} 100 = 2$
    $10^2 = 100$

15. $\log_2 64 = 6$
    $2^6 = 64$

17. $\log_8 1 = 0$
    $8^0 = 1$

19. $\log_{10} 0.001 = -3$
    $10^{-3} = 0.001$

21. $\log_6 36 = 2$
    $6^2 = 36$

23. $\log_5 \dfrac{1}{25} = -2$
    $5^{-2} = \dfrac{1}{25}$

25. $\log_3 x = 2$
    $3^2 = x$
    $9 = x$

27. $\log_5 x = -3$
    $5^{-3} = x$
    $\dfrac{1}{125} = x$

29. $\log_2 16 = x$
    $2^x = 16$
    $2^x = 2^4$
    $x = 4$

31. $\log_8 2 = x$
    $8^x = 2$
    $2^{3x} = 2^1$
    $3x = 1$
    $x = \dfrac{1}{3}$

33. $\log_x 4 = 2$

    $x^2 = 4$

    $x^2 = 2^2$

    $x = 2$

35. $\log_x 5 = 3$

    $x^3 = 5$

    $x^3 = \left(5^{1/3}\right)^3$

    $x = \sqrt[3]{5}$

37. $y = \log_3 x$ is $3^y = x$

    See the graph in the back of the textbook.

39. $y = \log_{1/3} x$ is $\left(\dfrac{1}{3}\right)^y = x$

    See the graph in the back of the textbook.

41. $y = \log_5 x$ is $5^y = x$

    See the graph in the back of the textbook.

43. $y = \log_{10} x$ is $10^y = x$

    See the graph in the back of the textbook.

45. If $(2,9)$ is a point on the graph of $y = b^x$,

    then $9 = b^2$ or $b = 3$. Also, if $x = 0$,

    then $y = 3^x = 3^0 = 1$ and $(0,\ 1)$ is also

    a point on the graph and the equation is

    $y = 3^x$.

47. If $(9,-2)$ is a point on the graph of $y = \log_b x$,

    then $-2 = \log_b 9$ or $b^{-2} = 9 \rightarrow b = \dfrac{1}{3}$. Also,

    if $y = 0$, then $0 = \log_{1/3} 1$ or $\left(\dfrac{1}{3}\right)^0 = 1$ and $(1,\ 0)$

    is also a point on the graph and the equation is

    $y = \log_{1/3} x$.

49. $\log_2 16 = \log_2 2^4 = 4$

51. $\log_{25} 125 = \log_{25} 25^{3/2} = \dfrac{3}{2}$

53. $\log_{10} 1000 = \log_{10} 10^3 = 3$

55. $\log_3 3 = \log_3 3^1 = 1$

57. $\log_5 1 = \log_5 5^0 = 0$

59. $\log_3 \left(\log_6 6\right) = \log_3 1$

    $\qquad\qquad\quad = \log_3 3^0$

    $\qquad\qquad\quad = 0$

61. $\log_4 \left[\log_2 \left(\log_2 16\right)\right] = \log_4 \left[\log_2 \left(\log_2 2^4\right)\right]$

    $\qquad\qquad\qquad\quad = \log_4 \left(\log_2 4\right)$

    $\qquad\qquad\qquad\quad = \log_4 \left(\log_2 2^2\right)$

    $\qquad\qquad\qquad\quad = \log_4 2$

    $\qquad\qquad\qquad\quad = \dfrac{1}{2}$

63. See the table in the back of the textbook.

65. $M = \log_{10} T,\ T = 100$

$M = \log_{10} 100$

$M = \log_{10} 10^2$

$M = 2$

67. $M = \log_{10} T,\ M = 8$

$8 = \log_{10} T$

$T = 10^8 = 100{,}000{,}000$

69. $2x^2 + 4x - 3$

$$x = \frac{-4 \pm \sqrt{4^2 - 4(2)(-3)}}{2(2)} = \frac{-4 \pm \sqrt{16 + 24}}{4} = \frac{-4 \pm \sqrt{40}}{4} = \frac{-4 \pm 2\sqrt{10}}{4} = \frac{2\left(-2 \pm \sqrt{10}\right)}{4} = \frac{-2 \pm \sqrt{10}}{2}$$

71. $(2y - 3)(2y - 1) = -4$

$4y^2 - 8y + 7 = 0$

$$y = \frac{-(-8) \pm \sqrt{(-8)^2 - 4(4)(7)}}{2(4)} = \frac{8 \pm \sqrt{-48}}{8} = \frac{8 \pm 4i\sqrt{3}}{8} = \frac{2 \pm i\sqrt{3}}{2}$$

73. $t^3 - 125 = 0$

$(t - 5)(t^2 + 5t + 25) = 0$

$t - 5 = 0$ or $t^2 + 5t + 25 = 0$

$t = 5 \qquad t = \dfrac{-5 \pm \sqrt{5^2 - 4(1)(25)}}{2(1)} = \dfrac{-5 \pm \sqrt{25 - 100}}{2} = \dfrac{-5 \pm \sqrt{-75}}{2} = \dfrac{-5 \pm i\sqrt{75}}{2} = \dfrac{-5 \pm 5i\sqrt{3}}{2}$

The solutions are $5,\ \dfrac{-5 + 5i\sqrt{3}}{2},\ \dfrac{-5 - 5i\sqrt{3}}{2}$.

75. $4x^5 - 16x^4 = 20x^3$

$4x^5 - 16x^4 - 20x^3 = 0$

$4x^3 \left(x^2 - 4x - 5\right) = 0$

$4x^3 (x + 1)(x - 5) = 0$

$4x^3 = 0$ or $x + 1 = 0$ or $x - 5 = 0$

$x = 0 \qquad\quad x = -1 \qquad\quad x = 5$

77. $\dfrac{1}{x - 3} + \dfrac{1}{x + 2} = 1$

$x + 2 + x - 3 = (x - 3)(x + 2)$

$2x - 1 = x^2 - x - 6$

$0 = x^2 - 3x - 5$

$$x = \frac{-(-3) \pm \sqrt{(-3)^2 - 4(1)(-5)}}{2(1)}$$

$$= \frac{3 \pm \sqrt{9 + 20}}{2}$$

$$= \frac{3 \pm \sqrt{29}}{2}$$

79. (a) See the table in the back of the textbook.
    (b) See the table in the back of the textbook
    (c) $f(x) = 8^x$
    (d) $f^{-1}(x) = \log_8 x$

11.4 Properties of Logarithms

1. $\log_3 4x = \log_3 4 + \log_3 x$

3. $\log_6 \dfrac{5}{x} = \log_6 5 - \log_6 x$

5. $\log_2 y^5 = 5 \log_2 y$

7. $\log_9 \sqrt[3]{z} = \dfrac{1}{3} \log_9 z$

9. $\log_6 x^2 y^4 = \log_6 x^2 + \log_6 y^4$
   $= 2 \log_6 x + 4 \log_6 y$

11. $\log_5 \sqrt{x} \cdot y^4 = \dfrac{1}{2} \log_5 x + 4 \log_5 y$

13. $\log_b \dfrac{xy}{z} = \log_b xy - \log_b z$
    $= \log_b x + \log_b y - \log_b z$

15. $\log_{10} \dfrac{4}{xy} = \log_{10} 4 - \log_{10} x - \log_{10} y$

17. $\log_{10} \dfrac{x^2 y}{\sqrt{z}} = \log_{10} \dfrac{x^2 y}{z^{1/2}}$
    $= \log_{10} x^2 y - \log_{10} z^{1/2}$
    $= \log_{10} x^2 + \log_{10} y - \log_{10} z^{1/2}$
    $= 2 \log_{10} x + \log_{10} y - \dfrac{1}{2} \log_{10} z$

19. $\log_{10} \dfrac{x^3 \cdot \sqrt{y}}{z^4} = \log_{10} \dfrac{x^3 y^{1/2}}{z^4}$
    $= \log_{10} x^3 + \log_{10} y^{1/2} - \log_{10} z^4$
    $= 3 \log_{10} x + \dfrac{1}{2} \log_{10} y - 4 \log_{10} z$

21. $\log_b \sqrt[3]{\dfrac{x^2 y}{z^4}} = \log_b \left(\dfrac{x^2 y}{z^4}\right)^{1/3}$

$\qquad = \dfrac{1}{3} \log_b \dfrac{x^2 y}{z^4}$

$\qquad = \dfrac{1}{3}\left(\log_b x^2 + \log_b y - \log_b z^4\right)$

$\qquad = \dfrac{1}{3}\left(2 \log_b x + \log_b y - 4 \log_b z\right)$

$\qquad = \dfrac{2}{3} \log_b x + \dfrac{1}{3} \log_b y - \dfrac{4}{3} \log_b z$

23. $\log_b x + \log_b z = \log_b xz$

25. $2 \log_3 x - 3 \log_3 y = \log_3 x^2 - \log_3 y^3$

$\qquad = \log_3 \dfrac{x^2}{y^3}$

27. $\dfrac{1}{2} \log_{10} x + \dfrac{1}{3} \log_{10} y = \log_{10} \sqrt{x} + \log_{10} \sqrt[3]{y}$

$\qquad = \log_{10} \sqrt{x} \sqrt[3]{y}$

29. $3 \log_2 x + \dfrac{1}{2} \log_2 y - \log_2 z$

$\qquad = \log_2 x^3 + \log_2 y^{1/2} - \log_2 z$

$\qquad = \log_2 x^3 y^{1/2} - \log_2 z$

$\qquad = \log_2 \dfrac{x^3 y^{1/2}}{z}$

$\qquad = \log_2 \dfrac{x^3 \sqrt{y}}{z}$

31. $\dfrac{1}{2} \log_2 x - 3 \log_2 y - 4 \log_2 z$

$\qquad = \log_2 \sqrt{x} - \log_2 y^3 - \log_2 z^4$

$\qquad = \log_2 \dfrac{\sqrt{x}}{y^3 z^4}$

33. $\dfrac{3}{2} \log_{10} x - \dfrac{3}{4} \log_{10} y - \dfrac{4}{5} \log_{10} z$

$\qquad = \log_{10} x^{3/2} - \log_{10} y^{3/4} - \log_{10} z^{4/5}$

$\qquad = \log_{10} \dfrac{x^{3/2}}{y^{3/4} z^{4/5}}$

35. $\log_2 x + \log_2 3 = 1$

$\qquad \log_2 x(3) = 1$

$\qquad 3x = 2^1$

$\qquad x = \dfrac{2}{3}$

37. $\log_3 x - \log_3 2 = 2$

$\qquad \log_3 \dfrac{x}{2} = 2$

$\qquad 3^2 = \dfrac{x}{2}$

$\qquad 9 = \dfrac{x}{2}$

$\qquad x = 18$

39. $\log_3 x + \log_3 (x-2) = 1$

$\qquad \log_3 x(x-2) = 1$

$\qquad x(x-2) = 3^1$

$\qquad x^2 - 2x - 3 = 0$

$\qquad (x-3)(x+1) = 0$

$\qquad x = 3 \quad \text{or} \quad x = -1$

Possible solutions: -1, 3; only 3 checks

41. $\log_3 (x+3) - \log_3 (x-1) = 1$

$$\log_3 \frac{x+3}{x-1} = 1$$

$$\frac{x+3}{x-1} = 3^1$$

$$(x-1)\frac{x+3}{x-1} = 3(x-1)$$

$$x+3 = 3x-3$$

$$6 = 2x$$

$$x = 3$$

43. $\log_2 x + \log_2 (x-2) = 3$

$$\log_2 x(x-2) = 3$$

$$x(x-2) = 2^3$$

$$x^2 - 2x - 8 = 0$$

$$(x+2)(x-4) = 0$$

$$x = -2 \quad \text{or} \quad x = 4$$

Possible solutions: -2 and 4; only 4 checks.

45. $\log_8 x + \log_8 (x-3) = \frac{2}{3}$

$$\log_8 \left[x(x-3) \right] = \frac{2}{3}$$

$$x(x-3) = 8^{2/3}$$

$$x(x-3) = 4$$

$$x^2 - 3x - 4 = 0$$

$$(x-4)(x+1) = 0$$

$$x-4 = 0 \quad \text{or} \quad x+1 = 0$$

$$x = 4 \qquad\qquad x = -1$$

Possible solutions: -1, 4; only 4 checks

47. $\log_5 \sqrt{x} + \log_5 \sqrt{6x+5} = 1$

$$\log_5 \sqrt{x}\ \sqrt{6x+5} = 1$$

$$\sqrt{x}\ \sqrt{6x+5} = 5^1$$

$$x(6x+5) = 25$$

$$6x^2 + 5x - 25 = 0$$

$$(2x+5)(3x-5) = 0$$

$$x = -\frac{5}{2} \quad \text{or} \quad x = \frac{5}{3}$$

Possible solutions: $-\frac{5}{2}$ and $\frac{5}{3}$; only $\frac{5}{3}$ checks.

49. $D = 10\log_{10}\left(\frac{I}{I_0}\right) = 10\left(\log_{10} I - \log_{10} I_0\right)$

51. $\log_{10} 8 = 0.903$, and $\log_{10} 5 = 0.699$

(a) $\log_{10} 40 = \log_{10}(5 \cdot 8)$

$$= \log_{10} 5 + \log_{10} 8$$

$$= 0.699 + 0.903$$

$$= 1.602$$

(b) $\log_{10} 320 = \log_{10}\left(5 \cdot 8^2\right)$

$$= \log_{10} 5 + 2\log_{10} 8$$

$$= 0.699 + 2(0.903)$$

$$= 2.505$$

(c) $\log_{10} 1600 = \log_{10}\left(5^2 \cdot 8^2\right)$

$$= 2\log_{10} 5 + 2\log_{10} 8$$

$$= 2(0.699) + 2(0.903)$$

$$= 3.204$$

53. $\text{pH} = 6.1 + \log_{10}\left(\frac{x}{y}\right) = 6.1 + \log_{10} x - \log_{10} y$

55.

$M = 0.21\left(\log_{10} a - \log_{10} b\right),$
$\quad a = 1 \text{ and } b = 10^{-12}$
$M = 0.21\left(\log_{10} 1 - \log_{10} 10^{-12}\right)$
$M = 0.21\left(\log_{10} \dfrac{1}{10^{-12}}\right)$
$M = 0.21\left(\log_{10} 10^{12}\right)$
$\dfrac{M}{0.21} = \log_{10} 10^{12}$

$10^{m/0.21} = 10^{12}$
$\dfrac{M}{0.21} = 12$
$M = 2.52$

$M = 0.21 \log_{10} \dfrac{a}{b}, \quad a = 1 \text{ and } b = 10^{-12}$
$M = 0.21 \log_{10} \dfrac{1}{10^{-12}}$
$M = 0.21 \log_{10} 10^{12}$
$\dfrac{M}{0.21} = \log_{10} 10^{12}$
$\dfrac{M}{0.21} = 12$
$M = 2.52$

57. $2x^2 - 5x + 4 = 0$

Use $a = 2$, $b = -5$ and $c = 4$ in the discriminant

$D = b^2 - 4ac = (-5)^2 - 4(2)(4) = 25 - 32 = -7$

Since $D = -7$, there are two complex solutions.

59. $x = -3 \quad \text{or} \quad x = 5$
$x + 3 = 0 \qquad x - 5 = 0$
$(x + 3)(x - 5) = 0$
$x^2 - 2x - 15 = 0$

61. $y = \dfrac{2}{3} \quad \text{or} \quad y = 3$
$3y = 2 \qquad y - 3 = 0$
$3y - 2 = 0$
$(3y - 2)(y - 3) = 0$
$3y^2 - 11y + 6 = 0$

11.5 Common Logarithms and Natural Logarithms

1. $\log 378 = 2.5775$

3. $\log 37.8 = 1.5775$

5. $\log 3{,}780 = 3.5775$

7. $\log 0.0378 = -1.4225$

9. $\log 37{,}800 = 4.5775$

11. $\log 600 = 2.7782$

13. $\log 2{,}010 = 3.3032$

15. $\log 0.00971 = -2.0128$

17. $\log 0.0314 = -1.5031$

19. $\log 0.399 = -0.3990$

21. $\log x = 2.8802$

$x = 10^{2.8802}$

$x = 759$

23. $\log x = -2.1198$

$x = 10^{-2.1198}$

$x = 0.00759$

25. $\log x = 3.1553$

$x = 10^{3.1553}$

$x = 1,430$

27. $\log x = -5.3497$

$x = 10^{-5.3497}$

$x = 0.00000447$

29. $\log x = -7.0372$

$x = 10^{-7.0372}$

$x = 0.0000000918$

31. $\log x = 10$

$x = 10^{10}$

33. $\log x = -10$

$x = 10^{-10}$

35. $\log x = 20$

$x = 10^{20}$

37. $\log x = -2$

$x = 10^{-2}$

$x = \dfrac{1}{100}$

39. $\log x = \log_2 8$

$\log x = \log_2 2^3$

$\log x = 3$

$x = 10^3$

$x = 1000$

41.

$\log x^2 = (\log x)^2$

$2 \log x = (\log x)^2$

$(\log x)^2 - 2 \log x = 0$

$\log x (\log x - 2) = 0$

$\log x - 2 = 0$ or $\log x = 0$

$\log x = 2$ $x = 10^0$

$x = 10^2$ $x = 1$

$x = 100$

Solutions are 1 and 100

43. The TI-83 says "ERR: NONREAL ANS", indicating that -10 is not in the domain of $f(x) = \log x$.

45. 1906 earthquake:

$\log T = 8.3$

$T = 10^{8.3}$

$= 2.00 \times 10^8$

Test earthquake:

$\log T = 5.0$

$T = 10^{5.0}$

$= 1.00 \times 10^5$

$\text{Ratio} = \dfrac{2.00 \times 10^8}{1.00 \times 10^5} = 2 \times 10^3$

The 1906 quake was 2,000 times larger

47. $f(x) = (1+x)^{1/x}$

See the table in the back of the textbook. $f(x)$ seems to approach 2.7183 as x gets closer and closer to zero.

49. Let $\dfrac{R^3}{T^2} = k,$ where k is a constant $\neq 0$

Take the logarithm of both sides to get

$\log \dfrac{R^3}{T^2} = \log k,$ where $\log k$ is a constant

$\log R^3 - \log T^2 = c$

$3 \log R - 2 \log T = c$

51. $pH = -\log \left[H^+ \right],\ \left[H^+ \right] = 6.5 \times 10^{-4}$

$pH = -\log \left[6.5 \times 10^{-4} \right]$

$pH \approx 3.19$

53. $pH = -\log \left[H^+ \right],$ $pH = 4.75$ becomes

$4.75 = -\log \left[H^+ \right]$

$\left[H^+ \right] = 10^{-4.75}$

$\left[H^+ \right] = 1.78 \times 10^{-5}$

55. $M = \log T,\ M = 5.5$

$5.5 = \log T$

$T = 10^{5.5}$

$T = 3.16 \times 10^5$

57. $M = \log T,\ M = 8.3$ becomes

$8.3 = \log T$

$T = 10^{8.3}$

$T = 2.00 \times 10^8$

59. $M = \log T,\ M_1 = 6.5,\ M_2 = 5.5$

$6.5 = \log T_1 \Rightarrow T_1 = 10^{6.5}$

$5.5 = \log T_2 \Rightarrow T_2 = 10^{5.5}$

$\dfrac{T_1}{T_2} = \dfrac{10^{6.5}}{10^{5.5}} = 10^1$

61. If $P = \$9000,\ t = 5,\ W = \$4,500$

$\log(1 - r) = \dfrac{1}{t} \log \dfrac{W}{P}$ becomes

$\log(1 - r) = \dfrac{1}{5} \log \dfrac{4500}{9000}$

$= 0.20 \log 0.5$

$= 0.20(-0.3010)$

$= -0.0602$

$1 - r = 10^{-0.0602}$

$1 - r = 0.871$

$-r = -0.129$

$r = 0.129 = 12.9\%$

63. If $P = \$7550,\ t = 5,$ and $W = \$5750$

$\log(1 - r) = \dfrac{1}{t} \log \dfrac{W}{P}$

$= \dfrac{1}{5} \log \dfrac{5750}{7550}$

$= -0.0237$

$1 - r = 10^{-0.0237}$

$1 - r = 0.947$

$-r = -0.053$

$r = 0.053$ or 5.3%

65. $\ln e = 1$

67. $\ln e^5 = 5$

69. $\ln e^x = x \ln e = x(1) = x$

71. $\ln 10e^{3t} = \ln 10 + \ln e^{3t}$

$= \ln 10 + 3t \ln e$

$= \ln 10 + 3t$

73. $\ln Ae^{-2t} = \ln A + \ln e^{-2t}$
$\qquad = \ln A - 2t \ln e$
$\qquad = \ln A - 2t$

75. $\ln 15 = \ln (3)(5)$
$\qquad = \ln 3 + \ln 5$
$\qquad = 1.0986 + 1.6094$
$\qquad = 2.7080$

77. $\ln \dfrac{1}{3} = \ln 1 - \ln 3$
$\qquad = 0 - 1.0986$
$\qquad = -1.0986$

79. $\ln 9 = \ln 3^2$
$\qquad = 2 \ln 3$
$\qquad = 2.1972$

81. $\ln 16 = \ln 2^4$
$\qquad = 4 \ln 2$
$\qquad = 4(0.6931)$
$\qquad = 2.7724$

83. $\qquad x^4 - 2x^2 - 8 = 0$
$\qquad (x^2 + 2)(x^2 - 4) = 0$
$\qquad x^2 + 2 = 0 \quad \text{or} \quad x^2 - 4 = 0$
$\qquad\qquad x^2 = -2 \qquad\qquad x^2 = 4$
$\qquad\qquad x = \pm i\sqrt{2} \qquad\qquad x = \pm 2$

85. $x^{2/3} - 5x^{1/3} + 6 = 0$
$\qquad y^2 - 5y + 6 = 0, \quad y = x^{1/3}$
$\qquad (y - 2)(y - 3) = 0$
$\qquad\qquad y - 2 = 0 \quad \text{or} \quad y - 3 = 0$
$\qquad\qquad\qquad y = 2 \qquad\qquad y = 3$
$\qquad\qquad\quad x^{1/3} = 2 \qquad\quad x^{1/3} = 3$
$\qquad\qquad\qquad x = 8 \qquad\qquad x = 27$

87. $\qquad 2x - 5\sqrt{x} + 3 = 0$
$\qquad 2y^2 - 5y + 3 = 0, \quad y = \sqrt{x}$
$\qquad (2y - 3)(y - 1) = 0$
$\qquad\qquad 2y - 3 = 0 \quad \text{or} \quad y - 1 = 0$
$\qquad\qquad\quad 2y = 3 \qquad\qquad y = 1$
$\qquad\qquad\qquad y = \dfrac{3}{2}$
$\qquad\qquad\quad \sqrt{x} = \dfrac{3}{2} \quad \text{or} \quad \sqrt{x} = 1$
$\qquad\qquad\qquad x = \dfrac{9}{4} \qquad\qquad x = 1$

89. $(3x + 1) - 6\sqrt{3x + 1} + 8 = 0$
$\qquad y^2 - 6y + 8 = 0, \quad y = \sqrt{3x + 1}$
$\qquad (y - 2)(y - 4) = 0$
$\qquad\qquad y = 2 \quad \text{or} \quad y = 4$
$\qquad\quad \sqrt{3x + 1} = 2 \quad \text{or} \quad \sqrt{3x + 1} = 4$
$\qquad\qquad 3x + 1 = 4 \qquad\quad 3x + 1 = 16$
$\qquad\qquad\quad x = 1 \qquad\qquad x = 5$

91. $kx^2 + 4x - k = 0$
$$x = \frac{-4 \pm \sqrt{4^2 - 4(k)(-k)}}{2k}$$
$$= \frac{-4 \pm \sqrt{4(4 + k^2)}}{2k}$$
$$= \frac{-4 \pm 2\sqrt{4 + k^2}}{2k}$$
$$= \frac{-2 \pm \sqrt{4 + k^2}}{k}$$

11.6 Exponential Equations and Change of Base

1. $3^x = 5$

    $\log 3^x = \log 5$

    $x \log 3 = \log 5$

    $x = \dfrac{\log 5}{\log 3}$

    $x = 1.4650$

3. $5^x = 3$

    $\log 5^x = \log 3$

    $x \log 5 = \log 3$

    $x = \dfrac{\log 3}{\log 5}$

    $x = 0.6826$

5. $5^{-x} = 12$

    $\log 5^{-x} = \log 12$

    $-x \log 5 = \log 12$

    $-x = \dfrac{\log 12}{\log 5}$

    $-x = 1.5440$

    $x = -1.5440$

7. $12^{-x} = 5$

    $\log 12^{-x} = \log 5$

    $-x \log 12 = \log 5$

    $-x = \dfrac{\log 5}{\log 12}$

    $x = -0.6477$

9. $8^{x+1} = 4$

    $\log 8^{x+1} = \log 4$

    $(x+1) \log 8 = \log 4$

    $x + 1 = \dfrac{\log 4}{\log 8}$

    $x = \dfrac{\log 4}{\log 8} - 1$

    $x = \dfrac{0.6021}{0.9031} - 1$

    $x = 0.6667 - 1$

    $x = -0.3333$

11. $4^{x-1} = 4$

    $\log 4^{x-1} = \log 4$

    $(x-1) \log 4 = \log 4$

    $x - 1 = \dfrac{\log 4}{\log 4}$

    $x = \dfrac{\log 4}{\log 4} + 1$

    $x = 2.0000$

13. $3^{2x+1} = 2$

    $\log 3^{2x+1} = \log 2$

    $(2x+1) \log 3 = \log 2$

    $2x + 1 = \dfrac{\log 2}{\log 3}$

    $2x = \dfrac{\log 2}{\log 3} - 1$

    $x = \dfrac{1}{2}\left(\dfrac{\log 2}{\log 3} - 1 \right)$

    $x = -0.1845$

15. $3^{1-2x} = 2$

    $\log 3^{1-2x} = \log 2$

    $(1-2x) \log 3 = \log 2$

    $(1-2x) = \dfrac{\log 2}{\log 3}$

    $-2x = \dfrac{\log 2}{\log 3} - 1$

    $x = -\dfrac{1}{2}\left(\dfrac{\log 2}{\log 3} - 1 \right)$

    $x = -\dfrac{1}{2}\left(\dfrac{0.3010}{0.4771} - 1 \right)$

    $x = 0.1845$

17.
$$15^{3x-4} = 10$$
$$\log 15^{3x-4} = \log 10$$
$$(3x-4)\log 15 = \log 10$$
$$3x-4 = \frac{\log 10}{\log 15}$$
$$3x = \frac{\log 10}{\log 15} + 4$$
$$x = \frac{1}{3}\left(\frac{\log 10}{\log 15} + 4\right)$$
$$x = \frac{1}{3}(4.8503)$$
$$x = 1.6168$$

19.
$$6^{5-2x} = 4$$
$$\log 6^{5-2x} = \log 4$$
$$(5-2x)\log 6 = \log 4$$
$$(5-2x) = \frac{\log 4}{\log 6}$$
$$x = -\frac{1}{2}\left(\frac{\log 4}{\log 6} - 5\right)$$
$$x = -\frac{1}{2}(-4.2262)$$
$$x = 2.1131$$

21. $\log_8 16 = \dfrac{\log 16}{\log 8} = 1.3333$

23. $\log_{16} 8 = \dfrac{\log 8}{\log 16} = 0.7500$

25. $\log_7 15 = \dfrac{\log 15}{\log 7} = 1.3917$

27. $\log_{15} 7 = \dfrac{\log 7}{\log 15} = 0.7186$

29. $\log_8 240 = \dfrac{\log 240}{\log 8} = 2.6356$

31. $\log_4 321 = \dfrac{\log 321}{\log 4} = 4.1632$

33. $\ln 345 = 5.8435$

35. $\ln 0.345 = -1.0642$

37. $\ln 10 = 2.3026$

39. $\ln 45,000 = 10.7144$

41.
$$A = P\left(1 + \frac{r}{n}\right)^{nt}, A = \$1,000 \ (\$500 \text{ doubled}),$$
$$r = 0.06, \quad n = 2 \text{ and } P = \$500$$
$$1,000 = 500\left(1 + \frac{0.06}{2}\right)^{2t}$$
$$2 = (1.03)^{2t}$$
$$\log 2 = \log (1.03)^{2t}$$
$$\log 2 = 2t \log (1.03)$$
$$\frac{\log 2}{2 \log (1.03)} = t$$
$$11.72 = t$$
It will take about 11.72 years to double.

43.
$$A = \$3000 \ (\$1000 \text{ tripled}), r = 0.12, \ n = 6$$
$$P = \$1000$$
$$3000 = 1000\left(1 + \frac{0.12}{6}\right)^{6t}$$
$$3 = 1.02^{6t}$$
$$\log 3 = \log 1.02^{6t}$$
$$\log 3 = 6t \log 1.02$$
$$t = \frac{\log 3}{6 \log 1.02}$$
$$t = 9.25 \text{ years}$$
It will take about 9.25 years to triple.

45. $\qquad A = 2P, \quad r = 0.08, \quad n = 4$

$$2P = P\left(1 + \frac{0.08}{4}\right)^{4t}$$

$$2 = (1.02)^{4t}$$

$$\log 2 = \log (1.02)^{4t}$$

$$\log 2 = 4t \log (1.02)$$

$$\frac{\log 2}{4 \log (1.02)} = t$$

$$8.75 = t$$

It will take about 8.75 years to double.

47. $\qquad A = \$75 \ (3 \cdot 25), r = 0.06, n = 2, P = \$25$

$$75 = 25\left(1 + \frac{0.06}{2}\right)^{2t}$$

$$3 = 1.03^{2t}$$

$$\log 3 = \log 1.03^{2t}$$

$$\log 3 = 2t \log 1.03$$

$$t = \frac{\log 3}{2 \log 1.03}$$

$$t = 18.58 \text{ years}$$

It will take about 18.58 years to triple.

49. $\qquad A = \$1,000 \ (\$500 \text{ doubled}), \quad r = 0.06, \text{ and } P = \$500$

$$A = Pe^{rt}$$

$$1,000 = 500e^{0.06t}$$

$$2 = e^{0.06t}$$

$$\ln 2 = \ln e^{0.06t}$$

$$\ln 2 = 0.06t \ \ln e$$

$$\ln 2 = 0.06t$$

$$\frac{\ln 2}{0.06} = t$$

$$11.55 = t$$

It will take about 11.55 years to double.

51. $\qquad A = \$1500 \ (\$500 \text{ tripled}),$

$$r = 0.06, P = \$500$$

$$1500 = 500e^{0.06t}$$

$$3 = e^{0.06t}$$

$$\ln 3 = \ln e^{0.06t}$$

$$\ln 3 = 0.06t \ \ln e$$

$$0.06t = \ln 3$$

$$t = \frac{\ln 3}{0.06}$$

$$= 18.1 \text{ years}$$

It will take about 18.1 years to triple

53. $\qquad A = \$2,500, \quad r = 0.08, \text{ and } P = \$1,000$

$$2,500 = 1,000e^{0.08t}$$

$$2.5 = e^{0.08t}$$

$$\ln 2.5 = \ln e^{0.08t}$$

$$\ln 2.5 = 0.08t \ \ln e$$

$$\ln 2.5 = 0.08t$$

$$\frac{\ln 2.5}{0.08} = t$$

$$11.45 = t : \text{ It will take about 11.45 years.}$$

55. $\quad P = 32,000 \ e^{0.05t}, \quad P = 64,000$

$$64,000 = 32,000 \ e^{0.05t}$$

$$2 = e^{0.05t}$$

$$\ln 2 = \ln e^{0.05t}$$

$$\ln 2 = 0.05t \ \ln e$$

$$\ln 2 = 0.05t$$

$$t = \frac{\ln 2}{0.05} = 13.9 \text{ years,}$$

toward the end of 2007

57. $\quad P = 15{,}000\ e^{0.04t},\ P = 45{,}000$

$45{,}000 = 15{,}000\ e^{0.04t}$

$3 = e^{0.04t}$

$\ln 3 = \ln\ e^{0.04t}$

$\ln 3 = 0.04t\ \ln\ e$

$\ln 3 = 0.04t$

$t = \dfrac{\ln 3}{0.04} = 27.5$ years

59. $\quad y = 2x^2 + 8x - 15$

$x = \dfrac{-b}{2a} = \dfrac{-8}{2(2)} = -2$

$y = 2(-2)^2 + 8(-2) - 15 = -23$

Vertex: $(-2,\ -23)$, lowest point

61. $\quad y = 12x - 4x^2$

$x = -\dfrac{b}{2a} = -\dfrac{12}{2(-4)} = \dfrac{3}{2}$

$y = 12\left(\dfrac{3}{2}\right) - 4\left(\dfrac{3}{2}\right)^2 = 9$

Vertex: $\left(\dfrac{3}{2},\ 9\right)$. highest point

63. $\quad h(t) = 64t - 16t^2$

$0 = -16t^2 + 64t - h$

$t = \dfrac{-b}{2a} = \dfrac{-64}{2(-16)} = \dfrac{64}{32} = 2$ seconds

$h(2) = 64(2) - 16(2)^2 = 128 - 64 = 64$ feet

65. $\quad A = Pe^{rt}$

$\dfrac{A}{P} = e^{rt}$

$\ln\ \dfrac{A}{P} = \ln\ e^{rt}$

$\ln\ \dfrac{A}{P} = rt\ \ln\ e$

$\ln\ \dfrac{A}{P} = rt$

$\dfrac{1}{r}\ \ln\ \dfrac{A}{P} = t$

67. $\quad A = P2^{-kt}$

$\dfrac{A}{P} = 2^{-kt}$

$\log\ \dfrac{A}{P} = \log\ 2^{-kt}$

$\log A - \log P = -kt\ \log\ 2$

$t = \dfrac{1}{k}\ \dfrac{\log P - \log A}{\log 2}$

69. $\quad A = P(1-r)^t$

$\dfrac{A}{P} = (1-r)^t$

$\log\ \dfrac{A}{P} = \log\ (1-r)^t$

$\log\ \dfrac{A}{P} = t\ \log\ (1-r)$

$\log A - \log P = t\ \log\ (1-r)$

$\dfrac{\log A - \log P}{\log(1-r)} = t$

71. Let $x = \log_6 23$ or $6^x = 23$

$Y_1 = 6^\wedge x$

$Y_2 = 23$

$\log_6 23 = 1.75$

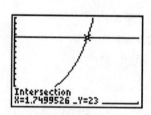

Intersection
X=1.7499526 Y=23

73. Let $x = \log_7 29$ or $7^x = 29$

$Y_1 = 7 \wedge x$

$Y_2 = 29$

$\log_7 29 = 1.73$

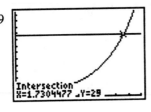
Intersection
X=1.7304477 _Y=29

75. $3^x = 5$

Let $Y_1 = 3^x$ and $Y_2 = 5$

$x = 1.465$

Approximately the same

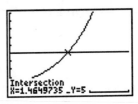
Intersection
X=1.4649735 _Y=5

Chapter 11 Review

1. $f(x) = 2^x$

$f(4) = 2^4 = 16$

3. $g(x) = \left(\dfrac{1}{3}\right)^x$

$g(2) = \left(\dfrac{1}{3}\right)^2 = \dfrac{1}{9}$

5. $f(-1) + g(1) = 2^{-1} + \left(\dfrac{1}{3}\right)^1$

$= \dfrac{1}{2} + \dfrac{1}{3}$

$= \dfrac{3}{6} + \dfrac{2}{6}$

$= \dfrac{5}{6}$

7. See the graph in the back of the textbook.

9. $y = 2x + 1$

$x = 2y + 1$

$2y = x - 1$

$y = \dfrac{x - 1}{2}$

See the graph in the back of the textbook.

11. $f(x) = 2x + 3$

$y = 2x + 3$

$x = 2y + 3$

$y = \dfrac{x - 3}{2}$

$f^{-1}(x) = \dfrac{x - 3}{2}$

13. $f(x) = \dfrac{1}{2}x + 2$

$y = \dfrac{1}{2}x + 2$

$x = \dfrac{1}{2}y + 2$

$y = 2x - 4$

$f^{-1}(x) = 2x - 4$

15. $3^4 = 81$

$\log_3 81 = 4$

17. $0.01 = 10^{-2}$
        $\log_{10} \ 0.01 = -2$

19. $\log_2 \ 8 = 3$
        $2^3 = 8$

21. $\log_4 \ 2 = \dfrac{1}{2}$
        $4^{1/2} = 2$

23. $\log_5 \ x = 2$
        $5^2 = x$
        $x = 25$

25. $\log_x \ 0.01 = -2$
        $x^{-2} = 0.01$
        $x^{-2} = 10^{-2}$
        $x = 10$

27. $y = \log_{1/2} \ x$ See the graph in the back of the textbook.

29. Let $x = \log_{27} \ 9$
        $27^x = 9$
        $3^{3x} = 3^2$
        $3x = 2$
        $x = \dfrac{2}{3}$

31. $\log_2 \ 5x = \log_2 \ 5 + \log_2 \ x$

33. $\log_a \ \dfrac{\sqrt{x}y^3}{z} = \log_a \ \dfrac{x^{1/2}y^3}{z} = \dfrac{1}{2} \ \log_a \ x + 3 \ \log_a \ y - \log_a \ z$

35. $\log_2 \ x + \log_2 \ y = \log_2 \ xy$

37. $2 \ \log_a \ 5 - \dfrac{1}{2} \ \log_a \ 9 = \log_a \ \dfrac{5^2}{9^{1/2}} = \log_a \ \dfrac{25}{3}$

39. $\log_2 \ x + \log_2 \ 4 = 3$
        $\log_2 \ 4x = 3$
        $4x = 2^3$
        $4x = 8$
        $x = 2$

41. $\log_3 \ x + \log_3 \ (x-2) = 1$
        $\log_3 \ x(x-2) = 1$
        $x(x-2) = 3^1$
        $x^2 - 2x - 3 = 0$
        $(x+1)(x-3) = 0$
        $x+1 = 0$ or $x-3 = 0$
        $x = -1$ $x = 3$

Possible solutions are 3 and -1; only 3 checks.

43. $\log_6 (x-1) + \log_6 x = 1$

$\qquad \log_6 (x-1)x = 1$

$\qquad\qquad x(x-1) = 6^1$

$\qquad\quad x^2 - x - 6 = 0$

$\qquad (x+2)(x-3) = 0$

$\qquad\qquad x+2 = 0 \quad \text{or} \quad x-3 = 0$

$\qquad\qquad\qquad x = -2 \qquad\qquad x = 3$

Possible solutions are 3 and −2; only 3 checks.

45. $\log 346 = 2.5391$

47. If $\log x = 3.9652$

$\quad$ then $x = 10^{3.9652} = 9,230$

49. $\ln e = 1$

51. $\ln e^2 = 2$

53. $pH = -\log \left[H^+ \right]$

$\quad pH = -\log 7.9 \times 10^{-3}, \quad \text{when } \left[H^+ \right] = 7.9 \times 10^{-3}$

$\qquad\quad = 2.1024$

$\quad pH = 2.1$

55. $pH = -\log \left[H^+ \right]$

$\quad 2.7 = -\log \left[H^+ \right], \quad \text{when } pH = 2.7$

$\quad \left[H^+ \right] = 10^{-2.7}$

$\quad \left[H^+ \right] = 2 \times 10^{-3}$

57. $\qquad 4^x = 8$

$\qquad \log 4^x = \log 8$

$\qquad x \log 4 = \log 8$

$\qquad\qquad x = \dfrac{\log 8}{\log 4}$

$\qquad\qquad x = \dfrac{0.9031}{0.6021}$

$\qquad\qquad x = \dfrac{3}{2}$

59. $\log_{16} 8 = \dfrac{\log 8}{\log 16} = \dfrac{0.9031}{1.2041} = 0.75$

61. Substituting $A = \$10,000, P = \$5,000, n = 1$
and $r = 16\%$ into the formula

$$A = P\left(1 + \frac{r}{n}\right)^{nt}$$

The equation becomes

$$10,000 = 5,000\left(1 + \frac{0.16}{1}\right)^{1(t)}$$

$$2 = (1.16)^t$$

$$\log 2 = t \log 1.16$$

$$\frac{\log 2}{\log 1.16} = t$$

$$\frac{0.30103}{0.06446} = t$$

$t =$ about 4.67 years

63. Substituting $P_0 = \$100,000, r = 4\% = 0.04$,
and $t = 8$ years into the formula for inflation

$$P = P_0(1 + r)^t$$

we get

$$P = 100,000(1 + 0.04)^8$$

$$= 100,000(1.04)^8$$

$$= 136,856.91$$

It will sell for $136,856.91

Cumulative Review: Chapters 1-11

1. $-8+2\left[5-3(-2-3)\right]=-8+2\left[5-3(-5)\right]=-8+2\left[5+15\right]=-8+2\left[20\right]\ =-8+40=32$

3. $\left(\dfrac{3}{5}\right)^{-2}-\left(\dfrac{3}{13}\right)^{-2}=\left(\dfrac{5}{3}\right)^{2}-\left(\dfrac{13}{3}\right)^{2}=\dfrac{25}{9}-\dfrac{169}{9}=-\dfrac{144}{9}=-16$

5. $\sqrt[3]{27x^{4}y^{3}}=\sqrt[3]{27x^{3}xy^{3}}=3xy\sqrt[3]{x}$

7. $\left[(6+2i)-(3-4i)\right]-(5-i)=\left[6+2i-3+4i\right]-5+i=3+6i-5+i=-2+7i$

9. $1+\dfrac{x}{1+\dfrac{1}{x}}=1+\dfrac{x}{\dfrac{x+1}{x}}=1+\dfrac{x^{2}}{x+1}=\dfrac{x+1+x^{2}}{x+1}=\dfrac{x^{2}+x+1}{x+1}$

11. $\left(3t^{2}+\dfrac{1}{4}\right)\left(4t^{2}-\dfrac{1}{3}\right)=12t^{4}-t^{2}+t^{2}-\dfrac{1}{12}=12t^{4}-\dfrac{1}{12}$

13.
$$
\begin{array}{r}
3x+7+\dfrac{10}{3x-4} \\
3x-4\overline{)9x^{2}+9x-18} \\
-\ \ \ + \\
\underline{9x^{2}-12x} \\
21x-18 \\
-\ \ \ + \\
\underline{21x-28} \\
10
\end{array}
$$

15. $\dfrac{7}{4x^{2}-x-3}-\dfrac{1}{4x^{2}-7x+3}$

$=\dfrac{7}{(4x+3)(x-1)}-\dfrac{1}{(x-1)(4x-3)}$

$=\dfrac{7}{(4x+3)(x-1)}\cdot\dfrac{4x-3}{4x-3}-\dfrac{1}{(x-1)(4x-3)}\cdot\dfrac{4x+3}{4x+3}$

$=\dfrac{28x-21-4x-3}{(x-1)(4x+3)(4x-3)}$

$=\dfrac{24x-24}{(x-1)(4x+3)(4x-3)}$

$=\dfrac{24(x-1)}{(x-1)(4x+3)(4x-3)}$

$=\dfrac{24}{(4x+3)(4x-3)}$

17. $\dfrac{2}{3}(6x-5)+\dfrac{1}{3}=13$

$2(6x-5)+1=39$

$12x-10+1=39$

$12x-9=39$

$12x=48$

$x=4$

19. $|3x-5|+6=2$

$|3x-5|=-4$

The absolute value can't be negative.

$\varnothing$

21. $\dfrac{1}{x+3} + \dfrac{1}{x-2} = 1$

Multiply both sides by $(x+3)(x-2)$

$x - 2 + x + 3 = (x+3)(x-2)$

$2x + 1 = x^2 + x - 6$

$x^2 - x - 7 = 0$

Let $a = 1$, $b = -1$ and $c = -7$

$x = \dfrac{-(-1) \pm \sqrt{(-1)^2 - 4(1)(-7)}}{2(1)}$

$= \dfrac{1 \pm \sqrt{29}}{2}$

The solution set is $\left\{ \dfrac{1+\sqrt{29}}{2} \text{ and } \dfrac{1-\sqrt{29}}{2} \right\}$.

23. $2x - 1 = x^2$

$x^2 - 2x + 1 = 0$

$(x-1)^2 = 0$

$x - 1 = \pm\sqrt{0}$

$x = 1$

25. $\sqrt{7x-4} = -2$

$\left(\sqrt{7x-4}\right)^2 = (-2)^2$

$7x - 4 = 4$

$7x = 8$

$x = \dfrac{8}{7}$

$\frac{8}{7}$ does not check in equation, $\emptyset$.

27. $x - 3\sqrt{x} + 2 = 0$

Replacing $\sqrt{x}$ with y, we have

$y^2 - 3y + 2 = 0$

$(y-1)(y-2) = 0$

$y - 1 = 0$ or $y - 2 = 0$

$y = 1$ $y = 2$

Replacing y with $\sqrt{x}$, we have

$\sqrt{x} = 1$ or $\sqrt{x} = 2$

$x = 1$ $x = 4$

The solution set is $\{1, 4\}$.

Both solutions check.

29. $\log_3 x = 3$

$x = 3^3$

$x = 27$

31. $\log_3 (x-3) - \log_3 (x+2) = 1$

$\log_3 \dfrac{x-3}{x+2} = 1$

$\dfrac{x-3}{x+2} = 3^1$

$x - 3 = 3(x+2)$

$x - 3 = 3x + 6$

$-9 = 2x$

$x = -\dfrac{9}{2}$ doesn't check

$\emptyset$

33. $4x + 7y = -3$

$x = -2y - 2$

Substitute $x = -2y - 2$ into the first equation

$4(-2y - 2) + 7y = -3$

$-8y - 8 + 7y = -3$

$-y = 5$

$y = -5$

Substitute $y = -5$ into the second equation

$x = -2(-5) - 2$

$x = 8$

The solution to the system is $(8, -5)$.

35. $3y - 6 \geq 3$ or $3y - 6 \leq -3$

$3y \geq 9$ or $3y \leq 3$

$y \geq 3$ or $y \leq 1$

See the graph in the back of the textbook.

37. $x^3 + 2x^2 - 9x - 18 < 0$

$(x + 2)(x + 3)(x - 3) < 0,$ either 3 or 1 negative

$x < -3$ or $-2 < x < 3$

See the graph in the back of the textbook.

Sign of $x - 3$

Sign of $x + 2$

Sign of $x + 3$

-3   -2   3

39. $m = -\dfrac{5}{3}$ and $(3, -3)$

$y - y_1 = m(x - x_1)$

$y - (-3) = -\dfrac{5}{3}(x - 3)$

$y + 3 = -\dfrac{5}{3}x + 5$

$y = -\dfrac{5}{3}x + 2$

41. $0.0000972 = 9.72 \times 10^{-5}$

43. $f(x) = \dfrac{1}{2}x + 3$ means $y = \dfrac{1}{2}x + 3$

The inverse is $x = \dfrac{1}{2}y + 3$

Solve for y in terms of x

$2x = y + 6$

$2x - 6 = y$

$f^{-1}(x) = 2x - 6$

45. $\{(2, -3), (2, -1), (-3, 3)\}$

Domain: $\{-3, 2\}$

Range: $\{-3, -1, 3\}$

Not a function

47. $2, -6, 18, \ldots$

$$r = \frac{a_2}{a_1} = \frac{-6}{2} = -3$$

$$r = \frac{a_3}{a_2} = \frac{18}{-6} = -3$$

Next term $= a_3 r = 18(-3) = -54$

Geometric sequence

49. Let x = number of gallons of 50% alcohol solution and y = number of gallons of 25% alcohol solution

The system of equations is

$$x + y = 20 \qquad (1)$$
$$0.50x + 0.25y = 20(0.425) \qquad (2)$$

$$\begin{array}{ll} 50x + 50y = 1,000 & 50 \text{ times } (1) \\ \underline{-50x - 25y = -850} & -100 \text{ times } (2) \\ 25y = 150 & \\ y = 6 & \end{array}$$

Substitute $y = 6$ into equation (1)

$$x + 6 = 20$$
$$x = 14$$

It will take 14 gallons of 50% alcohol solution and 6 gallons of 25% alcohol solution.

Chapter 11 Test

1. See the graph in the back of the textbook.

2. See the graph in the back of the textbook.

3.
$$f(x) = 2x - 3$$
$$y = 2x - 3$$
$$x = 2y - 3$$
$$x + 3 = 2y$$
$$y = \frac{x + 3}{2}$$
$$f^{-1}(x) = \frac{x + 3}{2}$$

See the graph in the back of the textbook.

4.
$$f(x) = x^2 - 4$$
$$y = x^2 - 4$$
$$x = y^2 - 4$$
$$y^2 = x + 4$$
$$y = \pm\sqrt{x + 4}$$

See the graph in the back of the textbook.

5.
$$\log_4 x = 3$$
$$4^3 = x$$
$$64 = x$$

6.
$$\log_x 5 = 2$$
$$x^2 = 5$$
$$x = \sqrt{5}$$

7. $y = \log_2 x$

See the graph in the back of the textbook.

8. $y = \log_{1/2} x$

See the graph in the back of the textbook.

9. Let $x = \log_8 4$

$$8^x = 4$$

$$2^{3x} = 2^2$$

$$3x = 2$$

$$x = \frac{2}{3}$$

10. $\log_7 21 = \dfrac{\log 21}{\log 7} = 1.5646$

11. $\log 23,400 = 4.3692$

12. $\log 0.0123 = -1.9101$

13. $\ln 46.2 = 3.8330$

14. $\ln 0.0462 = -3.0748$

15. $\log_2 \dfrac{8x^2}{y} = \log_2 8 + 2 \log_2 x - \log_2 y = 3 + 2 \log_2 x - \log_2 y$

16. $\log \dfrac{\sqrt{x}}{y^4 \sqrt[5]{z}} = \dfrac{1}{2} \log x - 4 \log y - \dfrac{1}{5} \log z$

17. $2 \log_3 x - \dfrac{1}{2} \log_3 y = \log_3 \dfrac{x^2}{\sqrt{y}}$

18. $\dfrac{1}{3} \log x - \log y - 2 \log z = \log \dfrac{\sqrt[3]{x}}{yz^2}$

19. $\log x = 4.8476$

$$x = 7.04 \times 10^4$$

20. $\log x = -2.6478$

$$x = 2.25 \times 10^{-3}$$

21. $5 = 3^x$

$$x = \log_3 5$$

$$x = \dfrac{\log 5}{\log 3}$$

$$x = 1.46$$

22. $4^{2x-1} = 8$

$$2x - 1 = \log_4 8$$

$$2x = \dfrac{\log 8}{\log 4} + 1$$

$$x = \dfrac{1}{2} \left[\dfrac{\log 8}{\log 4} + 1 \right]$$

$$x = \dfrac{1}{2} \left[\dfrac{0.9031}{0.6021} + 1 \right]$$

$$x = 1.25 \quad \text{or} \quad \dfrac{5}{4}$$

23. $\log_5 x - \log_5 3 = 1$

$$\log_5 \dfrac{x}{3} = 1$$

$$\dfrac{x}{3} = 5^1$$

$$x = 15$$

24. $\log_2 x + \log_2 (x-7) = 3$

$\qquad \log_2 x(x-7) = 3$

$\qquad\qquad x(x-7) = 2^3$

$\qquad\qquad x^2 - 7x - 8 = 0$

$\qquad\qquad (x+1)(x-8) = 0$

$\qquad\qquad x+1 = 0 \quad \text{or} \quad x-8 = 0$

$\qquad\qquad\qquad x = -1 \qquad\qquad x = 8$

Possible solutions are -1 and 8; only 8 checks.

25. $\text{pH} = -\log\left[H^+\right]$

$\quad \text{pH} = -\log\, 6.6 \times 10^{-7}, \;\text{when}\; \left[H^+\right] = 6.6 \times 10^{-7}$

$\qquad\quad = 6.1805$

$\quad \text{pH} = 6.2$

26. Substituting $P = \$400$, $n = 2$, $t = 5$

and $r = 10\%$ into the formula:

$$A = P\left(1 + \frac{r}{n}\right)^{nt}$$

The equation becomes

$$A = 400\left(1 + \frac{0.10}{2}\right)^{2(5)}$$

$\log A = \log\, 400(1.05)^{10}$

$\qquad = \log\, 400 + 10\,\log\, 1.05$

$\log A = 2.813953$

$\qquad A = \$651.56$

27. Substituting $A = \$1,800$, $P = \$600$, $n = 4$

and $r = 8\%$ into the formula:

$$A = P\left(1 + \frac{r}{n}\right)^{nt}$$

The equation becomes

$$1,800 = 600\left(1 + \frac{0.08}{4}\right)^{4t}$$

$\qquad 3 = (1.02)^{4t}$

$\qquad \log\, 3 = 4t\,\log\, 1.02$

$\qquad \dfrac{\log\, 3}{\log\, 1.02} = 4t$

$\qquad \dfrac{0.4771}{0.0086} = 4t$

$\qquad 55.4767 = 4t$

$\qquad t \approx 13.9 \text{ years}$

28. Substituting $P_0 = \$18,000$, $r = 20\% = 0.2$,

and $t = 4$ years into the formula for deprecation

$V(t) = P_0(1-r)^t$

we get

$V(4) = 18,000(1-0.2)^4$

$\qquad = 18,000(0.8)^4$

$\qquad = 7,372.8$

It' s value will be $\$7,372.80$

CHAPTER 12

Sequences and Series

12.1 Sequences

1. $a_n = 3n + 1$

$a_1 = 3(1) + 1 = 4$

$a_2 = 3(2) + 1 = 7$

$a_3 = 3(3) + 1 = 10$

$a_4 = 3(4) + 1 = 13$

$a_5 = 3(5) + 1 = 16$

3. $a_n = 4n - 1$

$a_1 = 4(1) - 1 = 3$

$a_2 = 4(2) - 1 = 7$

$a_3 = 4(3) - 1 = 11$

$a_4 = 4(4) - 1 = 15$

$a_5 = 4(5) - 1 = 19$

5. $a_n = n$

$a_1 = 1$

$a_2 = 2$

$a_3 = 3$

$a_4 = 4$

$a_5 = 5$

7. $a_n = n^2 + 3$

$a_1 = 1^2 + 3 = 4$

$a_2 = 2^2 + 3 = 7$

$a_3 = 3^2 + 3 = 12$

$a_4 = 4^2 + 3 = 19$

$a_5 = 5^2 + 3 = 28$

9. $a_n = \dfrac{n}{n+3}$

$a_1 = \dfrac{1}{1+3} = \dfrac{1}{4}$

$a_2 = \dfrac{2}{2+3} = \dfrac{2}{5}$

$a_3 = \dfrac{3}{3+3} = \dfrac{3}{6}$

$a_4 = \dfrac{4}{4+3} = \dfrac{4}{7}$

$a_5 = \dfrac{5}{5+3} = \dfrac{5}{8}$

11. $a_n = \dfrac{n+1}{n+2}$

$a_1 = \dfrac{1+1}{1+2} = \dfrac{2}{3}$

$a_2 = \dfrac{2+1}{2+2} = \dfrac{3}{4}$

$a_3 = \dfrac{3+1}{3+2} = \dfrac{4}{5}$

$a_4 = \dfrac{4+1}{4+2} = \dfrac{5}{6}$

$a_5 = \dfrac{5+1}{5+2} = \dfrac{6}{7}$

13. $a_n = \dfrac{1}{n^2}$

$a_1 = \dfrac{1}{1^2} = 1$

$a_2 = \dfrac{1}{2^2} = \dfrac{1}{4}$

$a_3 = \dfrac{1}{3^2} = \dfrac{1}{9}$

$a_4 = \dfrac{1}{4^2} = \dfrac{1}{16}$

$a_5 = \dfrac{1}{5^2} = \dfrac{1}{25}$

15. $a_n = 2^n$

$a_1 = 2^1 = 2$

$a_2 = 2^2 = 4$

$a_3 = 2^3 = 8$

$a_4 = 2^4 = 16$

$a_5 = 2^5 = 32$

17. $a_n = 3^{-n}$

$a_1 = 3^{-1} = \dfrac{1}{3}$

$a_2 = 3^{-2} = \dfrac{1}{9}$

$a_3 = 3^{-3} = \dfrac{1}{27}$

$a_4 = 3^{-4} = \dfrac{1}{81}$

$a_5 = 3^{-5} = \dfrac{1}{243}$

19. $a_n = 1 + \dfrac{1}{n}$

$a_1 = 1 + \dfrac{1}{1} = 2$

$a_2 = 1 + \dfrac{1}{2} = \dfrac{3}{2}$

$a_3 = 1 + \dfrac{1}{3} = \dfrac{4}{3}$

$a_4 = 1 + \dfrac{1}{4} = \dfrac{5}{4}$

$a_5 = 1 + \dfrac{1}{5} = \dfrac{6}{5}$

21. $a_n = n - \dfrac{1}{n}$

$a_1 = 1 - \dfrac{1}{1} = 0$

$a_2 = 2 - \dfrac{1}{2} = \dfrac{3}{2}$

$a_3 = 3 - \dfrac{1}{3} = \dfrac{8}{3}$

$a_4 = 4 - \dfrac{1}{4} = \dfrac{15}{4}$

$a_5 = 5 - \dfrac{1}{5} = \dfrac{24}{5}$

23. $a_n = (-2)^n$

$a_1 = (-2)^1 = -2$

$a_2 = (-2)^2 = 4$

$a_3 = (-2)^3 = -8$

$a_4 = (-2)^4 = 16$

$a_5 = (-2)^5 = -32$

25. $a_1 = 3$

$a_2 = -3a_{2-1} = -3a_1 = -3(3) = -9$

$a_3 = -3a_{3-1} = -3a_2 = -3(-9) = 27$

$a_4 = -3a_{4-1} = -3a_3 = -3(27) = -81$

$a_5 = -3a_{5-1} = -3a_4 = -3(-81) = 243$

27. $a_n = a_{n-1} - 3$

$a_1 = 3$

$a_2 = a_1 - 3 = 3 - 3 = 0$

$a_3 = a_2 - 3 = 0 - 3 = -3$

$a_4 = a_3 - 3 = -3 - 3 = -6$

$a_5 = a_4 - 3 = -6 - 3 = -9$

29. $a_1 = 1$

$a_2 = 2a_{2-1} + 3 = 2a_1 + 3 = 2(1) + 3 = 5$

$a_3 = 2a_{3-1} + 3 = 2a_2 + 3 = 2(5) + 3 = 13$

$a_4 = 2a_{4-1} + 3 = 2a_3 + 3 = 2(13) + 3 = 29$

$a_5 = 2a_{5-1} + 3 = 2a_4 + 3 = 2(29) + 3 = 61$

31. $a_n = a_{n-1} + n$

$a_1 = 1$

$a_2 = a_1 + 2 = 1 + 2 = 3$

$a_3 = a_2 + 3 = 3 + 3 = 6$

$a_4 = a_3 + 4 = 6 + 4 = 10$

$a_5 = a_4 + 5 = 10 + 5 = 15$

33. $2, 3, 4, 5, \ldots$

$a_1 = 2 = 1 + 1$

$a_2 = 3 = 2 + 1$

$a_3 = 4 = 3 + 1$

$a_4 = 5 = 4 + 1$

$a_n = n + 1$

35. $4, 8, 12, 16, 20, \ldots$

$a_1 = 4(1)$

$a_2 = 4(2)$

$a_3 = 4(3)$

$a_4 = 4(4)$

$a_n = 4n$

37. $7, 10, 13, 16, \ldots$

$a_1 = 7 = 3(1) + 4$

$a_2 = 10 = 3(2) + 4$

$a_3 = 13 = 3(3) + 4$

$a_4 = 16 = 3(4) + 4$

$a_n = 3(n) + 4$

or recursively as

$a_1 = 7, \ a_n = a_{n-1} + 3$

39. $1, 4, 9, 16, \ldots$

$a_1 = 1^2$

$a_2 = 2^2$

$a_3 = 3^2$

$a_4 = 4^2$

$a_n = n^2$

41. $3, 12, 27, 48, \ldots$

$a_1 = 3 = 3(1)^2$

$a_2 = 12 = 3(2)^2$

$a_3 = 27 = 3(3)^2$

$a_4 = 48 = 3(4)^2$

$a_n = 3n^2$

43. $4, 8, 16, 32, \ldots$

$a_1 = 2^{1+1}$

$a_2 = 2^{2+1}$

$a_3 = 2^{3+1}$

$a_4 = 2^{4+1}$

$a_n = 2^{n+1}$

or recursively as

$a_1 = 4, \ a_n = 2a_{n-1}$

45. $-2, 4, -8, 16, \ldots$

$a_1 = -2 = (-2)^1$

$a_2 = 4 = (-2)^2$

$a_3 = -8 = (-2)^3$

$a_4 = 16 = (-2)^4$

$a_n = (-2)^n$

or recursively as

$a_1 = -2, \ a_n = -2a_{n-1}$

47. $\dfrac{1}{4}, \dfrac{1}{8}, \dfrac{1}{16}, \dfrac{1}{32}, \ldots$

$a_1 = \dfrac{1}{2^{1+1}}$

$a_2 = \dfrac{1}{2^{2+1}}$

$a_3 = \dfrac{1}{2^{3+1}}$

$a_4 = \dfrac{1}{2^{4+1}}$

$a_n = \dfrac{1}{2^{n+1}}$

or recursively as

$a_1 = \dfrac{1}{4}, \ a_n = \dfrac{1}{2} a_{n-1}$

49. $\dfrac{1}{4}, \dfrac{2}{9}, \dfrac{3}{16}, \dfrac{4}{25}, \ldots$

$a_1 = \dfrac{1}{4} = \dfrac{1}{2^2} = \dfrac{1}{(1+1)^2}$

$a_2 = \dfrac{2}{9} = \dfrac{2}{3^2} = \dfrac{2}{(2+1)^2}$

$a_3 = \dfrac{3}{16} = \dfrac{3}{4^2} = \dfrac{3}{(3+1)^2}$

$a_4 = \dfrac{4}{25} = \dfrac{4}{5^2} = \dfrac{4}{(4+1)^2}$

$a_n = \dfrac{n}{(n+1)^2}$

51. (a) $28,000

$\quad 28,000 + .04(28,000) = \$29,120$

$\quad 29,120 + .04(29,120) = \$30,284.80$

$\quad 30,284.80 + .04(30,284.80) = \$31,496.19$

$\quad 31,496.19 + .04(31,496.19) = \$32,756.04$

(b) $a_n = 28,000(1.04)^{n-1}$

53. (a) $a_1 = 100(12) = 1,200$

   $a_2 = 100(24) = 2,400$

   $a_3 = 100(36) = 3,600$

   $a_4 = 100(48) = 4,800$

   $a_5 = 100(60) = 6,000$

   (b) $a_m = 100(m)$

   (c) $a_y = 100(12y)$

55. $\log_9 x = \dfrac{3}{2}$

   $x = 9^{3/2}$

   $= \left(9^{1/2}\right)^3$

   $= 3^3$

   $= 27$

57. $\log_2 32 = \dfrac{\log_{10} 32}{\log_{10} 2}$

   $= \dfrac{1.5051}{0.3010}$

   $= 5$

59. $\log_3 \left[\log_2 8\right] = \log_3 \left[\log_2 2^3\right]$

   $= \log_3 \left[3 \log_2 2\right]$

   $= \log_3 \left[(3)(1)\right]$

   $= \log_3 3$

   $= 1$

61. $a_{100} = \left(1 + \dfrac{1}{100}\right)^{100} = (1.01)^{100} \approx 2.7048$

   $a_{1000} = \left(1 + \dfrac{1}{1000}\right)^{1000} = (1.001)^{1000} \approx 2.7169$

   $a_{10,000} = \left(1 + \dfrac{1}{10,000}\right)^{10,000} = (1.0001)^{10,000} \approx 2.7181$

   $a_{100,000} = \left(1 + \dfrac{1}{100,000}\right)^{100,000} = (1.00001)^{100,000} \approx 2.7183$

   This is the same as the approx $e \approx 2.7183$

63. $a_n = a_{n-1} + a_{n-2},\ n > 2$

   $a_1 = 1,\ a_2 = 1$

   $a_3 = a_2 + a_1 = 1 + 1 = 2$

   $a_4 = a_3 + a_2 = 2 + 1 = 3$

   $a_5 = a_4 + a_3 = 3 + 2 = 5$

   $a_6 = a_5 + a_4 = 5 + 3 = 8$

   $a_7 = a_6 + a_5 = 8 + 5 = 13$

   $a_8 = a_7 + a_6 = 13 + 8 = 21$

   $a_9 = a_8 + a_7 = 21 + 13 = 34$

   $a_{10} = a_9 + a_8 = 34 + 21 = 55$

65. $1 + \dfrac{1}{1+1} = 1 + \dfrac{1}{2} = \dfrac{3}{2}$

   $1 + \dfrac{1}{1 + \dfrac{1}{1+1}} = 1 + \dfrac{1}{\dfrac{3}{2}} = 1 + \dfrac{2}{3} = \dfrac{5}{3}$

   $1 + \dfrac{1}{1 + \dfrac{1}{1 + \dfrac{1}{1+1}}} = 1 + \dfrac{1}{\dfrac{5}{3}} = 1 + \dfrac{3}{5} = \dfrac{8}{5}$

12.2 Series

1. $\displaystyle\sum_{i=1}^{4}(2i+4) = (2\cdot1+4)+(2\cdot2+4)+(2\cdot3+4)+(2\cdot4+4) = 6+8+10+12 = 36$

3. $\displaystyle\sum_{i=1}^{3}(2i-1) = 2(1)-1+2(2)-1+2(3)-1 = 1+3+5 = 9$

5. $\displaystyle\sum_{i=2}^{3}(i^2-1) = (2^2-1)+(3^2-1) = 3+8 = 11$

7. $\displaystyle\sum_{i=1}^{4}\frac{i}{1+i} = \frac{1}{1+1}+\frac{2}{1+2}+\frac{3}{1+3}+\frac{4}{1+4} = \frac{1}{2}+\frac{2}{3}+\frac{3}{4}+\frac{4}{5} = \frac{163}{60}$

9. $\displaystyle\sum_{i=1}^{3}\frac{i^2}{2i-1} = \frac{1^2}{2\cdot1-1}+\frac{2^2}{2\cdot2-1}+\frac{3^2}{2\cdot3-1} = \frac{1}{1}+\frac{4}{3}+\frac{9}{5} = \frac{15}{15}+\frac{20}{15}+\frac{27}{15} = \frac{62}{15}$

11. $\displaystyle\sum_{i=1}^{4}(-3)^i = (-3)^1+(-3)^2+(-3)^3+(-3)^4 = -3+9-27+81 = 60$

13. $\displaystyle\sum_{i=3}^{6}(-2)^i = (-2)^3+(-2)^4+(-2)^5+(-2)^6 = -8+16-32+64 = 40$

15. $\displaystyle\sum_{i=1}^{5}(x+i) = (x+1)+(x+2)+(x+3)+(x+4)+(x+5) = 5x+15$

17. $\displaystyle\sum_{i=2}^{7}(x+1)^i = (x+1)^2+(x+1)^3+(x+1)^4+(x+1)^5+(x+1)^6+(x+1)^7$

19. $\displaystyle\sum_{i=1}^{5}\frac{x+i}{x-1} = \frac{x+1}{x-1}+\frac{x+2}{x-1}+\frac{x+3}{x-1}+\frac{x+4}{x-1}+\frac{x+5}{x-1}$

21. $\displaystyle\sum_{i=3}^{8}(x+i)^i = (x+3)^3+(x+4)^4+(x+5)^5+(x+6)^6+(x+7)^7+(x+8)^8$

23. $\displaystyle\sum_{i=1}^{5}(x+i)^{i+1} = (x+1)^{1+1} + (x+2)^{2+1} + (x+3)^{3+1} + (x+4)^{4+1} + (x+5)^{5+1}$

$\qquad = (x+1)^2 + (x+2)^3 + (x+3)^4 + (x+4)^5 + (x+5)^6$

25. $2+4+8+16 = 2^1 + 2^2 + 2^3 + 2^4 = \displaystyle\sum_{i=1}^{4} 2^i$

27. $4+8+16+32+64 = 2^2 + 2^3 + 2^4 + 2^5 + 2^6 = \displaystyle\sum_{i=2}^{6} 2^i$

29. $5+10+17+26+37 = \left(2^2+1\right) + \left(3^2+1\right) + \left(4^2+1\right) + \left(5^2+1\right) + \left(6^2+1\right) = \displaystyle\sum_{i=2}^{6}\left(i^2+1\right)$

31. $\dfrac{3}{4} + \dfrac{4}{5} + \dfrac{5}{6} + \dfrac{6}{7} + \dfrac{7}{8} = \dfrac{3}{1+3} + \dfrac{4}{1+4} + \dfrac{5}{1+5} + \dfrac{6}{1+6} + \dfrac{7}{1+7} = \displaystyle\sum_{i=3}^{7} \dfrac{i}{1+i}$

33. $\dfrac{1}{3} + \dfrac{2}{5} + \dfrac{3}{7} + \dfrac{4}{9} = \dfrac{1}{2\cdot 1+1} + \dfrac{2}{2\cdot 2+1} + \dfrac{3}{2\cdot 3+1} + \dfrac{4}{2\cdot 4+1} = \displaystyle\sum_{i=1}^{4} \dfrac{i}{2i+1}$

35. $(x-3)+(x-4)+(x-5)+(x-6) = \displaystyle\sum_{i=3}^{6}(x-i)$

37. $\dfrac{x}{x+3} + \dfrac{x}{x+4} + \dfrac{x}{x+5} = \displaystyle\sum_{i=3}^{5} \dfrac{x}{x+i}$

39. $x^2(x+2) + x^3(x+3) + x^4(x+4) = \displaystyle\sum_{i=2}^{4} x^i(x+i)$

41. (a) $1/3 = 0.3 + 0.03 + 0.003 + 0.0003 + \cdots$

    (b) $2/9 = 0.2 + 0.02 + 0.002 + 0.0002 + \cdots$

    (c) $3/11 = 0.27 + 0.0027 + 0.000027 + \cdots$

43. First second $= a_1 = 16 = 1\cdot 16 = (1+0)16$

    Second second $= a_2 = 48 = 3\cdot 16 = (2+1)16$

    Third second $= a_3 = 80 = 5\cdot 16 = (3+2)16$

    Fourth second $= a_4 = 112 = 7\cdot 16 = (4+3)16$

    Fifth second $= a_5 = 144 = 9\cdot 16 = (5+4)16$

    Sixth second $= a_6 = 176 = 11\cdot 16 = (6+5)16$

    Seventh second $= a_7 = 208 = 13\cdot 16 = (7+6)16$

    nth $= a_n = \left[n+(n-1)\right]16$

    The skydiver will fall 208 feet in the seventh second. The total distance he will fall in seven seconds is 784 feet.

45. (a) $\dfrac{1}{3} + \dfrac{1}{15} + \dfrac{1}{35} + \dfrac{1}{63} + \dfrac{1}{99} + \dfrac{1}{143}$

    (b) $6/13$

    (c) $1/2$

47. $\log_2 x^3 y = \log_2 x^3 + \log_2 y$

$\qquad = 3\log_2 x + \log_2 y$

49. $\log_{10} \dfrac{\sqrt[3]{x}}{y^2} = \log_{10} \dfrac{x^{1/3}}{y^2}$

$\quad\quad\quad\quad = \log_{10} x^{1/3} - \log_{10} y^2$

$\quad\quad\quad\quad = \dfrac{1}{3} \log_{10} x - 2 \log_{10} y$

51. $\log_{10} x - \log_{10} y^2 = \log_{10} \dfrac{x}{y^2}$

53. $2 \log_3 x - 3 \log_3 y - 4 \log_3 z$

$\quad = \log_3 x^2 - \log_3 y^3 - \log_3 z^4$

$\quad = \log_3 \dfrac{x^2}{y^3 z^4}$

55. $\log_4 x - \log_4 5 = 2$

$\quad\quad \log_4 \dfrac{x}{5} = 2$

$\quad\quad\quad \dfrac{x}{5} = 4^2$

$\quad\quad\quad x = 80$

57. $\log_2 x + \log_2 (x-7) = 3$

$\quad\quad \log_2 x(x-7) = 3$

$\quad\quad\quad x(x-7) = 2^3$

$\quad\quad\quad x^2 - 7x = 8$

$\quad\quad\quad x^2 - 7x - 8 = 0$

$\quad\quad (x-8)(x+1) = 0$

$\quad\quad\quad x - 8 = 0 \ \text{ or } \ x + 1 = 0$

$\quad\quad\quad\quad x = 8 \quad\quad\quad x = -1$

Possible solutions are 8 and -1 but only 8 checks.

59. $\displaystyle\sum_{i=1}^{3} \left(2^i \cdot x - 4\right) = 20$

$\left(2^1 \cdot x - 4\right) + \left(2^2 \cdot x - 4\right) + \left(2^3 \cdot x - 4\right) = 20$

$\quad\quad 2x - 4 + 4x - 4 + 8x - 4 = 20$

$\quad\quad\quad\quad\quad\quad\quad 14x = 32$

$\quad\quad\quad\quad\quad\quad\quad x = \dfrac{16}{7}$

12.3 Arithmetic Sequences

1. 1, 2, 3, 4, ...

$2 - 1 = 1, \ 3 - 2 = 1, \ 4 - 3 = 1$

Arithmetic $d = 1$

3. 1, 2, 4, 7, ...

$2 - 1 = 1, \ 4 - 2 = 2$

Not arithmetic

5. 50, 45, 40, ...

$45 - 50 = -5, \ 40 - 45 = -5$

Arithmetic $d = -5$

7. 1, 4, 9, 16, ...

$4 - 1 = 3, \ 9 - 4 = 5$

Not arithmetic

9. $\dfrac{1}{3}, 1, \dfrac{5}{3}, \dfrac{7}{3}, \ ...$

$1 - \dfrac{1}{3} = \dfrac{2}{3}, \ \dfrac{5}{3} - 1 = \dfrac{2}{3}, \ \dfrac{7}{3} - \dfrac{5}{3} = \dfrac{2}{3}$

Arithmetic $d = \dfrac{2}{3}$

Use $a_n = a_1 + (n-1)d$ and $S_n = \dfrac{n}{2}(a_1 + a_n)$ in Problems 11-24.

11. $a_1 = 3,\ d = 4$

$a_n = 3 + (n-1)4$

$\quad = 4n - 1$

$a_{24} = 4(24) - 1 = 95$

13. $a_1 = 6,\ d = -2$

$a_n = 6 + (n-1)(-2)$

$a_{10} = 6 + (10-1)(-2)$

$\quad = 6 + (9)(-2)$

$\quad = 6 - 18$

$a_{10} = -12$

$S_{10} = \dfrac{10}{2}(6 - 12)$

$\quad = 5(-6)$

$S_{10} = -30$

15. $a_6 = 17,\ a_{12} = 29$

$17 = a_1 + (6-1)d$

$17 = a_1 + 5d \qquad (1)$

$29 = a_1 + (12-1)d$

$29 = a_1 + 11d \qquad (2)$

$\ \ 187 = 11a_1 + 55d \quad (1)\text{ times }11$

$\underline{-145 = -5a_1 - 55d \quad (2)\text{ times } -5}$

$\quad\ \ 42 = 6a_1$

$\quad\ \ \ \ 7 = a_1$

$17 = 7 + 5d \qquad\quad (1),\ a_1 = 7$

$\ \ 2 = d$

$a_{30} = 7 + (30-1)2 = 65$

17. $\qquad a_3 = 16,\ a_8 = 26$

$a_3 = a_1 + (3-1)d$

$16 = a_1 + 2d \qquad (1)$

$a_8 = a_1 + (8-1)d$

$26 = a_1 + 7d \qquad\quad (2)$

$\qquad$ Add (1) to (2)

$5d = 10$

$\ \ d = 2$

$a_1 + 2(2) = 16 \qquad\quad (1),\ d = 2$

$a_1 = 12$

$a_n = 12 + (n-1)2$

$a_n = 2n + 10$

$a_{20} = 2(20) + 10 = 50$

$S_{20} = \dfrac{20}{2}(12 + 50) = 10(62) = 620$

19. 5, 9, 13, 17, ...

$a_1 = 5,\ d = 9 - 5 = 4$

$a_{100} = 5 + 99(4) = 401$

$S_{100} = \dfrac{100}{2}(5 + 401) = 20{,}300$

21. 12, 7, 2, −3, ...

$a_1 = 12$, $d = 7 - 12 = -5$.

$a_{35} = 12 + (35 - 1)(-5)$

$\quad = 12 + (34)(-5)$

$\quad = 12 - 170$

$a_{35} = -158$

23. $\dfrac{1}{2}$, 1, $\dfrac{3}{2}$, 2, ...

$a_1 = \dfrac{1}{2}$, $d = 1 - \dfrac{1}{2} = \dfrac{1}{2}$

$a_{10} = \dfrac{1}{2} + 9\left(\dfrac{1}{2}\right) = 5$

$S_{10} = \dfrac{10}{2}\left(\dfrac{1}{2} + 5\right) = \dfrac{55}{2}$

25. (a) 18,000, 14,700, 11,400, 8,100, 4,800

(b) $d = -3,300$

(c) See the graph in the back of the textbook.

(d) $9,750

(e) $a_0 = 18,000$, $a_n = a_{n-1} - 3,300$, $n \geq 1$

27. (a) 1,500, 1,460, 1,420, 1,380, 1,340, 1,300

(b) It is arithmetic as the same amount is subtracted from each succeeding term

(c) $a_n = 1,500 - (n-1)40$

29. (a) 1, 3, 6, 10, 15, 21, 28, 36, 45, 66, 78, 91, 105, 120

(b) $a_1 = 1$, $a_n = a_{n-1} + n$, $n \geq 2$

(c) No, it is not arithmetic because the same amount is not added to each term.

31. 28,000, 28,850, 29,700, 30,550, 31,400

$a_n = a_1 + (n-1)d$

$\quad = 28,000 + (n-1)(850)$

$\quad = 28,000 + 850n - 850$

$a_n = 27,150 + 850n$

$a_{10} = 27,150 + 850(10)$

$\quad = 27,150 + 8500$

$\quad = 35,650$

33. $\log 576 = 2.7604$

35. $\log 0.0576 = -1.2396$

37. $\log x = 2.6484$

$x = 10^{2.6484}$

≈ 445

39. $\log x = -7.3516$

$x = 4.45 \times 10^{-8}$

12.4 Geometric Sequences

1. 1, 5, 25, 125, ...

$\frac{5}{1} = 5, \quad \frac{25}{5} = 5, \quad \frac{125}{25} = 5$

Geometric, $r = 5$

3. $\frac{1}{2}, \frac{1}{6}, \frac{1}{18}, \frac{1}{54}, \ldots$

$\frac{1/6}{1/2} = \frac{1}{3}, \quad \frac{1/18}{1/6} = \frac{1}{3}, \quad \frac{1/54}{1/18} = \frac{1}{3}$

Geometric, $r = \frac{1}{3}$

5. 4, 9, 16, 25, ...

$\frac{9}{4}, \frac{16}{9}, \frac{25}{16} 2$

Not geometric

7. $-2, 4, -8, 16, \ldots$

$\frac{4}{-2} = -2, \quad \frac{-8}{4} = -2, \quad \frac{16}{-8} = -2$

Geometric, $r = -2$

9. 4, 6, 8, 10, ...

$\frac{6}{4} = \frac{3}{2}, \quad \frac{8}{6} = \frac{4}{3}$

Not geometric

Use $a_n = a_1 r^{n-1}$ and $S_n = \dfrac{a_1\left(r^n - 1\right)}{r - 1}$ in Problems 11-30.

11. $a_1 = 4, \ r = 3$

$a_n = 4(3)^{n-1}$

13. $a_1 = -2, \ r = -\dfrac{1}{2}$

$a_6 = -2\left(-\dfrac{1}{2}\right)^{6-1}$

$= -2\left(-\dfrac{1}{2}\right)^5$

$= -2\left(-\dfrac{1}{32}\right)$

$a_6 = \dfrac{1}{16}$

15. $a_1 = 3, \ r = -1$

$a_{20} = 3(-1)^{20-1} = -3$

17. $a_1 = 10, \ r = 2$

$S_{10} = \dfrac{10\left(2^{10} - 1\right)}{2 - 1}$

$= \dfrac{10\left(1024 - 1\right)}{1}$

$= 10\left(1023\right)$

$S_{10} = 10,230$

19. $a_1 = 1, \ r = -1$

$S_{20} = \dfrac{1\left[(-1)^{20} - 1\right]}{-1 - 1} = 0$

21. $a_1 = \dfrac{1}{5}, \ r = \dfrac{1}{2}$

$a_8 = \dfrac{1}{5}\left(\dfrac{1}{2}\right)^{8-1}$

$= \dfrac{1}{5}\left(\dfrac{1}{2}\right)^7$

$= \dfrac{1}{5}\left(\dfrac{1}{128}\right)$

$a_8 = \dfrac{1}{640}$

23. $-\dfrac{1}{2}, -\dfrac{1}{4}, -\dfrac{1}{8}, \ldots$

$a_1 = -\dfrac{1}{2}, \quad r = \dfrac{-1/4}{-1/2} = \dfrac{1}{2}$

$S_5 = \dfrac{-\dfrac{1}{2}\left[\left(\dfrac{1}{2}\right)^5 - 1\right]}{\dfrac{1}{2} - 1} = -\dfrac{31}{32}$

25. $\sqrt{2}, 2, 2\sqrt{2}, \ldots$

$a_1 = \sqrt{2}, \quad r = \dfrac{2}{\sqrt{2}} = \sqrt{2}$

$a_{10} = \sqrt{2}\left(\sqrt{2}\right)^{10-1} = \left(\sqrt{2}\right)^{10} = 32$

$S_{10} = \dfrac{\sqrt{2}\left[\left(\sqrt{2}\right)^{10} - 1\right]}{\sqrt{2} - 1}$

$= \dfrac{31\sqrt{2}}{\sqrt{2} - 1}$

$= \dfrac{31\sqrt{2}}{\sqrt{2} - 1} \cdot \dfrac{\sqrt{2} + 1}{\sqrt{2} + 1}$

$= \dfrac{31(2) + 31\sqrt{2}}{2 - 1}$

$= 62 + 31\sqrt{2}$

27. $100, 10, 1, \ldots$

$a_1 = 100, \quad r = \dfrac{10}{100} = \dfrac{1}{10}$

$a_6 = 100\left(\dfrac{1}{10}\right)^{6-1} = \dfrac{1}{1000}$

$S_6 = \dfrac{100\left[\left(\dfrac{1}{10}\right)^6 - 1\right]}{\dfrac{1}{10} - 1}$

$= \dfrac{100\left(\dfrac{1}{1,000,000} - 1\right)}{-\dfrac{9}{10}}$

$= 111.111$

29. $a_4 = 40, \ a_5 = ?, \ a_6 = 160$

$a_4 = a_1 r^3 = 40$

$a_6 = a_1 r^5 = 160$

$\dfrac{a_6}{a_4} = \dfrac{a_1 r^5}{a_1 r^3} = \dfrac{160}{40}$

$r^2 = 4$

$r = \pm 2$

Use $S = \dfrac{a_1}{1-r}$ in Problems 31 - 42.

31. $\dfrac{1}{2} + \dfrac{1}{4} + \dfrac{1}{8} + \cdots$

$a_1 = \dfrac{1}{2}, \quad r = \dfrac{1/4}{1/2} = \dfrac{1}{2}$

$S = \dfrac{1/2}{1 - 1/2}$

$= 1$

33. $4 + 2 + 1 + \ldots,$

$a_1 = 4, \quad r = \dfrac{2}{4} = \dfrac{1}{2}$

$S = \dfrac{4}{1 - \dfrac{1}{2}} = \dfrac{4}{\dfrac{1}{2}} = 8$

35. $\dfrac{2}{5} + \dfrac{4}{25} + \dfrac{8}{125} + \cdots$

$a_1 = \dfrac{2}{5}, \quad r = \dfrac{4/25}{2/5} = \dfrac{2}{5}$

$S = \dfrac{2/5}{1 - 2/5} = \dfrac{2}{3}$

37. $\dfrac{3}{4} + \dfrac{1}{4} + \dfrac{1}{12} + \ldots,$ then $a_1 = \dfrac{3}{4}, \quad r = \dfrac{1}{3}$

$S = \dfrac{\dfrac{3}{4}}{1 - \dfrac{1}{3}} = \dfrac{\dfrac{3}{4}}{\dfrac{2}{3}} = \dfrac{9}{8}$

39. $0.444 \ldots = 0.4 + 0.04 + 0.004 + \ldots$

$= 0.4 + 0.4\left(\dfrac{1}{10}\right) + 0.4\left(\dfrac{1}{10}\right)^2 + \ldots$

$a_1 = 0.4, \quad r = \dfrac{1}{10}$

$S = \dfrac{0.4}{1 - 1/10} = \dfrac{4/10}{9/10} = \dfrac{4}{9}$

43. (a) 450,000, 315,000, 220,500, 154,350, 108,045

(b) 0.70

(c) See the graph in the back of the textbook.

(d) $90,396.93

(e) $a_0 = 450,000$

$a_n = 0.7a_{n-1}$

$n \geq 1$

41. $0.272727\ldots = 0.27 + 0.0027 + 0.000027 + \ldots$

$= \dfrac{27}{100} + \dfrac{27}{10,000} + \dfrac{27}{1,000,000} + \ldots$

$= \dfrac{27}{100} + \dfrac{27}{100}\left(\dfrac{1}{100}\right) + \dfrac{27}{100}\left(\dfrac{1}{100}\right)^2 + \ldots$

$a_1 = \dfrac{27}{100}, \quad r = \dfrac{1}{100}$

$S = \dfrac{a_1}{1-r} = \dfrac{\dfrac{27}{100}}{1 - \dfrac{1}{100}} = \dfrac{\dfrac{27}{100}}{\dfrac{99}{100}} = \dfrac{27}{99} = \dfrac{3}{11}$

45. (a) $\dfrac{1}{3} + \dfrac{1}{9} + \dfrac{1}{27} + \ldots,$ then $a_1 = \dfrac{1}{3}, \quad r = \dfrac{1}{3}$

$S = \dfrac{\dfrac{1}{3}}{1 - \dfrac{1}{3}} = \dfrac{\dfrac{1}{3}}{\dfrac{2}{3}} = \dfrac{1}{2}$

(b) $S_6 = \dfrac{\dfrac{1}{3}\left[\left(\dfrac{1}{3}\right)^6 - 1\right]}{\dfrac{1}{3} - 1} = \dfrac{\dfrac{1}{3}\left[\dfrac{1}{729} - \dfrac{729}{729}\right]}{-\dfrac{2}{3}} = \dfrac{364}{729}$

(c) $S - S_6 = \dfrac{1}{2} - \dfrac{364}{729} = \dfrac{729 - 728}{1458} = \dfrac{1}{1458}$

47. $20 + 2\left(\dfrac{7}{8}\right)20 + 2\left(\dfrac{7}{8}\right)^2 20 + 2\left(\dfrac{7}{8}\right)^3 20 + \ \ldots$

$40 + \left(\dfrac{7}{8}\right)40 + \left(\dfrac{7}{8}\right)^2 40 + \left(\dfrac{7}{8}\right)^3 40 + \ \ldots$

$a_1 = 40, \ \ r = \dfrac{7}{8}$

$S = \dfrac{40}{1 - 7/8} = 320$

Total vertical distance=320-20=300ft

49. $(x + 5)^2 = (x + 5)(x + 5)$

$\qquad\qquad = x^2 + 10x + 25$

51. $(x + y)^3 = (x + y)(x + y)(x + y)$

$\qquad\quad = \left(x^2 + 2xy + y^2\right)(x + y)$

$\qquad\quad = x^3 + 3x^2 y + 3xy^2 + y^3$

53. $(x + y)^4 = (x + y)(x + y)(x + y)(x + y)$

$\qquad\quad = \left(x^2 + 2xy + y^2\right)\left(x^2 + 2xy + y^2\right)$

$\qquad\quad = x^4 + 4x^3 y + 6x^2 y^2 + 4xy^3 + y^4$

55. $a_1 = 100, \ \ r = 2$

$a_{20} = 100(2)^{20-1}$

$\qquad = 100(2)^{19}$

$\qquad = 100(524{,}288)$

$\qquad = 52{,}428{,}800$

57. $a_1 = 100, \ \ r = \dfrac{1}{2}$

$S_{18} = \dfrac{100\left[\left(\dfrac{1}{2}\right)^{18} - 1\right]}{\dfrac{1}{2} - 1} = 199.99924$

59. $a_1 = a, \ \ r = \dfrac{1}{2}$

$S = \dfrac{a_1}{1 - r} = \dfrac{a}{1 - \dfrac{1}{2}} = \dfrac{a}{\dfrac{1}{2}} = 2a$

61. $\dfrac{a}{b} + \dfrac{a^2}{b^2} + \dfrac{a^3}{b^3} + \ \cdots \ \ \text{if } \left|\dfrac{a}{b}\right| < 1$

$a_1 = \dfrac{a}{b}, \ \ r = \dfrac{a^2 / b^2}{a / b} = \dfrac{a}{b}$

$S = \dfrac{a / b}{1 - a / b} = \dfrac{a}{b - a}$

12.5 The Binomial Expansion

1. $(x+2)^4 = \binom{4}{0}(x)^4(2)^0 + \binom{4}{1}(x)^3(2)^1 + \binom{4}{2}(x)^2(2)^2 + \binom{4}{3}(x)^1(2)^3 + \binom{4}{4}(x)^0(2)^4$

$= 1(x^4)(1) + 4(x^3)(2) + 6(x^2)(4) + 4(x)(8) + 1(1)(16)$

$= x^4 + 8x^3 + 24x^2 + 32x + 16$

3. $(x+y)^6 = \binom{6}{0}x^6y^0 + \binom{6}{1}x^5y^1 + \binom{6}{2}x^4y^2 + \binom{6}{3}x^3y^3 + \binom{6}{4}x^2y^4 + \binom{6}{5}x^1y^5 + \binom{6}{6}x^0y^6$

$= x^6 + 6x^5y + 15x^4y^2 + 20x^3y^3 + 15x^2y^4 + 6xy^5 + y^6$

5. $(2x+1)^5 = \binom{5}{0}(2x)^5(1)^0 + \binom{5}{1}(2x)^4(1)^1 + \binom{5}{2}(2x)^3(1)^2 + \binom{5}{3}(2x)^2(1)^3 + \binom{5}{4}(2x)^1(1)^4 + \binom{5}{5}(2x)^0(1)^5$

$= 1(32x^5)(1) + 5(16x^4)(1) + 10(8x^3)(1) + 10(4x^2)(1) + 5(2x)(1) + 1(1)(1)$

$= 32x^5 + 80x^4 + 80x^3 + 40x^2 + 10x + 1$

7. $(x-2y)^5 = \binom{5}{0}x^5(-2y)^0 + \binom{5}{1}x^4(-2y)^1 + \binom{5}{2}x^3(-2y)^2 + \binom{5}{3}x^2(-2y)^3 + \binom{5}{4}x^1(-2y)^4 + \binom{5}{5}x^0(-2y)^5$

$= 1x^5(1) + 5x^4(-2y) + 10x^3(4y^2) + 10x^2(-8y^3) + 5x(16y^4) + 1(1)(-32y^5)$

$= x^5 - 10x^4y + 40x^3y^2 - 80x^2y^3 + 80xy^4 - 32y^5$

9. $(3x-2)^4 = \binom{4}{0}(3x)^4(-2)^0 + \binom{4}{1}(3x)^3(-2)^1 + \binom{4}{2}(3x)^2(-2)^2 + \binom{4}{3}(3x)^1(-2)^3 + \binom{4}{4}(3x)^0(-2)^4$

$= 1(81x^4)(1) + 4(27x^3)(-2) + 6(9x^2)(4) + 4(3x)(-8) + 1(1)(16)$

$= 81x^4 - 216x^3 + 216x^2 - 96x + 16$

11. $(4x-3y)^3 = \binom{3}{0}(4x)^3(-3y)^0 + \binom{3}{1}(4x)^2(-3y)^1 + \binom{3}{2}(4x)^1(-3y)^2 + \binom{3}{3}(4x)^0(-3y)^3$

$= 1(64x^3)(1) + 3(16x^2)(-3y) + 3(4x)(9y^2) + 1(1)(-27y^3)$

$= 64x^3 - 144x^2y + 108xy^2 - 27y^3$

13. $(x^2+2)^4 = \binom{4}{0}(x^2)^4(2)^0 + \binom{4}{1}(x^2)^3(2)^1 + \binom{4}{2}(x^2)^2(2)^2 + \binom{4}{3}(x^2)^1(2)^3 + \binom{4}{4}(x^2)^0(2)^4$

$= 1(x^8)(1) + 4(x^6)(2) + 6(x^4)(4) + 4(x^2)(8) + 1(1)(16)$

$= x^8 + 8x^6 + 24x^4 + 32x^2 + 16$

15. $\left(x^2+y^2\right)^3 = \binom{3}{0}\left(x^2\right)^3\left(y^2\right)^0 + \binom{3}{1}\left(x^2\right)^2\left(y^2\right)^1 + \binom{3}{2}\left(x^2\right)^1\left(y^2\right)^2 + \binom{3}{3}\left(x^2\right)^0\left(y^2\right)^3$

$\qquad = x^6 + 3x^4 y^2 + 3x^2 y^4 + y^6$

17. $\left(\dfrac{x}{2}-4\right)^3 = \binom{3}{0}\left(\dfrac{x}{2}\right)^3(-4)^0 + \binom{3}{1}\left(\dfrac{x}{2}\right)^2(-4)^1 + \binom{3}{2}\left(\dfrac{x}{2}\right)^1(-4)^2 + \binom{3}{3}\left(\dfrac{x}{2}\right)^0(-4)^3$

$\qquad = 1\left(\dfrac{x^3}{8}\right)(1) + 3\left(\dfrac{x^2}{4}\right)(-4) + 3\left(\dfrac{x}{2}\right)(16) + 1(1)(-64)$

$\qquad = \dfrac{1}{8}x^3 - 3x^2 + 24x - 64$

19. $\left(\dfrac{x}{3}+\dfrac{y}{2}\right)^4 = \binom{4}{0}\left(\dfrac{x}{3}\right)^4\left(\dfrac{y}{2}\right)^0 + \binom{4}{1}\left(\dfrac{x}{3}\right)^3\left(\dfrac{y}{2}\right)^1 + \binom{4}{2}\left(\dfrac{x}{3}\right)^2\left(\dfrac{y}{2}\right)^2 + \binom{4}{3}\left(\dfrac{x}{3}\right)^1\left(\dfrac{y}{2}\right)^3 + \binom{4}{4}\left(\dfrac{x}{3}\right)^0\left(\dfrac{y}{2}\right)^4$

$\qquad = 1\left(\dfrac{x^4}{81}\right)(1) + 4\left(\dfrac{x^3}{27}\right)\left(\dfrac{y}{2}\right) + 6\left(\dfrac{x^2}{9}\right)\left(\dfrac{y^2}{4}\right) + 4\left(\dfrac{x}{3}\right)\left(\dfrac{y^3}{8}\right) + 1(1)\left(\dfrac{y^4}{16}\right)$

$\qquad = \dfrac{x^4}{81} + \dfrac{2x^3 y}{27} + \dfrac{x^2 y^2}{6} + \dfrac{xy^3}{6} + \dfrac{y^4}{16}$

21. The coefficients of the first three terms have been calculated in example 4, page 776, of your textbook. The fourth term is calculated as follows:

$\binom{9}{3} = \dfrac{9!}{3!6!} = \dfrac{9\cdot 8\cdot 7\cdot 6\cdot 5\cdot 4\cdot 3\cdot 2\cdot 1}{(3\cdot 2\cdot 1)(6\cdot 5\cdot 4\cdot 3\cdot 2\cdot 1)} = \dfrac{504}{6} = 84$

$(x+2)^9 = 1\left(x^9\right)(1) + 9\left(x^8\right)(2) + 36\left(x^7\right)(2)^2 + 84\left(x^6\right)(2)^3$

$\qquad = x^9 + 18x^8 + 144x^7 + 672x^6$

23. $\binom{10}{0} = \dfrac{10!}{0!10!} = \dfrac{10!}{(1)10!} = \dfrac{1}{1} = 1,\qquad \binom{10}{1} = \dfrac{10!}{1!9!} = \dfrac{10\cdot 9!}{(1)9!} = \dfrac{10}{1} = 10$

$\binom{10}{2} = \dfrac{10!}{2!8!} = \dfrac{10\cdot 9\cdot 8!}{(2\cdot 1)8!} = \dfrac{90}{2} = 45,\qquad \binom{10}{3} = \dfrac{10!}{3!7!} = \dfrac{10\cdot 9\cdot 8\cdot 7!}{(3\cdot 2\cdot 1)7!} = \dfrac{720}{6} = 120$

$(x-y)^{10} \Rightarrow \binom{10}{0}\left(x^{10}\right)(-y)^0 + \binom{10}{1}\left(x^9\right)(-y)^1 + \binom{10}{2}\left(x^8\right)(-y)^2 + \binom{10}{3}\left(x^7\right)(-y)^3$

$\qquad = 1\left(x^{10}\right)(1) + 10\left(x^9\right)(-y) + 45\left(x^8\right)\left(y^2\right) + 120\left(x^7\right)\left(-y^3\right)$

$\qquad = x^{10} - 10x^9 y + 45x^8 y^2 - 120x^7 y^3$

25. $\dbinom{10}{0} = \dfrac{10!}{0!10!} = \dfrac{10!}{(1)10!} = \dfrac{1}{1} = 1,$ $\qquad$ $\dbinom{10}{1} = \dfrac{10!}{1!9!} = \dfrac{10\cdot 9!}{(1)9!} = \dfrac{10}{1} = 10$

$\dbinom{10}{2} = \dfrac{10!}{2!8!} = \dfrac{10\cdot 9\cdot 8!}{(2\cdot 1)8!} = \dfrac{90}{2} = 45,$ $\qquad$ $\dbinom{10}{3} = \dfrac{10!}{3!7!} = \dfrac{10\cdot 9\cdot 8\cdot 7!}{(3\cdot 2\cdot 1)7!} = \dfrac{720}{6} = 120$

$(x+2y)^{10} \Rightarrow 1\left(x^{10}\right)(2y)^{0} + 10\left(x^{9}\right)(2y)^{1} + 45\left(x^{8}\right)(2y)^{2} + 120\left(x^{7}\right)(2y)^{3}$

$\qquad = x^{10} + 20x^{9}y + 180x^{8}y^{2} + 960x^{7}y^{3}$

27. $\dbinom{15}{0} = \dfrac{15!}{0!15!} = \dfrac{15!}{(1)15!} = \dfrac{1}{1} = 1,$ $\dbinom{15}{1} = \dfrac{15!}{1!14!} = \dfrac{15\cdot 14!}{(1)14!} = \dfrac{15}{1} = 15,$ $\dbinom{15}{2} = \dfrac{15!}{2!13!} = \dfrac{15\cdot 14\cdot 13!}{(2\cdot 1)13!} = \dfrac{210}{2} = 105$

$(x+1)^{15} \Rightarrow \dbinom{15}{0}\left(x^{15}\right)\left(1^{0}\right) + \dbinom{15}{1}\left(x^{14}\right)\left(1^{1}\right) + \dbinom{15}{2}\left(x^{13}\right)\left(1^{2}\right)$

$\qquad = x^{15} + 15x^{14} + 105x^{13}$

29. $\dbinom{12}{0} = \dfrac{12!}{0!12!} = \dfrac{12!}{(1)12!} = \dfrac{1}{1} = 1,$ $\dbinom{12}{1} = \dfrac{12!}{1!11!} = \dfrac{12\cdot 11!}{(1)11!} = \dfrac{12}{1} = 12,$ $\dbinom{12}{2} = \dfrac{12!}{2!10!} = \dfrac{12\cdot 11\cdot 10!}{(2\cdot 1)10!} = \dfrac{132}{2} = 66$

$(x-y)^{12} \Rightarrow \dbinom{12}{0}\left(x^{12}\right)(-y)^{0} + \dbinom{12}{1}\left(x^{11}\right)(-y)^{1} + \dbinom{12}{2}\left(x^{10}\right)(-y)^{2}$

$\qquad = x^{12} - 12x^{11}y + 66x^{10}y^{2}$

31. $\dbinom{20}{0} = \dfrac{20!}{0!20!} = \dfrac{20!}{(1)20!} = \dfrac{1}{1} = 1,$ $\dbinom{20}{1} = \dfrac{20!}{1!19!} = \dfrac{20\cdot 19!}{(1)19!} = \dfrac{20}{1} = 20,$ $\dbinom{20}{2} = \dfrac{20!}{2!18!} = \dfrac{20\cdot 19\cdot 18!}{(2\cdot 1)18!} = \dfrac{380}{2} = 190$

$(x+2)^{20} \Rightarrow \dbinom{20}{0}\left(x^{20}\right)\left(2^{0}\right) + \dbinom{20}{1}\left(x^{19}\right)\left(2^{1}\right) + \dbinom{20}{2}\left(x^{18}\right)\left(2^{2}\right)$

$\qquad = x^{20} + 40x^{19} + 760x^{18}$

33. $\dbinom{100}{0} = \dfrac{100!}{0!100!} = \dfrac{100!}{1\cdot 100!} = \dfrac{1}{1} = 1,$ $\dbinom{100}{1} = \dfrac{100!}{1!99!} = \dfrac{100\cdot 99!}{1\cdot 99!} = \dfrac{100}{1} = 100$

$(x+2)^{100} \Rightarrow 1\left(x^{100}\right)(2)^{0} + 100\left(x^{99}\right)(2)^{1}$

$\qquad = x^{100} + 200x^{99}$

35. $\dbinom{50}{0} = \dfrac{50!}{0!50!} = \dfrac{50!}{1\cdot 50!} = \dfrac{1}{1} = 1,$ $\dbinom{50}{1} = \dfrac{50!}{1!49!} = \dfrac{50\cdot 49!}{1\cdot 49!} = \dfrac{50}{1} = 50$

$(x+y)^{50} \Rightarrow \dbinom{50}{0}\left(x^{50}\right)\left(y^{0}\right) + \dbinom{50}{1}\left(x^{49}\right)\left(y^{1}\right)$

$\qquad = x^{50} + 50x^{49}y$

37. $\dbinom{12}{8} = \dfrac{12!}{8!\,4!} = \dfrac{12\cdot11\cdot10\cdot9\cdot8!}{8!(4\cdot3\cdot2\cdot1)} = \dfrac{11,880}{24} = 495$

$\begin{aligned} a_9 &= 495(2x)^4(3y)^8 \\ &= 495\left(16x^4\right)\left(6561y^8\right) \\ &= 51,963,120x^4y^8 \end{aligned}$

39. $\dbinom{10}{4} = \dfrac{10!}{4!\,6!} = \dfrac{10\cdot9\cdot8\cdot7\cdot6!}{(4\cdot3\cdot2\cdot1)6!} = \dfrac{5040}{24} = 210$

$\begin{aligned} (x-2)^{10} &\Rightarrow \dbinom{10}{4}\left(x^6\right)(-2)^4 \\ &= 210\left(x^6\right)(16) \\ &= 3360x^6 \end{aligned}$

41. $\dbinom{9}{3} = \dfrac{9!}{3!\,6!} = \dfrac{9\cdot8\cdot7\cdot6!}{(3\cdot2\cdot1)6!} = \dfrac{504}{6} = 84$

$a_4 = 84(x)^6(3)^3 = 84x^6(27) = 2,268x^6$

43. $(2x+5y)^{20} \Rightarrow \dbinom{20}{11}(2x)^9(5y)^{11}$

$\qquad = \dfrac{20!}{11!\,9!}(2x)^9(5y)^{11}$

45. The third term from the binomial formula is

$\dbinom{7}{2}\left(\dfrac{1}{2}\right)^5\left(\dfrac{1}{2}\right)^2 = \dfrac{7!}{2!5!}\left(\dfrac{1}{2}\right)^7$

$\qquad = \dfrac{7\cdot6\cdot5\cdot4\cdot3\cdot2\cdot1}{(2\cdot1)(5\cdot4\cdot3\cdot2\cdot1)}\left(\dfrac{1}{2}\right)^7$

$\qquad = \dfrac{42}{2}\cdot\dfrac{1}{128}$

$\qquad = \dfrac{21}{128}$

47. $(1+.01)^4 = \dbinom{4}{0}(1)^4(.01)^0 + \dbinom{4}{1}(1)^3(.01)^1 + \dbinom{4}{2}(1)^2(.01)^2 + \dbinom{4}{3}(1)^1(.01)^3 + \dbinom{4}{4}(1)^0(.01)^4$

$\qquad = 1(1) + 4(.01) + 6(.0001) + 4(.000001) + 1(.00000001)$

$\qquad = 1 + .04 + .0006 + .000004 + .00000001$

$\qquad = 1.04060401$

49. $5^x = 7$

$\qquad x = \dfrac{\log 7}{\log 5} \approx 1.21$

51. $\qquad 8^{2x+1} = 16$

$\qquad \left(2^3\right)^{2x+1} = 2^4$

$\qquad 2^{6x+3} = 2^4$

$\qquad 6x+3 = 4$

$\qquad 6x = 1$

$\qquad x = \dfrac{1}{6} \text{ or } 0.17$

53.　$A = 800,\ P = 400,\ r = 0.10,\ n = 4$

$$A = P\left(1 + \frac{r}{n}\right)^{nt}$$

$$800 = 400\left(1 + \frac{0.10}{4}\right)^{4t}$$

$$2 = (1.025)^{4t}$$

$$4t = \frac{\log 2}{\log 1.025}$$

$$t = \frac{\log 2}{4\ \log 1.025} \approx 7 \text{ years}$$

55.　$\log_4 20 = \dfrac{\log 20}{\log 4}$

$$= \frac{1.3010}{0.6021}$$

$$= 2.16$$

57.　$\ln 576 \approx 6.36$

59.
$$A = 10e^{5t}$$
$$\ln A = \ln 10 + \ln e^{5t}$$
$$\ln A = \ln 10 + 5t \ln e$$
$$\ln A = \ln 10 + 5t$$
$$\ln A - \ln 10 = 5t$$
$$t = \frac{1}{5} \cdot \left(\ln A - \ln 10\right)$$
$$t = \frac{1}{5} \ln \frac{A}{10}$$

61.　$\dbinom{8}{5} = \dfrac{8!}{5!\ 3!} = \dfrac{8 \cdot 7 \cdot 6 \cdot 5 \cdot 4 \cdot 3!}{(5 \cdot 4 \cdot 3 \cdot 2 \cdot 1)3!} = \dfrac{6720}{120} = 56$

$\dbinom{8}{3} = \dfrac{8!}{3!\ 5!} = \dfrac{8 \cdot 7 \cdot 6 \cdot 5!}{(3 \cdot 2 \cdot 1)5!} = \dfrac{336}{6} = 56$

63.　$\dbinom{20}{12} = \dfrac{20!}{12!\,8!}$

$$= \frac{20 \cdot 19 \cdot 18 \cdot 17 \cdot 16 \cdot 15 \cdot 14 \cdot 13 \cdot 12!}{12\,!(8 \cdot 7 \cdot 6 \cdot 5 \cdot 4 \cdot 3 \cdot 2 \cdot 1)} = 125{,}970$$

$\dbinom{20}{8} = \dfrac{20!}{8!12!}$

$$= \frac{20 \cdot 19 \cdot 18 \cdot 17 \cdot 16 \cdot 15 \cdot 14 \cdot 13 \cdot 12!}{(8 \cdot 7 \cdot 6 \cdot 5 \cdot 4 \cdot 3 \cdot 2 \cdot 1)12!} = 125{,}970$$

65.　$\dbinom{n}{r} = \dfrac{n!}{(n-r)!\ r!}$

$\dbinom{n}{n-r} = \dfrac{n!}{r!(n-r)!}$

67. $\binom{3}{0}+\binom{3}{1}+\binom{3}{2}+\binom{3}{3}=1+3+3+1=8=2^3$

$\binom{5}{0}+\binom{5}{1}+\binom{5}{2}+\binom{5}{3}+\binom{5}{4}+\binom{5}{5}=1+5+10+10+5+1=32=2^5$

$\binom{7}{0}+\binom{7}{1}+\binom{7}{2}+\binom{7}{3}+\binom{7}{4}+\binom{7}{5}+\binom{7}{6}+\binom{7}{7}=1+7+21+35+35+21+7+1=128=2^7$

Chapter 12 Review

1. General term $=a_n=2n+5$
   First term $=a_1=2(1)+5=7$
   Second term $=a_2=2(2)+5=9$
   Third term $=a_3=2(3)+5=11$
   Fourth term $=a_4=2(4)+5=13$

3. General term $=a_n=n^2-1$
   First term $=a_1=1^2-1=0$
   Second term $=a_2=2^2-1=3$
   Third term $=a_3=3^2-1=8$
   Fourth term $=a_4=4^2-1=15$

5. General term $=a_n=4a_{n-1}$, $n>1$
   First term $=a_1=4$
   Second term $=a_2=4(4)=16$
   Third term $=a_3=4(16)=64$
   Fourth term $=a_4=4(64)=256$

7. $a_n=3n-1$ or recursively as
   $a_1=2$, $a_n=a_{n-1}+3$

9. $a_n=n^4$

11. $a_n=2^{-n}$ or $\dfrac{1}{2^n}$ or recursively as
    $a_1=\dfrac{1}{2}$, $a_n=\dfrac{1}{2}a_{n-1}$

13. $\displaystyle\sum_{i=1}^{4}(2i+3)=[2(1)+3]+[2(2)+3]+[2(3)+3]+[2(4)+3]=32$

15. $\displaystyle\sum_{i=2}^{3}\frac{i^2}{i+2}=\frac{2^2}{2+2}+\frac{3^2}{3+2}=\frac{14}{5}$

17. $\displaystyle\sum_{i=3}^{5}\left(4i+i^2\right)=\left[4(3)+3^2\right]+\left[4(4)+4^2\right]+\left[4(5)+5^2\right]=98$

19. $\displaystyle\sum_{i=1}^{4}3i$

21. $\displaystyle\sum_{i=1}^{5}(2i+3)$

23. $\displaystyle\sum_{i=1}^{4}\frac{1}{i+2}$

25. $\displaystyle\sum_{i=1}^{3}(x-2i)$

27. Geometric

29. Arithmetic

31. Geometric

33. Arithmetic

35. $a_1 = 2,\ d = 3$
$$a_n = 2 + (n-1)3$$
$$= 3n - 1$$
$$a_{20} = 3(20) - 1$$
$$= 59$$

37. $a_1 = -2,\ d = 4$
$$a_n = -2 + (n-1)4$$
$$= 4n - 6$$
$$a_{10} = 4(10) - 6$$
$$= 34$$
$$S_{10} = \frac{10}{2}(-2 + 34)$$
$$= 160$$

39. $a_5 = 21,\ a_8 = 33$
$$a_5 = a_1 + (5-1)d$$
$$21 = a_1 + 4d$$

$$a_8 = a_1 + (8-1)d$$
$$33 = a_1 + 7d$$

$$a_1 + 4d = 21$$
$$a_1 + 7d = 33$$

Solve the system to get
$$a_1 = 5,\ d = 4$$
$$a_{10} = 5 + (10-1)(4)$$
$$= 41$$

41. $a_4 = -10,\ a_8 = -18$ Solve the system to get
$$a_4 = a_1 + (4-1)d \qquad a_1 = -4,\ d = -2$$
$$-10 = a_1 + 3d \qquad a_{20} = -4 + (20-1)(-2)$$
$$= -42$$
$$a_8 = a_1 + (8-1)d \qquad S_{20} = \frac{20}{2}\big[-4 + (-42)\big]$$
$$-18 = a_1 + 7d \qquad = -460$$

$$a_1 + 3d = -10$$
$$a_1 + 7d = -18$$

43. $a_1 = 100,\ d = -5$
$$a_{40} = 100 + (40-1)(-5)$$
$$= -95$$

45. $a_n = a_1 r^{n-1}$
$$= 5(-2)^{n-1}$$
$$a_{16} = 5(-2)^{16-1}$$
$$= 5(-2)^{15}$$

47. $S = \dfrac{a_1}{1-r} = \dfrac{-2}{1 - \dfrac{1}{3}} = \dfrac{-2}{\dfrac{2}{3}} = -3$

49. $a_3 = a_1 r^2 = 12$ $a_1 r^2 = 12$

   $a_4 = a_1 r^3 = 24$ $4a_1 = 12$

                          $a_1 = 3$

   $\dfrac{a_4}{a_3} = \dfrac{a_1 r^3}{a_1 r^2} = \dfrac{24}{12}$ $a_1 = 3, \ r = 2$

   $r = 2$ $a_6 = a_1 r^{6-1}$

   $\qquad\qquad = 3(2)^{6-1}$

   $\qquad\qquad = 3(2)^5$

   $\qquad\qquad = 96$

51. $\dbinom{8}{2} = \dfrac{8!}{2!6!} = 28$

53. $\dbinom{6}{3} = \dfrac{6!}{3!3!} = 20$

55. $\dbinom{10}{8} = \dfrac{10!}{8!2!} = 45$

57. $(x-2)^4 = \dbinom{4}{0}(x)^4(-2)^0 + \dbinom{4}{1}(x)^3(-2)^1 + \dbinom{4}{2}(x)^2(-2)^2 + \dbinom{4}{3}(x)^1(-2)^3 + \dbinom{4}{4}(x)^0(-2)^4$

   $= 1(x^4)(1) + 4(x^3)(-2) + 6(x^2)(4) + 4(x)(-8) + 1(1)(16)$

   $= x^4 - 8x^3 + 24x^2 - 32x + 16$

59. $(3x+2y)^3 = \dbinom{3}{0}(3x)^3(2y)^0 + \dbinom{3}{1}(3x)^2(2y)^1 + \dbinom{3}{2}(3x)^1(2y)^2 + \dbinom{3}{3}(3x)^0(2y)^3$

   $= 1(27x^3)(1) + 3(9x^2)(2y) + 3(3x)(4y^2) + 1(1)(8y^3)$

   $= 27x^3 + 54x^2 y + 36xy^2 + 8y^3$

61. $\left(\dfrac{x}{2}+3\right)^4 = \dbinom{4}{0}\left(\dfrac{x}{2}\right)^4(3)^0 + \dbinom{4}{1}\left(\dfrac{x}{2}\right)^3(3)^1 + \dbinom{4}{2}\left(\dfrac{x}{2}\right)^2(3)^2 + \dbinom{4}{3}\left(\dfrac{x}{2}\right)^1(3)^3 + \dbinom{4}{4}\left(\dfrac{x}{2}\right)^0(3)^4$

   $= 1\left(\dfrac{x^4}{16}\right)(1) + 4\left(\dfrac{x^3}{8}\right)(3) + 6\left(\dfrac{x^2}{4}\right)(9) + 4\left(\dfrac{x}{2}\right)(27) + 1(1)(81)$

   $= \dfrac{1}{16}x^4 + \dfrac{3}{2}x^3 + \dfrac{27}{2}x^2 + 54x + 81$

63. $\dbinom{10}{0} = \dfrac{10!}{0!10!} = 1, \ \dbinom{10}{1} = \dfrac{10!}{1!9!} = 10, \ \dbinom{10}{2} = \dfrac{10!}{2!8!} = 45$

   From the binomial formula, we write the first three terms:

   $(x+3y)^{10} = 1(x)^{10}(3y)^0 + 10(x)^9(3y)^1 + 45(x)^8(3y)^2$

   $\qquad\qquad = x^{10} + 30x^9 y + 405x^8 y^2$

65. $\binom{11}{0} = \frac{11!}{0!11!} = 1,\quad \binom{11}{1} = \frac{11!}{1!10!} = 11,\quad \binom{11}{2} = \frac{11!}{2!9!} = 55$

From the binomial formula, we write the first three terms:

$$(x+y)^{11} = 1(x)^{11}(y)^0 + 11(x)^{10}(y)^1 + 55(x)^9(y)^2$$
$$= x^{11} + 11x^{10}y + 55x^9y^2$$

67.

$\binom{16}{0} = \frac{16!}{0!16!} = 1,\quad \binom{16}{1} = \frac{16!}{1!15!} = 16$

From the binomial formula, we write the first two terms:

$$(x-2y)^{16} = 1(x)^{16}(-2y)^0 + 16(x)^{15}(-2y)^1$$
$$= x^{16} - 32x^{15}y$$

69. $\binom{50}{0} = \frac{50!}{0!50!} = 1,\quad \binom{50}{1} = \frac{50!}{1!49!} = 50$

From the binomial formula, we write the first two terms:

$$(x-1)^{50} = 1(x)^{50}(-1)^0 + 50(x)^{49}(-1)^1$$
$$= x^{50} - 50x^{49}$$

71. $a_6 = \binom{10}{5}(x)^5(-3)^5$

$$= \frac{10!}{5!5!}x^5(-243)$$

$$= -61{,}236x^5$$

Cumulative Review: Chapters 1-12

1. $\dfrac{5(-6)+3(-2)}{4(-3)+3} = \dfrac{-30+(-6)}{-12+3} = \dfrac{-36}{-9} = 4$

3. $\dfrac{18a^7b^{-4}}{36a^2b^{-8}} = \dfrac{a^{7-2}b^{-4-(-8)}}{2} = \dfrac{a^5b^4}{2}$

5. $8^{-2/3} = \left(2^3\right)^{-2/3} = 2^{3(-2/3)} = 2^{-2} = \dfrac{1}{2^2} = \dfrac{1}{4}$

7. $ab^3 + b^3 + 6a + 6 = b^3(a+1) + 6(a+1)$
$$= (a+1)(b^3+6)$$

9. $6 - 2(5x-1) + 4x = 20$
$$6 - 10x + 2 + 4x = 20$$
$$-6x + 8 = 20$$
$$-6x = 12$$
$$x = -2$$

11. $(x+1)(x+2) = 12$

$\qquad x^2 + 3x + 2 = 12$

$\qquad x^2 + 3x - 10 = 0$

$\qquad (x+5)(x-2) = 0$

$\qquad\qquad x+5 = 0 \quad \text{or} \quad x-2 = 0$

$\qquad\qquad\qquad x = -5 \qquad\qquad x = 2$

The solutions are $-5, 2$

13. $\qquad t - 6 = \sqrt{t-4}$

$\qquad t^2 - 12t + 36 = t - 4$

$\qquad t^2 - 13t + 40 = 0$

$\qquad (t-8)(t-5) = 0$

$\qquad\qquad t - 8 = 0 \quad \text{or} \quad t - 5 = 0$

$\qquad\qquad\qquad t = 8 \qquad\qquad t = 5$

Only 8 checks: The solution is 8

15. $\qquad 8t^3 - 27 = 0$

$\qquad (2t)^3 - (3)^3 = 0$

$(2t-3)(4t^2 + 6t + 9) = 0$

$4t^2 + 6t + 9 = 0 \qquad \text{or} \quad 2t - 3 = 0$

Let $a = 4, b = 6, c = 9 \qquad\qquad 2t = 3$

$t = \dfrac{-6 \pm \sqrt{6^2 - 4(4)(9)}}{2(4)} \qquad t = \dfrac{3}{2}$

$= \dfrac{-6 \pm \sqrt{-108}}{8}$

$= \dfrac{-6 \pm 6i\sqrt{3}}{8}$

$= \dfrac{-3 \pm 3i\sqrt{3}}{4}$

The solutions are $\dfrac{3}{2}, \dfrac{-3 \pm 3i\sqrt{3}}{4}$

17. $-3y - 2 < 7$

$\qquad -3y < 9$

$\qquad\quad y > -3$

See the graph in the back of the textbook.

19. See the graph in the back of the textbook.

21. See the graph in the back of the textbook.

23. $5x - 3y = -4 \qquad$ No change $\rightarrow \qquad 5x \ -3y = -4$

$\quad x + 2y = \ 7 \qquad$ Multiply by $-5 \rightarrow \ \underline{-5x - 10y = -35}$

$\qquad\qquad\qquad\qquad\qquad\qquad\qquad\qquad -13y = -39$

$\qquad\qquad\qquad\qquad\qquad\qquad\qquad\qquad\qquad y = 3$

Substitute $y = 3$ into the second equation to solve

for the x - coordinate.

$\qquad x + 2(3) = 7$

$\qquad\qquad\quad x = 1$

The solution to the system is $(1, 3)$.

25. $\dfrac{3y^2 - 3y}{3y - 12} \cdot \dfrac{y^2 - 2y - 8}{y^2 + 3y + 2} = \dfrac{3y(y-1)}{3(y-4)} \cdot \dfrac{(y-4)(y+2)}{(y+1)(y+2)}$

$\qquad\qquad\qquad\qquad\qquad\quad = \dfrac{y(y-1)}{y+1}$

27. $(2+3i)(1-4i) = 2+3i-8i-12i^2$
$$= 2-5i-12(-1)$$
$$= 14-5i$$

29. $4\sqrt{50} + 3\sqrt{8} = 4\sqrt{25(2)} + 3\sqrt{4(2)}$
$$= 4(5)\sqrt{2} + 3(2)\sqrt{2}$$
$$= 20\sqrt{2} + 6\sqrt{2}$$
$$= 26\sqrt{2}$$

31. $f(x) = 4x+1$ means $y = 4x+1$

The inverse is $x = 4y+1$

Solve for y in terms of x

$4y = x-1$

$$y = -\frac{1}{4}x - \frac{1}{4} = \frac{x-1}{4}$$

$$f^{-1}(x) = \frac{x-1}{4}$$

33. $\log_6 14 = \dfrac{\ln 14}{\ln 6} \approx 1.47$

35. $16, 8, 4, \cdots$

$$\frac{a_2}{a_1} = \frac{8}{16} = \frac{1}{2}$$

$$\frac{a_3}{a_2} = \frac{4}{8} = \frac{1}{2}$$

The common ratio $r = 1/2$

The general term is $a_n = \dfrac{1}{2}a_{n-1}$, $n \geq 2$

37. $\displaystyle\sum_{i=1}^{5} (2i-1) = (2\cdot1-1)+(2\cdot2-1)+(2\cdot3-1)+(2\cdot4-1)+(2\cdot5-1) = 1+3+5+7+9 = 25$

39. $(2,5), (6,-3)$

$$m = \frac{y_2-y_1}{x_2-x_1} = \frac{-3-5}{6-2} = \frac{-8}{4} = -2$$

Using $(x_1,y_1) = (6,-3)$ and $m = -2$

in $y-y_1 = m(x-x_1)$ yields

$y-(-3) = -2(x-6)$

$y+3 = -2x+12$

$y = -2x+9$

41. $t = -\dfrac{3}{4}$ and $t = \dfrac{1}{5}$

$t+\dfrac{3}{4} = 0 \qquad t-\dfrac{1}{5} = 0$

$4t+3 = 0 \qquad 5t-1 = 0$

$(4t+3)(5t-1) = 0$

$20t^2 + 11t - 3 = 0$

43. $3x - 5y = 2$ (1)

$2x + 4y = 1$ (2)

$6x - 10y = 4$ 2 times (1)

$-6x - 12y = -3$ -3 times (2)

$-22y = 1$

$y = -\dfrac{1}{22}$

$3x - 5\left(-\dfrac{1}{22}\right) = 2$ (1), $y = -\dfrac{1}{22}$

$3x + \dfrac{5}{22} = 2$

$3x = \dfrac{39}{22}$

$x = \dfrac{13}{22}$

The solution is $\left(\dfrac{13}{22}, -\dfrac{1}{22}\right)$.

47. Let x = one number, $5x$ = the other number.

$x + 5x = 180$ Supplementary angles

$6x = 180$

$x = 30$

$5x = 150$

The two supplementary angles are $30°$ and $150°$.

45. $-7 - 4 + (-8) = -7 + (-4) + (-8) = -19$

49. Let x = the smaller number. The larger number is $3x$.

Their reciprocals are $\dfrac{1}{x}$ and $\dfrac{1}{3x}$.

$3x \cdot \dfrac{1}{x} + 3x \cdot \dfrac{1}{3x} = 3x \cdot \dfrac{4}{3}$

$3 + 1 = 4x$

$4 = 4x$

$1 = x$

$x = 1$

The smaller number is 1 and the larger number is 3.

Chapter 12 Test

1. General term $= a_n = 3n - 5$
   First term $= a_1 = 3(1) - 5 = -2$
   Second term $= a_2 = 3(2) - 5 = 1$
   Third term $= a_3 = 3(3) - 5 = 4$
   Fourth term $= a_4 = 3(4) - 5 = 7$
   Fifth term $= a_5 = 3(5) - 5 = 10$

2. General term $= a_n = a_{n-1} + 4, \ n > 1$
   First term $= a_1 = 3$
   Second term $= a_2 = 3 + 4 = 7$
   Third term $= a_3 = 7 + 4 = 11$
   Fourth term $= a_4 = 11 + 4 = 15$
   Fifth term $= a_5 = 15 + 4 = 19$

3. General term $= a_n = n^2 + 1$
   First term $= a_1 = 1^2 + 1 = 2$
   Second term $= a_2 = 2^2 + 1 = 5$
   Third term $= a_3 = 3^2 + 1 = 10$
   Fourth term $= a_4 = 4^2 + 1 = 17$
   Fifth term $= a_5 = 5^2 + 1 = 26$

4. General term $= a_n = 2n^3$
   First term $= a_1 = 2(1)^3 = 2$
   Second term $= a_2 = 2(2)^3 = 16$
   Third term $= a_3 = 2(3)^3 = 54$
   Fourth term $= a_4 = 2(4)^3 = 128$
   Fifth term $= a_5 = 2(5)^3 = 250$

5. General term $= a_n = \dfrac{n+1}{n^2}$
   First term $= a_1 = \dfrac{1+1}{1^2} = 2$
   Second term $= a_2 = \dfrac{2+1}{2^2} = \dfrac{3}{4}$
   Third term $= a_3 = \dfrac{3+1}{3^2} = \dfrac{4}{9}$
   Fourth term $= a_4 = \dfrac{4+1}{4^2} = \dfrac{5}{16}$
   Fifth term $= a_5 = \dfrac{5+1}{5^2} = \dfrac{6}{25}$

6. General term $= a_n = -2a_{n-1}, \ n > 1$
   First term $= a_1 = 4$
   Second term $= a_2 = -2(4) = -8$
   Third term $= a_3 = -2(-8) = 16$
   Fourth term $= a_4 = -2(16) = -32$
   Fifth term $= a_5 = -2(-32) = 64$

7. $a_n = 4n + 2$ or recursively as
   $a_1 = 6, \ a_n = a_{n-1} + 4$

8. $a_n = 2^{n-1}$ or recursively as
   $a_1 = 1, \ a_n = 2a_{n-1}$

9. $a_n = \left(\dfrac{1}{2}\right)^n$ or recursively as
   $= \dfrac{1}{2^n}$ or 2^{-n}
   $a_1 = \dfrac{1}{2}, \ a_n = \dfrac{1}{2}a_{n-1}$

10. $a_n = (-3)^n$ or recursively as
    $a_1 = -3, \ a_n = -3a_{n-1}$

11a. $\displaystyle\sum_{i=1}^{5}(5i+3)=[5(1)+3]+[5(2)+3]+[5(3)+3]+[5(4)+3]+[5(5)+3]=90$

11b. $\displaystyle\sum_{i=3}^{5}(2^i-1)=(2^3-1)+(2^4-1)+(2^5-1)=53$

11c. $\displaystyle\sum_{i=2}^{6}(i^2+2i)=[2^2+2(2)]+[3^2+2(3)]+[4^2+2(4)]+[5^2+2(5)]+[6^2+2(6)]=130$

12. $a_5=a_1+4d=11$
$a_9=a_1+8d=19$
$4d=8$
$d=2$
$a_1=3$

13. $\dfrac{a_5}{a_3}=\dfrac{a_1r^4}{a_1r^2}=\dfrac{162}{18}$
$r^2=9$
$r=\pm3$
$a_3=a_1(\pm3)^2=18$
$a_1=2$
$a_2=a_1r^1=2(\pm3)=\pm6$

14. $a_{10}=5+(10-1)6$
$=59$
$S_{10}=\dfrac{10}{2}(5+59)$
$=320$

15. $a_{10}=25+(10-1)(-5)$
$=-20$
$S_{10}=\dfrac{10}{2}[25+(-20)]$
$=25$

16. $S_{50}=\dfrac{3(2^{50}-1)}{2-1}$
$=3(2^{50}-1)$

17. $S=\dfrac{\frac{1}{2}}{1-\frac{1}{3}}=\dfrac{\frac{1}{2}}{\frac{2}{3}}=\left(\dfrac{1}{2}\right)\left(\dfrac{3}{2}\right)=\dfrac{3}{4}$

18. $\dbinom{4}{0}=\dfrac{4!}{0!4!}=1,\ \dbinom{4}{1}=\dfrac{4!}{1!3!}=4,\ \dbinom{4}{2}=\dfrac{4!}{2!2!}=6,\ \dbinom{4}{3}=\dfrac{4!}{3!1!}=4,\ \dbinom{4}{4}=\dfrac{4!}{4!0!}=1$

$(x-3)^4=1(x)^4(-3)^0+4(x)^3(-3)^1+6(x)^2(-3)^2+4(x)^1(-3)^3+1(x)^0(-3)^4$
$=x^4-12x^3+54x^2-108x+81$

19. $\dbinom{5}{0}=\dfrac{5!}{0!5!}=1,\ \dbinom{5}{1}=\dfrac{5!}{1!4!}=5,\ \dbinom{5}{2}=\dfrac{5!}{2!3!}=10,\ \dbinom{5}{3}=\dfrac{5!}{3!2!}=10,\ \dbinom{5}{4}=\dfrac{5!}{4!1!}=5,\ \dbinom{5}{5}=\dfrac{5!}{5!0!}=1$

$(2x-1)^5=1(2x)^5(-1)^0+5(2x)^4(-1)^1+10(2x)^3(-1)^2+10(2x)^2(-1)^3+5(2x)^1(-1)^4+1(2x)^0(-1)^5$
$=32x^5-80x^4+80x^3-40x^2+10x-1$

20. $\binom{20}{0} = \dfrac{20!}{0!20!} = 1$, $\binom{20}{1} = \dfrac{20!}{1!19!} = 20$, $\binom{20}{2} = \dfrac{20!}{2!18!} = 190$

From the binomial formula, we write the first three terms:

$$(x-1)^{20} = 1(x)^{20}(-1)^0 + 20(x)^{19}(-1)^1 + 190(x)^{18}(-1)^2$$
$$= x^{20} - 20x^{19} + 190x^{18}$$

21. $a_6 = \binom{8}{5}(2x)^3(-3y)^5$

$$= \dfrac{8!}{5!3!}(8x^3)(-243y^5)$$
$$= -108{,}864x^3y^5$$

CHAPTER 13

Conic Sections

13.1 The Circle

Use $d = \sqrt{(x_2 - x_1)^2 + (y_2 - y_1)^2}$ in Problems 1-11 and 59-61.

1. $(3, 7)$ and $(6, 3)$

   $d = \sqrt{(6-3)^2 + (3-7)^2} = \sqrt{(3)^2 + (-4)^2} = 5$

3. $(0, 9)$ and $(5, 0)$

   $d = \sqrt{(5-0)^2 + (0-9)^2} = \sqrt{106}$

5. $(3, -5)$ and $(-2, 1)$

   $d = \sqrt{(-2-3)^2 + [1-(-5)]^2} = \sqrt{61}$

7. $(-1, -2)$ and $(-10, 5)$

   $d = \sqrt{(-10+1)^2 + (5+2)^2}$
   $= \sqrt{130}$

9. $(x, 2)$ and $(1, 5)$, $d = \sqrt{13}$

   $\sqrt{13} = \sqrt{(1-x)^2 + (5-2)^2}$

   $13 = (1-x)^2 + 3^2$

   $13 = 1 - 2x + x^2 + 9$

   $0 = x^2 - 2x - 3$

   $0 = (x+1)(x-3)$

   $x = -1$ or $x = 3$

11. $(7, y)$ and $(8, 3)$, $d = 1$

    $1 = \sqrt{(8-7)^2 + (3-y)^2}$

    $1 = \sqrt{1 + 9 - 6y + y^2}$

    $1 = y^2 - 6y + 10$

    $0 = y^2 - 6y + 9$

    $0 = (y-3)(y-3)$

    $y = 3$

Use $(x-a)^2 + (y-b)^2 = r^2$ in Problems 13-47, and 55-57.

13. $(a, b) = (2, 3)$, $r = 4$.

    $(x-2)^2 + (y-3)^2 = 4^2$

    $(x-2)^2 + (y-3)^2 = 16$

15. $(a, b) = (3, -2)$; $r = 3$

    $(x-3)^2 + (y+2)^2 = 3^2$

    $(x-3)^2 + (y+2)^2 = 9$

17. $(a, b) = (-5, -1)$, $r = \sqrt{5}$.

    $[x-(-5)]^2 + [y-(-1)]^2 = (\sqrt{5})^2$

    $(x+5)^2 + (y+1)^2 = 5$

19. $(a, b) = (0, -5)$; $r = 1$

    $(x-0)^2 + (y+5)^2 = 1$

    $x^2 + (y+5)^2 = 1$

21. $(a, b) = (0, 0)$, $r = 2$.

$$(x-0)^2 + (y-0)^2 = 2^2$$
$$x^2 + y^2 = 4$$

23.
$$x^2 + y^2 = 4$$
$$(x-0)^2 + (y-0)^2 = 2^2$$
Center $= (0, 0)$, Radius $= 2$
See the graph in the back of the textbook

25. $(x-1)^2 + (y-3)^2 = 25$

$$(x-1)^2 + (y-3)^2 = 5^2$$
Center $= (1, 3)$, Radius $= 5$.
See the graph in the back of the textbook.

27. $(x+2)^2 + (y-4)^2 = 8$

$$\left[x-(-2)\right]^2 + (y-4)^2 = \left(2\sqrt{2}\right)^2$$
Center $= (-2, 4)$, Radius $= 2\sqrt{2}$
See the graph in the back of the textbook.

29.
$$(x+1)^2 + (y+1)^2 = 1$$
$$\left[x-(-1)\right]^2 + \left[y-(-1)^2\right] = 1^2$$
Center $= (-1, -1)$, Radius $= 1$.
See the graph in the back of the textbook.

31.
$$x^2 + y^2 - 6y = 7$$
$$x^2 + y^2 - 6y + 9 = 7 + 9$$
$$(x-0)^2 + (y-3)^2 = 4^2$$
Center $= (0, 3)$, Radius $= 4$
See the graph in the back of the textbook.

33.
$$x^2 + y^2 + 2x = 1$$
$$x^2 + 2x + y^2 = 1$$
$$x^2 + 2x + 1 + y^2 = 1 + 1$$
$$(x+1)^2 + (y-0)^2 = 2$$
$$\left[x-(-1)\right]^2 + (y-0)^2 = \left(\sqrt{2}\right)^2$$
Center $= (-1, 0)$, $= \sqrt{2}$.

See the graph in the back of the textbook.

35.
$$x^2 + y^2 - 4x - 6y = -4$$
$$x^2 - 4x + 4 + y^2 - 6y + 9 = -4 + 4 + 9$$
$$(x-2)^2 + (y-3)^2 = 3^2$$
Center $= (2, 3)$, Radius $= 3$
See the graph in the back of the textbook.

37.
$$x^2 + y^2 + 2x + y = \frac{11}{4}$$
$$x^2 + 2x + y^2 + y = \frac{11}{4}$$
$$x^2 + 2x + 1 + y^2 + y + \frac{1}{4} = \frac{11}{4} + 1 + \frac{1}{4}$$
$$(x+1)^2 + \left(y + \frac{1}{2}\right)^2 = 4$$
$$\left[x-(-1)\right]^2 + \left[y - \left(-\frac{1}{2}\right)\right]^2 = 2^2$$
Center $= \left(-1, -\frac{1}{2}\right)$, Radius $= 2$.

See the graph in the back of the textbook.

39.
$$\text{Center} = (3, 4)$$
$$(x-3)^2 + (y-4)^2 = r^2$$
$$\text{Point} = (0, 0)$$
$$(0-3)^2 + (0-4)^2 = r^2$$
$$25 = r^2$$
$$(x-3)^2 + (y-4)^2 = 25$$

41. A: $a = 1/2,\ b = 1,\ r = 1/2$

$$(x - 1/2)^2 + (y - 1)^2 = (1/2)^2$$

$$(x - 1/2)^2 + (y - 1)^2 = 1/4$$

B: $a = 1,\ b = 1,\ r = 1$

$$(x - 1)^2 + (y - 1)^2 = 1$$

C: $a = 2,\ b = 1,\ r = 2$

$$(x - 2)^2 + (y - 1)^2 = 2^2$$

$$(x - 2)^2 + (y - 1)^2 = 4$$

43. Center $= (0, 0)$, Point $= (3, 4)$

$$d = \sqrt{(3 - 0)^2 + (4 - 0)^2}$$

$$d = \sqrt{25}$$

$$d = 5$$

$$(x - 0)^2 + (y - 0)^2 = 5^2$$

$$x^2 + y^2 = 25$$

45. The distance from the origin to 3 or -3 is 3.

This is the radius of the circle with center $(0, 0)$.

$$(x - a)^2 + (y - b)^2 = r^2$$

$$(x - 0)^2 + (y - 0)^2 = 3^2$$

$$x^2 + y^2 = 9$$

47. Center $= (-1, 3)$, Point $= (4, 3)$

$$d = \sqrt{(4 + 1)^2 + (3 - 3)^2}$$

$$d = \sqrt{25}$$

$$d = 5$$

$$(x + 1)^2 + (y - 3)^2 = 25$$

49. Look at the first four terms:

$$a_1 = 5 = 4(1) + 1$$

$$a_2 = 9 = 4(2) + 1$$

$$a_3 = 13 = 4(3) + 1$$

$$a_4 = 17 = 4(4) + 1$$

The general term is $a_n = 4(n) + 1$

51. $\displaystyle\sum_{i=2}^{5} \left(\frac{1}{2}\right)^i = \left(\frac{1}{2}\right)^2 + \left(\frac{1}{2}\right)^3 + \left(\frac{1}{2}\right)^4 + \left(\frac{1}{2}\right)^5$

$$= \frac{1}{4} + \frac{1}{8} + \frac{1}{16} + \frac{1}{32}$$

$$= \frac{15}{32}$$

53. $1 + 3 + 5 + 7 + 9 = \left[2(1) - 1\right] + \left[2(2) - 1\right] + \left[2(3) - 1\right] + \left[2(4) - 1\right] + \left[2(5) - 1\right] = \displaystyle\sum_{i=1}^{5} (2i - 1)$

55. Center $(2, 3)$

If the circle is tangent to the y-axis, the radius is 2.

$$(x - 2)^2 + (y - 3)^2 = 2^2$$

$$(x - 2)^2 + (y - 3)^2 = 4$$

57. If the circle is tangent to the vertical line $x = 4$,

and the center is $(2, 3)$ then the radius is $4 - 2 = 2$.

$$(x - 2)^2 + (y - 3)^2 = 2^2$$

$$(x - 2)^2 + (y - 3)^2 = 4$$

59. $\qquad x^2 + y^2 - 6x + 8y = 144$

$$x^2 - 6x + 9 + y^2 + 8y + 16 = 144 + 9 + 16$$

$$(x - 3)^2 + (y + 4)^2 = 169$$

$$(x - 3)^2 + (y + 4)^2 = 13^2$$

Center $= (3, -4)$, Radius $= 13$

$$d = \sqrt{3^2 + 4^2} = 5$$

61. $\qquad x^2 + y^2 - 6x - 8y = 144$

$$x^2 - 6x + 9 + y^2 - 8y + 16 = 144 + 9 + 16$$

$$(x - 3)^2 + (y - 4)^2 = \left(\sqrt{169}\right)^2$$

$$(x - 3)^2 + (y - 4)^2 = 13$$

Center $= (3, 4)$, Radius $= 13$

$$d = \sqrt{3^2 + 4^2} = 5$$

13.2 Ellipses and Hyperbolas

1. $\dfrac{x^2}{9} + \dfrac{y^2}{16} = 1$

$\dfrac{x^2}{3^2} + \dfrac{y^2}{4^2} = 1$

See the graph in the back of the textbook.

3. $\dfrac{x^2}{16} + \dfrac{y^2}{9} = 1$

$\dfrac{x^2}{4^2} + \dfrac{y^2}{3^2} = 1$

See the graph in the back of the textbook.

5. $\dfrac{x^2}{3} + \dfrac{y^2}{4} = 1$

$\dfrac{x^2}{\left(\sqrt{3}\right)^2} + \dfrac{y^2}{2^2} = 1$

See the graph in the back of the textbook.

7. $4x^2 + 25y^2 = 100$

$\dfrac{x^2}{25} + \dfrac{y^2}{4} = 1$

$\dfrac{x^2}{5^2} + \dfrac{y^2}{2^2} = 1$

See the graph in the back of the textbook

9. $x^2 + 8y^2 = 16$

$\dfrac{x^2}{16} + \dfrac{y^2}{2} = 1$

$\dfrac{x^2}{4^2} + \dfrac{y^2}{\left(\sqrt{2}\right)^2} = 1$

See the graph in the back of the textbook.

11. $\dfrac{x^2}{9} - \dfrac{y^2}{16} = 1$

$\dfrac{x^2}{3^2} - \dfrac{y^2}{4^2} = 1$

See the graph in the back of the textbook.

13. $\dfrac{x^2}{16} - \dfrac{y^2}{9} = 1$

$\dfrac{x^2}{4^2} - \dfrac{y^2}{3^2} = 1$

See the graph in the back of the textbook.

15. $\dfrac{y^2}{9} - \dfrac{x^2}{16} = 1$

$\dfrac{y^2}{3^2} - \dfrac{x^2}{4^2} = 1$

See the graph in the back of the textbook.

17. $\dfrac{y^2}{36} - \dfrac{x^2}{4} = 1$

$\dfrac{y^2}{6^2} - \dfrac{x^2}{2^2} = 1$

See the graph in the back of the textbook.

19. $x^2 - 4y^2 = 4$

$\dfrac{x^2}{4} - \dfrac{y^2}{1} = 1$

$\dfrac{x^2}{2^2} - \dfrac{y^2}{1^2} = 1$

See the graph in the back of the textbook.

21. $16y^2 - 9x^2 = 144$

$$\frac{y^2}{9} - \frac{x^2}{16} = 1$$

$$\frac{y^2}{3^2} - \frac{x^2}{4^2} = 1$$

See the graph in the back of the textbook.

23. $0.4x^2 + 0.9y^2 = 3.6$

$$\frac{x^2}{9} + \frac{y^2}{4} = 1$$

$$\frac{x^2}{3^2} + \frac{y^2}{2^2} = 1$$

x-intercepts $= \pm 3$

y-intercepts $= \pm 2$

25. $\dfrac{x^2}{0.04} - \dfrac{y^2}{0.09} = 1$

$$\frac{x^2}{(0.02)^2} - \frac{y^2}{(0.03)^2} = 1$$

x-intercepts $= \pm 0.2$.

No y-intercepts.

27. $\dfrac{25x^2}{9} + \dfrac{25y^2}{4} = 1$

$$\frac{x^2}{\frac{9}{25}} + \frac{y^2}{\frac{4}{25}} = 1$$

$$\frac{x^2}{\left(\frac{3}{5}\right)^2} + \frac{y^2}{\left(\frac{2}{5}\right)^2} = 1$$

x-intercepts $= \pm\dfrac{3}{5}$

y-intercepts $= \pm\dfrac{2}{5}$

29. $\dfrac{(x-4)^2}{4} + \dfrac{(y-2)^2}{9} = 1$

$$\frac{(x-4)^2}{2^2} + \frac{(y-2)^2}{3^2} = 1$$

Center $= (4, 2)$.

See the graph in the back of the textbook.

31. $4x^2 + y^2 - 4y - 12 = 0$

$$4x^2 + y^2 - 4y + 4 = 12 + 4$$

$$4x^2 + (y-2)^2 = 16$$

$$\frac{x^2}{4} + \frac{(y-2)^2}{16} = 1$$

$$\frac{x^2}{2^2} + \frac{(y-2)^2}{4^2} = 1$$

Center $= (0, 2)$

See the graph in the back of the textbook

33. See Example 3 on page 801 of the text.

35. $\dfrac{(x-2)^2}{16} - \dfrac{y^2}{4} = 1$

$$\frac{(x-2)^2}{4^2} - \frac{y^2}{2^2} = 1$$

Center $= (2, 0)$

See the graph in the back of the textbook.

37.
$$9y^2 - x^2 - 4x + 54y + 68 = 0$$
$$9y^2 + 54y - x^2 - 4x = -68$$
$$9(y^2 + 6y + 9) - 1(x^2 + 4x + 4) = -68 + 81 - 4$$
$$9(y+3)^2 - (x+2)^2 = 9$$
$$\frac{(y+3)^2}{1} - \frac{(x+2)^2}{9} = 1$$
$$\frac{(y+3)^2}{1^2} - \frac{(x+2)^2}{3^2} = 1$$
$$\text{Center} = (-2, -3)$$

See the graph in the back of the textbook.

39.
$$4y^2 - 9x^2 - 16y + 72x - 164 = 0$$
$$4(y^2 - 4y + 4) - 9(x^2 - 8x + 16) = 164 + 16 - 144$$
$$4(y-2)^2 - 9(x-4)^2 = 36$$
$$\frac{(y-2)^2}{9} - \frac{(x-4)^2}{4} = 1$$
$$\frac{(y-2)^2}{3^2} - \frac{(x-4)^2}{2^2} = 1$$
$$\text{Center} = (4, 2)$$

See the graph in the back of the textbook.

41. The center is $(2, -5)$.
$$a = \frac{-1-(-9)}{2} = 4,\ b = 2-(-1) = 3$$
$$\frac{(x-2)^2}{3^2} + \frac{[y-(-5)]^2}{4^2} = 1$$
$$\frac{(x-2)^2}{9} + \frac{(y+5)^2}{16} = 1$$

45. $\dfrac{y^2}{9} - \dfrac{x^2}{16} = 1$

The equations of the asymptotes are
$$y = \pm \frac{\sqrt{a}}{\sqrt{b}} x$$
$$y = \pm \frac{\sqrt{9}}{\sqrt{16}} x$$
$$y = \pm \frac{3}{4} x$$

43. Center $= (0,0)$, $a = 5 - 0 = 5$
Asymptote: $y = \dfrac{b}{5} x$, point: $\left(1, \dfrac{3}{5}\right)$
$$\frac{3}{5} = \frac{b}{5}(1) \rightarrow b = 3$$
$$\frac{x^2}{25} - \frac{y^2}{9} = 1$$

47. $\dfrac{x^2}{16} + \dfrac{y^2}{9} = 1$
$$|-4| + |4| = 8$$

49. The common difference is 6
and the first term is 5.
$$a_n = a_1 + (n-1)d$$
$$a_n = 5 + (n-1)6$$
$$a_n = 6n - 1$$

51. $a_1 = 4$ and $d = 5$
$$a_n = a_1 + (n-1)d$$
$$a_{20} = 4 + (20-1)5 = 99$$
$$S_n = \frac{n}{2}(a_1 + a_n)$$
$$S_{20} = \frac{20}{2}(4 + 99) = 1030$$

53. If $a_1 = 8$ and $r = \dfrac{1}{2}$

$$S_n = \frac{a_1\left(r^n - 1\right)}{r - 1}$$

$$S_6 = \frac{8\left[\left(\frac{1}{2}\right)^6 - 1\right]}{\frac{1}{2} - 1} = \frac{8\left(\frac{1}{64} - 1\right)}{-\frac{1}{2}} = \frac{8\left(-\frac{63}{64}\right)}{-\frac{1}{2}} = \frac{63}{4}$$

13.3 Second Degree Inequalities and Nonlinear Systems

1. $x^2 + y^2 \le 49$

$\le$, the boundary is included and
the area inside the circle is shaded.
See the graph in the back of the textbook.

3. $(x-2)^2 + (y+3)^2 < 16$

$<$, the boundary is not included and
the area inside the circle is shaded.
See the graph in the back of the textbook.

5. $y < x^2 - 6x + 7$

$<$, the boundary is not included and
the area below the parabola is shaded.
See the graph in the back of the textbook.

7. $\dfrac{x^2}{25} - \dfrac{y^2}{9} \ge 1$

$\ge$, the boundary is included and
the regions to the left and right are shaded.
See the graph in the back of the textbook.

9. $4x^2 + 25y^2 \le 100$

$\dfrac{x^2}{25} + \dfrac{y^2}{4} \le 1$

$\le$, the boundary is included and
the region inside is shaded.
See the graph in the back of the textbook.

11. $x^2 + y^2 < 9$

$y \ge x^2 - 1$

See the graph in the back of the textbook.

13. $\dfrac{x^2}{9} + \dfrac{y^2}{25} \le 1$

$\dfrac{x^2}{4} - \dfrac{y^2}{9} > 1$

See the graph in the back of the textbook.

15. $4x^2 + 9y^2 \le 36$

$\dfrac{x^2}{9} + \dfrac{y^2}{4} \le 1$

$y > x^2 + 2$

See the graph in the back of the textbook

17. $x + y \le 3$

$x - 3y \le 3$

$x \ge -2$

See the graph in the back of the textbook.

19. $x + y \le 2$

$-x + y \le 2$

$y \ge -2$

See the graph in the back of the textbook

21. $x + y \le 4$

    $x \ge 0$

    $y \ge 0$

See the graph in the back of the textbook.

23.
$$x^2 + y^2 = 9 \qquad (1)$$
$$2x + y = 3 \qquad (2)$$
$$y = 3 - 2x$$
$$x^2 + (3 - 2x)^2 = 9 \qquad (1), \ y = 3 - 2x$$
$$x^2 + 9 - 12x + 4x^2 = 9$$
$$5x^2 - 12x = 0$$
$$x(5x - 12) = 0$$

$$x = 0 \quad \text{or} \quad x = \frac{12}{5}$$

$$y = 3 - 2(0) \quad y = 3 - 2\left(\frac{12}{5}\right)$$

$$y = 3 \qquad y = -\frac{9}{5}$$

Solutions: $(0, \ 3), \ \left(\dfrac{12}{5}, \ -\dfrac{9}{5}\right)$.

25. $x^2 + y^2 = 16$

    $x + 2y = 8 \Rightarrow x = 8 - 2y$

Substitute $x = 8 - 2y$ first equation

$$(8 - 2y)^2 + y^2 = 16$$
$$64 - 32y + 4y^2 + y^2 = 16$$
$$5y^2 - 32y + 48 = 0$$
$$(y - 4)(5y - 12) = 0$$

$$y = 4 \quad \text{or} \quad y = \frac{12}{5}$$

Substitute $y = 4$ and $y = \dfrac{12}{5}$ in

the second equation

$$x + 2(4) = 8 \qquad x + 2\left(\frac{12}{5}\right) = 8$$

$$x = 0 \qquad\qquad x = -\frac{24}{5} + 8$$

$$x = \frac{16}{5}$$

The solutions are: $(0, \ 4)$ and $\left(\dfrac{16}{5}, \ \dfrac{12}{5}\right)$.

27. $x^2 + y^2 = 25 \quad (1)$

    $\dfrac{x^2 - y^2 = 25 \quad (2)}{}$

$$2x^2 \qquad = 50$$
$$x^2 \qquad = 25$$
$$x = \pm 5$$
$$x^2 + y^2 = 25 \qquad (1), \ x = \pm 5$$
$$25 + y^2 = 25$$
$$y = 0$$

Solutions: $(5, \ 0), \ (-5, \ 0)$.

29. $x^2 + y^2 = 9$ $\qquad$ $y = x^2 - 3$

$\qquad$ $x^2 = 9 - y^2$ $\qquad$ $y + 3 = x^2$

Substitute $9 - y^2$ for x^2 in the second equation,

$$9 - y^2 = y + 3$$
$$0 = y^2 + y - 6$$
$$0 = (y + 3)(y - 2)$$
$$y = -3 \quad \text{or} \quad y = 2$$

Substitute $y = -3$ and $y = 2$ into the second equation

$-3 = x^2 - 3$ $\qquad$ $2 = x^2 - 3$

$0 = x^2$ $\qquad\qquad$ $5 = x^2$

$0 = x$ $\qquad\qquad$ $\pm\sqrt{5} = x$

The solutions are: $(0, -3), \left(\sqrt{5}, 2\right), \left(-\sqrt{5}, 2\right)$.

31. $\qquad$ $x^2 + y^2 = 16$

$\qquad$ $x^2 = 16 - y^2$ $\quad$ (1)

$$y = x^2 - 4$$
$$y + 4 = x^2$$
$$16 - y^2 = y + 4 \qquad (1), \ x^2 = y + 4$$
$$y^2 + y - 12 = 0$$
$$(y + 4)(y - 3) = 0$$
$$y = -4 \quad \text{or} \quad y = 3$$

$-4 = x^2 - 4$ $\qquad$ $3 = x^2 - 4$

$0 = x^2$ $\qquad\qquad$ $7 = x^2$

$0 = x$ $\qquad\qquad$ $\pm\sqrt{7} = x$

Solutions: $(0, -4), \left(\sqrt{7}, 3\right), \left(-\sqrt{7}, 3\right)$.

33. $3x + 2y = 10$

$\qquad$ $y = x^2 - 5$

Substitute $x^2 - 5$ for y in the first equation

$$3x + 2\left(x^2 - 5\right) = 10$$
$$3x + 2x^2 - 10 = 10$$
$$2x^2 + 3x - 20 = 0$$
$$(x + 4)(2x - 5) = 0$$

$x = -4$ $\qquad$ or $\qquad$ $x = \dfrac{5}{2}$

then $y = x^2 - 5$ $\qquad$ and $\qquad$ $y = x^2 - 5$

becomes $y = (-4)^2 - 5$ $\qquad$ $y = \left(\dfrac{5}{2}\right)^2 - 5$

$y = 16 - 5$ $\qquad\qquad$ $y = \dfrac{25}{4} - 5$

$y = 11$ $\qquad\qquad\qquad$ $y = \dfrac{5}{4}$

Solutions: $(-4, 11)$ and $\left(\dfrac{5}{2}, \dfrac{5}{4}\right)$.

35. $\qquad$ $y = x^2 + 2x - 3 \quad (1)$

$\qquad$ $y = -x + 1 \qquad\quad (2)$

$$x^2 + 2x - 3 = -x + 1 \qquad (2), \ y = x^2 + 2x - 3$$
$$x^2 + 3x - 4 = 0$$
$$(x + 4)(x - 1) = 0$$

$x = -4$ $\qquad$ or $\qquad$ $x = 1$

$y = -(-4) + 1$ $\qquad$ $y = -1 + 1$

$y = 5$ $\qquad\qquad$ $y = 0$

Solutions: $(-4, 5), (1, 0)$.

37. $y = x^2 - 6x + 5$

$y = x - 5$

Substitute $x - 5$ for y in the first equation,

$$x - 5 = x^2 - 6x + 5$$

$$0 = x^2 - 7x + 10$$

$$0 = (x - 2)(x - 5)$$

$$x = 2 \quad \text{or} \quad x = 5$$

then $y = x - 5$ and $y = x - 5$

becomes $y = 2 - 5$ $\qquad y = 5 - 5$

$\qquad y = -3 \qquad\qquad y = 0$

Solutions: $(2, -3)$ and $(5, 0)$.

39. $4x^2 - 9y^2 = 36$ $\quad$ (1)

$\underline{4x^2 + 9y^2 = 36 \quad (2)}$

$8x^2 \qquad\quad = 72$

$x^2 \qquad\quad = 9$

$\qquad x = \pm 3$

$4(\pm 3)^2 - 9y^2 = 36 \quad$ (1), $x = \pm 3$

$\qquad -9y^2 = 0$

$\qquad\qquad y = 0$

Solutions: $(3, 0), (-3, 0)$.

41. $\quad x - y = 4$

$x^2 + y^2 = 16$

Substitute $y + 4$ for x in the second equation,

$$(y + 4)^2 + y^2 = 16$$

$$y^2 + 8y + 16 + y^2 = 16$$

$$2y^2 + 8y = 0$$

$$2y(y + 4) = 0$$

$$y = 0 \quad \text{or} \quad y = -4$$

then $x - y = 4$ and $x - y = 4$

becomes $x - 0 = 4$ $\qquad x - (-4) = 4$

$\qquad x = 4 \qquad\qquad x = 0$

Solutions: $(4, 0)$ and $(0, -4)$.

43. $\quad x^2 + y^2 = 89$ $\quad$ (1)

$\underline{x^2 - y^2 = 39 \quad (2)}$

$2x^2 \qquad\quad = 128$

$x^2 \qquad\quad = 64$

$\qquad x = \pm 8$

$(\pm 8)^2 + y^2 = 89 \quad$ (1), $x = \pm 8$

$\qquad\quad y^2 = 25$

$\qquad\quad y = \pm 5$

The numbers are $8, 5$ or $-8, 5$

or $8, -5$ or $-8, -5$.

45. $\quad x = y^2 - 3$

$x + y = 9$

Substitute $9 - y$ for x in the first equation,

$$9 - y = y^2 - 3$$

$$0 = y^2 + y - 12$$

$$0 = (y - 3)(y + 4)$$

$$y = 3 \quad \text{or} \quad y = -4$$

then $x + y = 9$ and $x + y = 9$

becomes $x + 3 = 9$ $\qquad x - 4 = 9$

$\qquad x = 6 \qquad\qquad x = 13$

The two sets of numbers are

6 and 3 or 13 and -4.

47. $(x+2)^4 = \binom{4}{0}x^4(2)^0 + \binom{4}{1}x^3(2)^1 + \binom{4}{2}x^2(2)^2 + \binom{4}{3}x^1(2)^3 + \binom{4}{4}x^0(2)^4$

$\qquad = 1x^4(1) + 4x^3(2) + 6x^2(4) + 4x(8) + 1(16)$

$\qquad = x^4 + 8x^3 + 24x^2 + 32x + 16$

49. $(2x+y)^3 = \binom{3}{0}(2x)^3(y)^0 + \binom{3}{1}(2x)^2(y)^1 + \binom{3}{2}(2x)^1(y)^2 + \binom{3}{3}(2x)^0(y)^3$

$(2x+y)^3 = 1(8x^3) + 3(4x^2)y + 3(2x)y^2 + 1(y^3)$

$\qquad = 8x^3 + 12x^2y + 6xy^2 + y^3$

51. $(x+3)^{50} \Rightarrow \binom{50}{0}x^{50}(3)^0 + \binom{50}{1}x^{49}(3)^1$

$\qquad = 1x^{50}(1) + 50x^{49}(3)$

$\qquad = x^{50} + 150x^{49}$

Chapter 13 Review

1. $d = \sqrt{(-1-2)^2 + (5-6)^2}$

$\quad d = \sqrt{10}$

3. $d = \sqrt{(-4-0)^2 + (0-3)^2}$

$\quad d = 5$

5. $\qquad 5 = \sqrt{(2-x)^2 + [-4-(-1)]^2}$

$\qquad 5 = \sqrt{(2-x)^2 + (-3)^2}$

$\qquad 25 = 4 - 4x + x^2 + 9$

$x^2 - 4x - 12 = 0$

$(x+2)(x-6) = 0$

$\qquad x = -2 \ \text{ or } \ x = 6$

7. $(x-3)^2 + (y-1)^2 = 2^2$

$\quad (x-3)^2 + (y-1)^2 = 4$

9. $[x-(-5)]^2 + (y-0)^2 = 3^2$

$\quad (x+5)^2 + y^2 = 9$

11. $(x-0)^2 + (y-0)^2 = 5^2$

$\qquad x^2 + y^2 = 25$

13.
$$d = \sqrt{[2-(-2)]^2 + (0-3)^2}$$
$$= \sqrt{4^2 + (-3)^2}$$
$$= \sqrt{25}$$
$$= 5$$
$$[x-(-2)]^2 + (y-3)^2 = 5^2$$
$$(x+2)^2 + (y-3)^2 = 25$$

15.
$$x^2 + y^2 = 4$$
$$(x-0)^2 + (y-0)^2 = 2^2$$
Center $(0,\ 0)$; radius $= 2$
See the graph in the back of the textbook.

17.
$$x^2 + y^2 - 6x + 4y = -4$$
$$x^2 - 6x + 9 + y^2 + 4y + 4 = -4 + 9 + 4$$
$$(x-3)^2 + (y+2)^2 = 9$$
Center $(3,\ -2)$ $r = 3$
See the graph in the back of the textbook.

19-23. See the graph in the back of the textbook.

25.
$$9y^2 - x^2 - 4x + 54y + 68 = 0$$
$$9y^2 + 54y - x^2 - 4x = -68$$
$$9(y^2 + 6y) - (x^2 + 4x) = -68$$
$$9(y^2 + 6y + 9 - 9) - (x^2 + 4x + 4 - 4) = -68$$
$$9(y+3)^2 - 81 - (x+2)^2 + 4 = -68$$
$$9(y+3)^2 - (x+2)^2 = 9$$
$$\frac{(y+3)^2}{1} - \frac{(x+2)^2}{9} = 1$$
See the graph in the back of the textbook.

27-31. See the graph in the back of the textbook.

33.
$$x^2 + y^2 = 16 \qquad (1)$$
$$2x + y = 4 \qquad (2)$$
$$y = -2x + 4 \qquad (2)$$
$$(1),\ y = -2x + 4$$
$$x^2 + (-2x+4)^2 = 16$$
$$x^2 + 4x^2 - 16x + 16 = 16$$
$$5x^2 - 16x = 0$$
$$x(5x - 16) = 0$$
$$x = 0 \quad \text{or}$$

$$5x - 16 = 0$$
$$5x = 16$$
$$x = \frac{16}{5}$$

$$2(0) + y = 4 \qquad (2),\ x = 0$$
$$y = 4$$
$$2\left(\frac{16}{5}\right) + y = 4 \qquad (2),\ x = \frac{16}{5}$$
$$\frac{32}{5} + y = 4$$
$$y = \frac{20}{5} - \frac{32}{5}$$
$$y = -\frac{12}{5}$$
The solutions are $(0,\ 4)$ and $\left(\frac{16}{5},\ -\frac{12}{5}\right)$.

35. $9x^2 - 4y^2 = 36$ (1)

    $\underline{9x^2 + 4y^2 = 36}$ (2)

    $18x^2 \qquad = 72$

        $x^2 = 4$

         $x = \pm 2$

   $9(4) + 4y^2 = 36$ (2), $x^2 = 4$

        $4y^2 = 0$

         $y = 0$

The solutions are $(-2, 0)$ and $(2, 0)$.

Cumulative Review: Chapters 1-13

1. $2^3 + 3(2 + 20 \div 4) = 2^3 + 3(2 + 5)$

                $= 2^3 + 3(7)$

                $= 8 + 3(7)$

                $= 8 + 21$

                $= 29$

3. $-5(2x + 3) + 8x = -10x - 15 + 8x$

                    $= -2x - 15$

5. $(3y + 2)^2 - (3y - 2)^2 = 9y^2 + 12y + 4 - (9y^2 - 12y + 4)$

                        $= 9y^2 + 12y + 4 - 9y^2 + 12y - 4$

                        $= 24y$

7. $x^{2/3} \cdot x^{1/5} = x^{\frac{2}{3} + \frac{1}{5}} = x^{\frac{10}{15} + \frac{3}{15}} = x^{13/15}$

9. $5y - 2 = -3y + 6$

    $8y - 2 = 6$

      $8y = 8$

        $y = 1$

11. $|3x - 1| - 2 = 6$

      $|3x - 1| = 8$

      $3x - 1 = 8$ or $3x - 1 = -8$

         $3x = 9$ $3x = -7$

          $x = 3$ $x = -\dfrac{7}{3}$

13. $x^3 - 3x^2 - 4x + 12 = 0$

    $x^2(x - 3) - 4(x - 3) = 0$

     $(x^2 - 4)(x - 3) = 0$

         $x^2 - 4 = 0$ or $x - 3 = 0$

            $x^2 = 4$ $x = 3$

             $x = \pm 2$

15. $x - 2 = \sqrt{3x + 4}$

        $(x - 2)^2 = 3x + 4$

    $x^2 - 4x + 4 = 3x + 4$

          $x^2 - 7x = 0$

          $x(x - 7) = 0$

          $x - 7 = 0$ or $x = 0$

             $x = 7$

Possible solutions 0 and 7, only 7 checks.

17. $4x^2 + 6x = -5$

$4x^2 + 6x + 5 = 0$

$$x = \frac{-6 \pm \sqrt{6^2 - 4(4)(5)}}{2(4)}$$

$$= \frac{-6 \pm \sqrt{-44}}{2(4)}$$

$$= \frac{-2(3) \pm 2i\sqrt{11}}{2(4)}$$

$$= -\frac{3}{4} \pm \frac{i\sqrt{11}}{4}$$

19. $\log_2 x + \log_2 5 = 1$

$\log_2 5x = 1$

$5x = 2^1$

$x = \frac{2}{5}$

21. $4x + 2y = 4$

$y = -3x + 1$

Substitute $y = -3x + 1$ into the first equation.

$4x + 2(-3x + 1) = 4$

$4x - 6x + 2 = 4$

$-2x = 2$

$x = -1$

Substitute $x = -1$ into the second equation.

$y = -3(-1) + 1$

$y = 4$

The solution to the system is $(-1, 4)$

23. $x + 2y - z = 4$ (1)

$2x - y - 3z = -1$ (2)

$-x + 2y + 2z = 3$ (3)

$x + 2y - z = 4$ (1)

$\underline{-x + 2y + 2z = 3}$ (3)

$4y + z = 7$ (4).

$2x - y - 3z = -1$ (2)

$\underline{-2x + 4y + 4z = 6}$ 2 times (3)

$3y + z = 5$ (5)

$4y + z = 7$ (4)

$\underline{-3y - z = -5}$ -1 times (5)

$y = 2$

Substitute $y = 2$ into equation (5)

$3(2) + z = 5$

$z = -1$

Substitute $y = 2$ and $z = -1$ into equation (1)

$x + 2(2) - (-1) = 4$

$x = -1$

The solution is $(-1, 2, -1)$.

25. $\left(x^{1/5} + 3\right)\left(x^{1/5} - 3\right) = \left(x^{1/5}\right)^2 - 3^2 = x^{2/5} - 9$

27. $\dfrac{3-2i}{1+2i} = \dfrac{3-2i}{1+2i} \cdot \dfrac{1-2i}{1-2i} = \dfrac{3-6i-2i+4i^2}{1-4i^2} = \dfrac{3-8i-4}{1+4} = \dfrac{-1-8i}{5} = -\dfrac{1}{5} - \dfrac{8}{5}i$

29. $3x - y < -2$

Boundary is a broken line.

Shade above the boundary.

See the graph in the back of the textbook.

31. $y = \log_2 x$

See the graph in the back of the textbook.

33. $9x^2 - 4y^2 = 36$

$\dfrac{9x^2}{36} - \dfrac{4y^2}{36} = \dfrac{36}{36}$

$\dfrac{x^2}{4} - \dfrac{y^2}{9} = 1$

The center is at $(0, 0)$

The vertices are at $(-2, 0)$ and $(2, 0)$

See the graph in the back of the textbook.

35. $y = (x-2)^2 - 3$

The vertex is at $(2, -3)$ and it opens upward.

See the graph in the back of the textbook.

37. $f(x) = 7x - 8$ means $y = 7x - 8$

Inverse: $x = 7y - 8$

$x + 8 = 7y$

$\dfrac{x+8}{7} = y$

$f^{-1}(x) = \dfrac{x+8}{7}$

39. $f(x) = -\dfrac{3}{2}x + 1$

$f(4) = -\dfrac{3}{2}(4) + 1 = -5$

41. $(-6, -1), (-3, -5)$

$m = \dfrac{y_2 - y_1}{x_2 - x_1} = \dfrac{-5 - (-1)}{-3 - (-6)} = \dfrac{-4}{3} = -\dfrac{4}{3}$

Using $(x_1, y_1) = (-6, -1)$ and $m = -\dfrac{4}{3}$

in $y - y_1 = m(x - x_1)$ yields

$y - (-1) = -\dfrac{4}{3}[x - (-6)]$

$y + 1 = -\dfrac{4}{3}x - 8$

$y = -\dfrac{4}{3}x - 8 - 1$

$y = -\dfrac{4}{3}x - 9$

43. $d = \sqrt{[4 - (-3)]^2 + (5-1)^2}$

$d = \sqrt{7^2 + 4^2}$

$d = \sqrt{49 + 16}$

$d = \sqrt{65}$

45. $f(x) = 4 - x^2,\ g(x) = 5x - 1$

$(g \circ f)(x) = g(4 - x^2)$

$= 5(4 - x^2) - 1$

$= 20 - 5x^2 - 1$

$= 19 - 5x^2$

47. When $s(t) = 48t - 16t^2$, and $s(t) = 20$

$20 = 48t - 16t^2$

$16t^2 - 48t + 20 = 0$

$4(4t^2 - 12t + 5) = 0$

$(2t - 5)(2t - 1) = 0$

$2t - 5 = 0$ or $2t - 1 = 0$

$2t = 5$ $2t = 1$

$t = \dfrac{5}{2}$ seconds $t = \dfrac{1}{2}$ seconds

49. $x^2 + x - 6 > 0$

$(x + 3)(x - 2) > 0$ same sign

$x < -3$ or $x > 2$

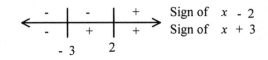

Sign of x - 2
Sign of x + 3

- 3   2

See the graph in the back of the textbook.

51. 5, 15, 45, 135, $\cdots$

$\dfrac{a_2}{a_1} = \dfrac{15}{5} = 3$

$\dfrac{a_3}{a_2} = \dfrac{45}{15} = 3$

$\dfrac{a_4}{a_3} = \dfrac{135}{45} = 3$

The common ratio $r = 3$

The general term is $a_n = ar^{n-1}$

$a_n = 5(3)^{n-1}$

Chapter 13 Test

1.
$$2\sqrt{5} = \sqrt{(-1-x)^2 + (4-2)^2}$$
$$20 = 1 + 2x + x^2 + 4$$
$$x^2 + 2x - 15 = 0$$
$$(x+5)(x-3) = 0$$
$$x = -5 \quad \text{or}$$
$$x = 3$$

2.
$$[x-(-2)]^2 + (y-4)^2 = 3^2$$
$$(x+2)^2 + (y-4)^2 = 9$$

3.
$$d = \sqrt{(-3-0)^2 + (-4-0)^2}$$
$$d = \sqrt{25}$$
$$d = 5$$
$$(x-0)^2 + (y-0)^2 = 5^2$$
$$x^2 + y^2 = 25$$

4.
$$x^2 + y^2 - 10x + 6y = 5$$
$$x^2 - 10x + 25 + y^2 + 6y + 9 = 5 + 25 + 9$$
$$(x-5)^2 + (y+3)^2 = 39$$
$$\text{Center} = (5, -3)$$
$$\text{Radius} = \sqrt{39}$$

5. $4x^2 - y^2 = 16$

$$\frac{x^2}{4} - \frac{y^2}{16} = 1$$

See the graph in the back of the textbook.

6. See the graph in the back of the textbook.

7. $(x-2)^2 + (y+1)^2 \le 9$

$$\frac{(x-2)^2}{9} + \frac{(y+1)^2}{9} \le 1$$

See the graph in the back of the textbook.

8.
$$9x^2 + 4y^2 - 72x - 16y + 124 = 0$$
$$9x^2 - 72x + 4y^2 - 16y = -124$$
$$9(x^2 - 8x) + 4(y^2 - 4y) = -124$$
$$9(x^2 - 8x + 16 - 16) + 4(y^2 - 4y + 4 - 4) = -124$$
$$9(x-4)^2 - 144 + 4(y-2)^2 - 16 = -124$$
$$9(x-4)^2 + 4(y-2)^2 = 36$$
$$\frac{(x-4)^2}{4} + \frac{(y-2)^2}{9} = 1$$

See the graph in the back of the textbook.

9.

$$x^2 + y^2 = 25 \quad (1)$$
$$2x + y = 5 \quad (2)$$
$$y = 5 - 2x \quad (2)$$

$$x^2 + (5 - 2x)^2 = 25 \quad (1), \ y = 5 - 2x$$
$$x^2 + 25 - 20x + 4x^2 = 25$$
$$5x^2 - 20x = 0$$
$$5x(x - 4) = 0$$
$$x = 0 \quad \text{or}$$
$$x = 4$$

$$y = 5 - 2(0) \quad (2), \ x = 0$$
$$y = 5 \quad \text{or}$$
$$y = 5 - 2(4) \quad (2), \ x = 4$$
$$y = -3$$

The solutions are $(0, \ 5), \ (4, \ -3)$.

10.

$$x^2 + y^2 = 16 \quad (1)$$
$$y = x^2 - 4 \quad (2)$$

$$x^2 + (x^2 - 4)^2 = 16 \quad (1), \ y = x^2 - 4$$
$$x^2 + x^4 - 8x^2 + 16 = 16$$
$$x^4 - 7x^2 = 0$$
$$x^2(x^2 - 7) = 0$$
$$x = 0 \quad \text{or}$$

$$x^2 - 7 = 0$$
$$x^2 = 7$$
$$x = \pm\sqrt{7}$$

$$y = 0^2 - 4 \quad (2), \ x = 0$$
$$y = -4 \quad \text{or}$$

$$y = \left(\pm\sqrt{7}\right)^2 - 4 \quad (2), \ x = \pm\sqrt{7}$$
$$y = 3$$

The solutions are $(0, \ -4), \ \left(-\sqrt{7}, \ 3\right), \ \left(\sqrt{7}, \ 3\right)$.

Introduction to Application Problems

As indicated in the Preface at the beginning of this manual, the following pages comprise the 50 real-world application problems by Utah Valley State College's Lori Palmer. These problems have been formatted in such a way to allow room for you to complete the problems and turn them in to your professor. On the next two pages, there are two correlation charts, which provide a breakdown of the problems. The first chart presents a correlation based upon mathematical concept. The second chart presents a correlation based upon the application topic.

Applications by Concept for *Math in Practice: An Applied Video Companion CD-ROM*

Math Concept	Application	Page	Corresponds to
Basics	Flying to Hawaii	A-1	Chapter 1
	Beetle Infestation	A-3	
	Population Growth	A-5	
	Premature Infant Formula	A-7	
	Aeronautics	A-9	
	World War II Bombers	A-11	
Linear Equations	Air Pollution	A-13	Chapter 2
	Investing and Debt	A-15	
	Falling Objects	A-17	
	Acoustics	A-19	
	Investing and Debt	A-21	
	Investigating Accidents	A-23	
	Optical Lab	A-25	
Linear Equations	Constructing Buildings	A-27	Chapter 3
Linear Systems	Map Evolution	A-29	Chapter 3
	Air Pollution	A-31	
	Metal Alloys	A-33	
	Premature Infant Formula	A-35	
	Living Soil Crust	A-37	
	Electronics	A-39	
Basics	Constructing Buildings	A-41	Chapter 4
Exponents and Roots	Line of Sight	A-43	Chapter 4
	Atlatl	A-45	
	Optical Lab	A-47	
Rational Equations	Electronics	A-49	Chapter 6
	Flying to Hawaii	A-51	
	Map Evolution	A-53	
	Beetle Infestation	A-55	
	Piano Pitch	A-57	
	Lighting a Stage	A-59	
Linear Equations - 2 Variables	Flying to Hawaii	A-61	Chapter 8
	Lighting a Stage	A-63	
	Investing and Debt	A-65	
	Metal Alloys	A-67	
	World War II Bombers	A-69	
	Acoustics	A-71	
Exponents and Roots	Piano Pitch	A-73	Chapter 8
Rational Equations	Dinosaur Motion	A-75	Chapter 9
Exponents and Roots	Fighting Fires	A-77	Chapter 9
	Pendulum	A-79	
	Pendulum	A-81	
	Dinosaur Motion	A-83	
Quadratic Equations	Line of Sight	A-85	Chapter 9
	Electronics	A-87	
Quadratic Equations	Population Growth	A-89	Chapter 10
	Fighting Fires	A-91	
Quadratic Equations	Atlatl	A-93	Chapter 10
	Optical Lab	A-95	
Exponents and Roots	Investigating Accidents	A-97	Chapter 11
	Population Growth	A-99	
Additional Topics	Earthquakes	A-101	Chapter 11
	Earthquakes	A-103	
Sequences and Series	Aeronautics	A-105	Chapter 12
	Piano Pitch	A-107	

Answers for these problems start on page A-109 in this manual.

Problems by Application Topic

Math in the Environment   Problems
    Beetle Infestation   2, 28
    Air Pollution   7, 16
    Earthquakes   51, 52
    Population Growth   3, 45, 50
    Living Soil Crust   19

Math Over Time
    World War II Bombers   6, 35
    Map Evolution   15, 27
    Atlatl   23, 47
    Dinosaur Motion   38, 42
    Premature Infant Formula 4, 18

Math of Matter and Motion
    Pendulum   40, 41
    Electronics   20, 25, 44
    Metal Alloys   17, 34
    Line of Sight   22, 43
    Falling Objects   9
    Optical Lab   13, 24, 48

Math on Stage
    Lighting a Stage   30, 32
    Piano Pitch   29, 37, 54
    Acoustics   10, 36

Math at Work
    Fighting Fires   39, 46
    Constructing Buildings   14, 21
    Investigating Accidents   12, 49
    Investing & Debt   8, 11, 33
    Flying to Hawaii   1, 26, 31
    Aeronautics   5, 53

#1

| Math Application by Topic: | Flying to Hawaii |
| Math Application by Concept: | Basics |

There are four time zones separating Salt Lake City, Utah and Honolulu, Hawaii. When it is 3:00 in the afternoon, in Salt Lake City, it is 11:00 in the morning in Honolulu. When Phyllis Upchurch boarded a plane at the Salt Lake City airport at 7:30 A.M. and piloted it west to Hawaii, the trip took 7 hours 40 minutes. What was the local time when she arrived in Honolulu?

#2

| Math Application by Topic: | Beetle Infestation |
| Math Application by Concept: | Basics |

The video indicates the costs involved with beetle infestation of forests. One cost is the loss of douglas-fir trees. Large douglas-fir trees are the most often attacked because they provide the best conditions for beetles to complete their life cycle.

Wood harvested from timber is measured in board feet. A board foot's value for a douglas-fir is about $0.25 The table below estimates the number of board feet which can be harvested from trees of varying height and diameter.

a. Determine the loss (-) in dollars of a single douglas-fir tree that is 70 feet tall, with a diameter of 19 inches, if the loss of one board foot is -$0.25.

b. One assessment is that the beetle infests 145 million acres in the northwest. If even 5 trees (of similar size to the one calculated above) per acre were infested, what would be a cost estimate of the loss?

Number of Board Feet Harvested
Total Height

Diameter (in inches)	40	50	60	70	80
11	44	57	69	82	95
12	58	72	87	101	115
13	73	89	104	120	136
14	88	105	123	140	158
15	103	122	142	161	181
16	118	139	161	182	204
17	134	157	181	204	227
18	150	175	201	227	253
19	166	194	222	251	279
20	182	213	244	275	306

#3

Math Application by Topic: Population Growth
Math Application by Concept: Basics

As indicated in the video, one important consideration for planners is identifying dense pockets of population. Drought, lack of employment and climate are all factors which cause groups of people to migrate from one place to another.

The measure of migration used by researchers is "net migration". It is calculated by subtracting the number of people who have moved out of an area from those who have moved into the area: Net Migration = In Migration − Out Migration

Between 1990 to 1991, 346,000 moved into the Northeast region of the U.S. and 932,000 moved out. Use the equation above to determine the Net Migration for the area.

#4

Math Application by Topic: Premature Infant Formula

Math Application by Concept: Basics

Valerie Johnson explained that the formula for the hyperalumuntation solution used for premature infants was based on the infant's changing weight. The infants weight may change daily or even hourly. The solution described in the video contained: dextrose, protein, and electrolytes A doctor prescribed 2.5 grams of protein kg/day, based on the number of kilograms (kg) of the baby's weight. How many grams of protein would a premature infant require per day who weighed 1.7 kg (almost $3\frac{3}{4}$ pounds)? Round to the nearest tenth of a gram.

#5

Math Application by Topic: Aeronautics

Math Application by Concept: Basics

As shown in the video, when pilots face unexpected events they have to react within seconds to save the aircraft. The pilot is the one ultimately responsible. For this reason, most late-model aircraft carry systems to assist pilots in dangerous situations such as wind shear. Wind shear is a sudden change in the direction and velocity of the wind and is most dangerous when approaching a runway.

If a pilot is approaching a runway from a height of 2,500 feet and descending at a normal rate of -150 feet per second when a wind shear forces the plane down at an additional rate of -472 feet per second, what would be the plane's altitude after three seconds?

Date_____ Name_____

Course_____ Section_____ Student Number_____

#6

Math Application by Topic: World War II Bombers

Math Application by Concept: Basics

The B-25 bomber shown in the video was named in honor of Billy Mitchell, a hero of World War I, who tried to persuade the United States government that World War II would be won in the air. He died before the government realized the truth of his argument.

The propellers of the B-25 bomber had a diameter of 12.5 feet. Mechanics felt the plane would lose efficiency if the tips of the propellers rotated faster than the speed of sound (about 761 mph) so they were constantly calculating the appropriate revolutions per minute to rotate the propellers as fast as possible, but under the speed of sound.

a. Determine how fast the tip of each propeller was moving (in feet per minute) when rotating at 1700 revolutions per minute (rpm). Using the diameter of the propeller, determine the circumference of the propeller's rotation ($C = \pi d$). Next multiply by the number of revolutions per minute.

b. Convert your answer (in feet per minute) to calculate how many miles per hour the propellers were rotating. Round to the nearest mile per hour. At 1700 rpm were the propeller tips moving more or less than the speed of sound?

#7

Math Application by Topic: Air Pollution

Math Application by Concept: Linear Equations

The video demonstrated the dramatic increase in carbon dioxide and pollutants in the air over Los Angeles from the 1950s to the 1980s. One automobile can deposit 7,200 pounds of carbon dioxide in the air each year.

Fortunately, carbon dioxide is also necessary for the growth of all plants. Thus, people living within Los Angeles have been urged to plant trees. Scientists have shown that the rate of carbon dioxide consumed annually by mature trees will increase linearly at a rate of 24 lb. per mature tree.

a. Write a linear equation showing the relationship between carbon dioxide consumed and the number of trees planted. Let C equal the total number of pounds of carbon dioxide consumed by the mature trees and T equal the number of trees.

b. Determine how many mature trees would be necessary to consume the carbon dioxide emitted by the automobile described.

#8

Math Application by Topic:	Investing and Debt
Math Application by Concept:	Linear Equations

As indicated by the banker, an individual's debt ratio should not exceed 35%. Debt ratio is determined by the total amount of debts divided by an individual's gross income (income before taxes and deductions). Debt ratio = $\dfrac{\text{Debts}}{\text{Gross Income}}$. Amanda is a full-time student, who works part time while attending school. She has exceeded the maximum recommended debt ratio of 35%. If her gross income is $975 for the month, and her debt ratio is 37%,

a. What is the monthly debt she carries?
b. What increase in her gross income must she earn to bring her debt ratio back to 35%?

Date_____ Name_____

Course_____ Section_____ Student Number_____

#9

Math Application by Topic: Falling Objects

Math Application by Concept: Linear Equations

The video showed a bull rider being thrown. He landed on his head and shoulders. We can determine the speed at which the rider hit the ground by using the linear equation described: $s = v + gt$, where s is the speed of the fall, v is the initial velocity that the rider left the bull's back, g is the pull of gravity, and t is the time it took the rider to hit the ground.

Determine what the cowboy's falling speed was when he hit the ground, if his initial downward velocity was 6 feet per second and it took him 0.75 seconds to hit the ground. The pull of gravity is 32 ft/sec^2.

A-18

#10

Math Application by Topic:	Acoustics
Math Application by Concept:	Linear Equations

As discussed in the video, the time it takes sound to travel can change the way an audience hears a presentation. In large cathedrals a choir director may need to stand half way between the choir at the front of the chapel and the organ at the back in order to conduct the voices and music together.

The length of the main nave at Westminster Abbey is 530 feet. Sounds travels through air at about 1,116 feet per second. (Sound travels somewhat faster on a hot day.)

On a cool day in the fall, how long would it take music played at the back of the chapel to be heard by the choir at the front? Use the relationship Distance = (Rate)(Time). Round your answer to the nearest hundredth of a second.

#11

Math Application by Topic: Investing and Debt
Math Application by Concept: Linear Equations

The investment banker shown in the video indicated that owning stock in a company may be unpredictable. This is because a company's worth is not measured by how many buildings it owns or its inventory but rather by how much it is worth on the stock market. The market value of a company is called market capitalization or market cap. The equation to calculate market cap of a company is:

Market Cap = (price per share) × (total number of shares outstanding)

The Excelsior-Henderson Motorcycle Manufacturing Company had 13.6 million shares outstanding on the day that it announced it was restructuring its business operations. At the opening of the stock market day, the price per share was $3.00. At the lowest point of the day the price per share had dropped to $2.25 per share. At the end of the trading day, the price per share was $2.375.

a. Use the Market Cap equation to determine the difference in the market value from the company's highest value to its lowest value during the day.
b. If a group of investors owned 12% of the company's stock, what was the market value of their stock at the end of the day?

#12

Math Application by Topic:	Investigating Accidents
Math Application by Concept:	Linear Equations

The weight and speed of a car determines the harm that is done to another vehicle at the point of a collision. That is one of the reasons that insurance companies are trying to require sport utility vehicles to pay higher premiums. The damage they inflict on smaller cars is significant.

An estimate of the weight of each car can be calculated with the formula: $W = \dfrac{APN}{2000}$,

where W is the vehicle's weight in tons, A is the average tire contact with a hard surface in square inches, P is the air pressure in the tires in pounds per square inch (psi or lb/in^2), and N is the number of tires.

a. In the video two cars were involved in the accident. What is the approximate weight of the Dodge if the average tire contact area is a rectangle 8.0 inches by 10.0 inches, and the air pressure is 38 psi? Round to the nearest tenth.

b. What is the approximate weight of the smaller car if the average tire contact area is a rectangle that is 6.0 inches by 5.0 inches and the tire pressure is 29 psi? Round to the nearest tenth.

Date_____ Name_____

Course_____ Section_____ Student Number_____

#13

Math Application by Topic: Optical Lab
Math Application by Concept: Linear Equations

The optical technician in the video stated that he uses math constantly in his job, especially when he sizes a lens for an individual. The point where a person looks through the lens needs to be placed correctly in the frame or else blurred vision will occur.

Technicians measure the width of the lens and the width of the frame's bridge. Then they factor in the distance from the center of a client's pupil to the center of the nose. They then work through an algebraic expression. If the value of this expression is less than 5, the glasses are too small. If the value is much greater than 8, the customer will have blurred vision.

The equation for optimizing vision is: $5 \le \dfrac{(D+B)-2p}{2} + .5 \le 8$, where D is the lens width, B is the bridge width and p is the distance from the center of the pupil to center of nose.

A customer came into the optical center to purchase a pair of glasses. The distance from the center of the pupil to the center of the individual's nose was 30 millimeters (mm). The width of the lens was 50 mm. And the width of the bridge was 18 mm. Would the glasses fit the customer's face correctly?

A-25

#14

Math Application by Topic: Constructing Buildings
Math Application by Concept: Linear Equations

The video describes the components of a roof. Because local building codes dictate the slope of the roof required, builders must be accurate in constructing the pitch.

What is the slope or pitch of a roof that rises 3 feet for every 5 feet of run? The equation $5y = 3x$ may be used to describe the relationship between the amount of rise (y feet) that corresponds with the run (x feet). The slope of the equation is the slope of the roof.

Place the equation in slope-intercept form to determine the slope of the roof.
(Solve for y).

#15

Math Application by Topic: Map Evolution

Math Application by Concept: Systems of Equations

As illustrated in the video, wave action constantly erodes coastal regions. Beach replenishment projects are costly. The cost of replenishing each mile of coastal beaches year after year, may be written as a function of time in the equation $C = \$6{,}000{,}000t$, where t is the continuing number of years that replenishment projects are necessary.

An alternative to sand replenishing projects is to erect barriers such as seawalls and jetties. These projects have a much larger initial cost, but the maintenance cost over time is less. An equation for the cost of building and maintaining a seawall could be written: $C = \$44{,}000{,}000 + \$500{,}000t$, where t is the continuing number of years that seawall maintenance is required.

Use these two equations to write a system of equations. Use graphing, substitution or elimination to determine at what point in time the seawall becomes the more cost efficient method to protect the coastline.

#16
Math Application by Topic: Air Pollution
Math Application by Concept: Systems of Equations

During a one week period in the 1940s, a small town in Pennsylvania had a temperature inversion resulting in extreme air stagnation. Thousands of people experienced acute illness. Symptoms included coughing, sore throat, chest constriction, shortness of breath, eye irritation, nausea and vomiting and many people died. Seventy percent of the people who died had some form of heart or lung disease.

A system of equations can be used to illustrate the deaths which occurred in this community as a result of polluted air: $x - 20 = 0$ and $7x - 10y = 0$, where x is the number of deaths and y represents the number of deaths who also had some form of heart and lung disease. Solve the system of equations.

#17

Math Application by Topic: Metal Alloys

Math Application by Concept: Systems of Equations

The development of materials specific to the need of a product is an important part of manufacturing. Metal workers solve systems of equations when forming metal alloys. The video described a 7-ingredient recipe for a new steel consisting of chromium, cobalt, and nickel and other substances.

The Walter Rolling Mills of Oklahoma developed a copper alloy consisting of copper, silver, magnesium and phosphorus. The alloy is very expensive to manufacture and an exact amount of each alloy must be produced to come within the dollar amount bid for the project. Using what is on hand at the mill, the team must create 350 pounds of a metal alloy that is 75% copper. In storage is an alloy which is 82% copper and another alloy is 60% copper. A system of equations may be written to determine the amount of each alloy necessary to develop the product. Write the system of equations and determine the amount.

#18
Math Application by Topic: Premature Infant Formula
Math Application by Concept: Systems of Equations

Valerie Johnson described the need to continually change the proportions of the intravenous solutions given to patients. One solution contained dextrose which was manufactured in categories of : 5%, 10%, 50%, 70% Sometimes patients don't fit into any of these manufactured categories of dextrose and a custom solution must be created.

It was determined that a patient needs a 30% dextrose solution. Using stock solutions of 70% and 10%, determine how much of each solution would be needed to make 3 kl of a 30% dextrose solution.

#19
Math Application by Topic: Living Soil Crust
Math Application by Concept: Systems of Equations

Lichens are an alliance between fungi and algae. Dr. Rosemary Pendleton described how a living crust of lichens keeps soil from eroding. She indicated that a dry inoculum was needed to rehabilitate damaged soil crust.

Studies are being conducted to find the optimal dry algae inoculum to spread on the soil to promote lichen growth. One study required a 48% algae solution. A lab assistant had produced both a 60% algae solution and a 30% solution. How much of each would be needed to produce 25 gallons of the proposed 48% solution?

A-37

#20

Math Application by Topic: Electronics
Math Application by Concept: Systems of Equations

Use the explanation in the video and the information below to find the amount of electric current flowing through the circuit described.

The following diagram is of an electrical circuit where x, y, and z represent the amount of current (in amperes) flowing across the 5-ohm, 20-ohm, and 10-ohm resistors, respectively. In circuit diagrams resistors are represented by —⋀⋀⋁ and potential differences by —| |—

The system of equations used to find the three currents x, y, and z in the diagram below is:

$x - y - z = 0$
$5x + 20y = 80$
$20y - 10z = 50$

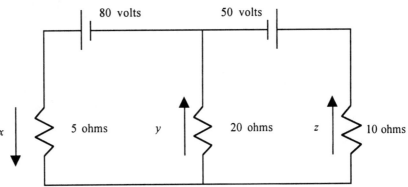

Solve the system for all variables

#21

Math Application by Topic: Constructing Buildings

Math Application by Concept: Basics

In the video, the construction worker was standing on a 12-foot ladder, which leaned against the house under construction. The base of the ladder was 3 feet from the base of the house. How high did the ladder reach along the side of the building? Round to the nearest tenth of a foot.

#22

Math Application by Topic:	Line of Sight
Math Application by Concept:	Exponents and Roots

The video discussed how large numbers can be written in scientific notation. Just as the distance from earth to the sun is a large number, the same can be said about the distances between the sun and each of the other planets.

The following table shows the mean distance each planet is from the sun. (The mean distance is listed because planet orbits are not perfectly circular.) Distances are shown in kilometers. Express each of the following distances in scientific notation.

Planet	Mean Distance from Sun (in kilometers)
Mercury	58,000,000
Venus	108,100,000
Earth	150,000,000
Mars	227,840,000
Jupiter	778,100,000
Saturn	1,427,000,000
Uranus	2,870,300,000
Neptune	4,500,000,000
Pluto	5,913,000,000

#23
Math Application by Topic: Atlatl
Math Application by Concept: Exponents and Roots

In the video, scientists were studying the additional amount of force and energy which the ancient device gave to a hunting spear. Since the atlatl multiplied the velocity of the spear when thrown, the energy force when the target was hit was greater than that when the spear was thrown by hand.

One scientist in the video made the statement that 15 times the spear's speed gave over 200 times the energy. Use the equation, $I = mr^2$, for calculating the Moment of Inertia (energy) to determine whether the statement was correct. The Moment of Inertia (energy) is I, m is the mass of the spear, and r is the velocity of the spear thrown without the atlatl ($15r$ would be the spear thrown with the atlatl.)

#24

Math Application by Topic: Optical Lab
Math Application by Concept: Exponents and Roots

The formula $f = \dfrac{ab}{a+b}$ is used in optics to find the focal length of a lens. For an optical lab technician, like that shown in the video, the formula might be used to determine how far the lens should be from the retina of the eye for correct vision. Show that the formula $f = (a^{-1} + b^{-1})^{-1}$ is equivalent to the preceding formula by rewriting it without the negative exponents and then simplifying the results.

#25
Math Application by Topic: Electronics
Math Application by Concept: Rational Equations

The electronics technician in the video explained the function of a resistor and demonstrated what form parallel resistors took in a circuit. The total resistance of three resistors R_1, R_2, and R_3 in parallel can be given by the equation: $\dfrac{1}{\dfrac{1}{R_1}+\dfrac{1}{R_2}+\dfrac{1}{R_3}}$. Simplify the complex fraction.

#26
Math Application by Topic: Flying to Hawaii
Math Application by Concept: Rational Equations

As indicated by the pilot in the video, air currents moving east to west affect the speed of aircraft as they travel between the mainland and Hawaii. The average speed for a round trip is given by the complex fraction: Average Speed $= \dfrac{2}{\dfrac{1}{v_1} + \dfrac{1}{v_2}}$, where v_1 is the average

speed against an air current west to Honolulu and v_2 is the average speed with an air current on the return trip east. Find the average speed, to the nearest mile per hour, for a round trip if heading toward Hawaii the speed was 270 mph while the plane's speed was 320 mph on the return flight.

#27

Math Application by Topic: Map Evolution

Math Application by Concept: Rational Equations

To compare the shorelines of past centuries to the current shoreline, a procedure similar to that described in the video is used. The actual distance between points on the historic map must be calculated. The scale on one map drawn in the late 1600s indicates that every 1.3 inches on the map corresponds to an actual distance of 65 miles. Two river outlets were 4 inches apart on the map. Use a proportion to determine the actual distance between the two points at the time the map was drawn.

#28

Math Application by Topic: Beetle Infestation

Math Application by Concept: Rational Equations

In 1999 a study was conducted to test the use of a beetle "scent" called a pheromone to fight a pine beetle infestation in the Idaho Panhandle National Forest. A pheromone is a blend of chemicals naturally produced by beetles. The chemical "scent" tricks insects into thinking the tree is already overrun by beetles, thus hanging out a "no vacancy" sign.

The economics of placing the packets in the forest areas was a concern. For the study approximately 75 pheromone packets were hung in trees on 2.5 acre test plots. If it took 2 hours for one forestry crew to place the pheromone packets on one test plot and it took 3 hours for a second crew to place packets on similar test plot, how long would it take both crews working together to continue placing the pheromone packets on each of the other 2.5 acre plots?

#29

Math Application by Topic:	Piano Pitch
Math Application by Concept:	Rational Expressions

As Dr. Pond explained, the faster a piano string vibrates, the higher the pitch. Frequency of vibration is measured in Hertz (Hz). A piano's middle C (a reference pitch) will vibrate at 263 times per second and thus has a frequency of 263 Hz. The A string above middle C has a frequency of 440 Hertz.

The equation in the video which explained this relationship between string length and frequency was $F = \dfrac{k}{S}$, where F is the frequency of vibration (pitch), S is the length of the string, and k is a constant of variation (determined by the type of piano). The frequency F varies inversely with the string length. The length of the middle C string on the Yamaha grand piano shown in the video is 60.7 inches. How long would a string, with the same mass, be to play an A note?

#30

Math Application by Topic: Lighting a Stage
Math Application by Concept: Rational Equations

The video described different types of lighting that are needed for a musical in a concert hall. The illumination from any of the light sources in the theater is inversely proportional to the square of the distance from the light. This dissipation of illumination must be taken into account when designing lighting for the production.

If the illumination of a spot light at the back of the hall (a distance of 130 feet) is 50 lumens on the stage actor, what is the illumination at a distance of 75 feet from the source? Use the equation: $I = \dfrac{k}{d^2}$, where I is the amount of illumination, d is the distance, and k is a constant of variation. Round to the nearest lumen.

Date_____ Name_____

Course_____ Section_____ Student Number_____

#31

Math Application by Topic: Flying to Hawaii

Math Application by Concept: Linear Equations-2 Variables

Ocean Wide Airlines is flying to LAX from Hawaii. The airplane has an altitude of 30,000 feet (approximately 5.7 miles), and is 65 miles from the airport when it begins its decent into LAX. The equation for the airplane's flight path during its decent is given by $y = -\dfrac{57}{650}x$. Use the equation to complete the table below. Round your answers to the nearest tenth of a mile. Then use your results to graph the equation. (The graph will appear in the third quadrant only because we are giving distance to the airport in negative numbers.

Distance to Airport (miles)	Altitude (miles)
-65	
-55	
-45	
-35	
-25	
-15	
-5	

#32
Math Application by Topic: Lighting a Stage
Math Application by Concept: Linear Equations-2 variables

The design and structure of a concert hall must be considered when lighting a stage production. When light comes into contact with impenetrable objects, such as costume fabric or parts of the set, it is reflected or absorbed, but not transmitted.

If we let R represent the percent of light reflected and A the percent of light absorbed by a surface, then the relationship between R and A is $R + A = 100$. Graph this equation on a coordinate system where the horizontal axis is the A-axis and the vertical axis is the R axis. Find the intercepts first, and limit your graph to the first quadrant.

#33

| Math Application by Topic: | Investing and Debt |
| Math Application by Concept: | Linear Equations |

As indicated in the video, an individual's debt burden needs to be carefully considered. Students taking out loans to finance their education are advised that the maximum monthly payment on the amount borrowed for student loans should not exceed 8% of their expected monthly starting salary upon graduation. The recommended maximum monthly payment (y) for such a student loan can be described with the formula: $y = 0.08x$, where x is their predicted monthly salary upon graduation.

a. What is the recommended maximum monthly payment for a journalism student if his/her predicted monthly starting salary is $2350 at a suburban paper in Detroit?

b. What is the needed monthly starting salary for a medical student who was expecting to make a student loan payment of $475?

#34
Math Application by Topic: Metal Alloys
Math Application by Concept: Linear Equations-2 Variables

The video described the mixing of metal alloys. Brasses are alloys which combine copper and zinc. The properties of brass depend primarily upon the proportion of zinc present but can be modified by the introducing of additional elements to further improve specific characteristics.

Alloys with a higher proportion of zinc cost less because zinc itself costs less. Brasses with a higher proportion of copper and a lower proportion of zinc (10-20%) may be used in decorative architecture or scientific equipment. Brasses with medium amounts of zinc (28-37%) combine the properties of good strength and pliability of copper with the corrosion resistance and a cheap metal price of zinc.

Below is a table of brass alloy prices charged by Mountain Metal.

Type of Brass	% Zinc	Cost/pound
Architectural Brass	15%	$2.55
Brass 260	30%	$2.10
Brass 360	39%	$1.83
Navel Brass 464	40%	$1.80

a. Use the paired data in the table above to create a line graph that shows the relationship between the percent of zinc in the brass alloy and the price per pound.

b. Use any two coordinate pairs to develop an equation which would describe the relationship between price and the copper content in any type of brass.

A-67

#35

| Math Application by Topic: | World War II Bombers |
| Math Application by Concept: | Linear Equations-2 Variables |

At the beginning of World War II the United States was at a great disadvantage in the air war. Germany had built up their air force prior to the war, but because the Wright brothers owned most patents for the design of aircraft, the US air corps had not been able to progress as far as the European military.

In 1940 the United States only assembled 6,026 aircraft annually. The U.S. government began to feel that the air war was so important that they had to overcome the issue of patents. By 1944 after a shift in priorities and while the aircraft assembly lines were mainly staffed by newly trained female riveters, 96,318 air craft were produced.

In 1939 Germany was already producing 8,300 aircraft annually. Germany steadily improved the quality and quantity of their air force. At the height of the war, the year 1944, the Third Reich produced 39,800 aircraft.

a. Use the points associated with US manufacturing of aircraft (1940, 6026) and (1944, 96318), to develop a linear equation (in slope-intercept form) to represent the growth in U.S. aircraft manufacturing.

b. Use the points associated with Germany's manufacturing of aircraft (1939, 8300) and (1944, 39800), to develop a linear equation (in slope-intercept form) to represent the growth in German aircraft manufacturing.

c. Use these two equations to determine approximately when U.S. aircraft production over took German aircraft production.

#36

Math Application by Topic: Acoustics

Math Application by Concept: Linear Equations-2 Variables

A concert hall becomes an extension of a musical instrument. The angles of the walls, materials in room, and size of the space all add subtle echoes and warmth to music. In a well designed music hall the sound and tones are rich and vibrant.

In a much more sterile setting, Dr. Pond demonstrated how an electronic keyboard can synthesize time delay and echo. Because the distance sound travels varies directly as the time it travels ($d = kt$), an electronic keyboard can be programmed to take advantage of the variation constant (k).

If the illusion of sound desired by the musician was that of a large concert hall where sound traveled 336 feet in 0.30 seconds, how far would the sound travel in 0.45 seconds?

#37

Math Application by Topic:	Piano Pitch
Math Application by Concept:	Exponents and Roots

Dr. Pond described the variables in tuning a grand piano. The tension on each string was one of the variables which determined the pitch of the note. The greater the tension, the higher the note. The tension of the strings is held by pegs which can be tightened or loosened.

As stated in the video, the pitch (or frequency) of the note varies directly with the square root of the tension of the string. The equation used was: $F = k\sqrt{T}$, where F is the frequency (measured in Hertz), T is the tension on the string (measured in pounds), and k is a constant of variation.

a. Suppose a string is tuned to a frequency of 233 Hertz (an A sharp) and there are 152 pounds of tension on the string. Use the equation of frequency: $F = k\sqrt{T}$ to find the value of the constant of variation (k). Round to the nearest whole number.

b. Considering the relationship between frequency and tension, raising any note an octave (doubling the frequency) would require how much more tension on a string?

#38

| Math Application by Topic: | Dinosaur Motion |
| Math Application by Concept: | Rational Equations |

The animators of the dinosaurs in the movie Jurassic Park used a computer program to build the dinosaurs on film. As the actors were running, the animated dinosaurs had to look as if they were moving within the same environment. The computer program used mathematical ratios similar to the following: $Dimensionless\ speed = \dfrac{speed}{\sqrt{lg}}$,

where l = leg length, g = gravitational acceleration and *speed* is the speed that the actor was running.

If an actor was moving at a *speed* of 15 feet per second, had a leg length l of 3 feet, with gravitational acceleration g of 16 feet per second, what would be the ratio for *Dimensionless speed* that the computer would use to animate the dinosaurs? Round to the nearest tenth.

#39

Math Application by Topic: Fighting Fires
Math Application by Concept: Exponents and Roots

The pressure of a stream of water from a fire hydrant can be calculated from the equation $G = 26.8D^2 P^{1/2}$ or $G = 26.8D^2 \sqrt{P}$, where G is the discharge of water (in gallons per minute); D is the diameter of the outlet (in inches); and P is the water pressure (in pounds per square inch).

A "feather-light" nozzle was advertised by the Cordoma Fire Equipment Co. as a heavy-duty nozzle with a weight of only 10 ounces. A fire fighter questions the ability of such a light nozzle to sustain the appropriate amount of water pressure. The nozzle was tested. The diameter was 1.5 inches and produced a stream of 350 gallons per minute. What was the water pressure at the nozzle? Round to the nearest tenth.

Date_____ Name_____

Course_____ Section_____ Student Number_____

#40

Math Application by Topic: Pendulum

Math Application by Concept: Exponents and Roots

The Foucault pendulum shown in the video has a large weight attached to the end of a long cable. The cable is attached to the ceiling of the second story of the structure. The length of the wire cable is 30 feet to the bottom of the weighted bob.

a.   Use the equation described in the video, $T = 2\pi \dfrac{\sqrt{L}}{\sqrt{32}}$, to determine the cycle of the pendulum. Let T be the time of a cycle in seconds and L the length of the pendulum in feet. Remember that a cycle is a swing in one direction and back again.

b.   Rewrite the formula so it may be used to determine the Length (L) of the cable. (Solve for L).

A-79

#41

Math Application by Topic: Pendulum

Math Application by Concept: Exponents and Roots

The antique pendulum clock shown in the video should have a cycle of one second long to keep accurate time. Sometimes when the clock's time is too slow or fast, the length of the pendulum must be adjusted. This can be done by raising or lowering the bob at the bottom thus making the pendulum length shorter or longer.

Using the equation described, $T = 2\pi \dfrac{\sqrt{L}}{\sqrt{32}}$, where T is the time of a cycle and L is the length of the pendulum, how many feet long should the pendulum be to keep accurate time?

#42

| Math Application by Topic: | Dinosaur Motion |
| Math Application by Concept: | Exponents and Roots |

The scientist in the video described how the speed of a dinosaur's movements is determined by the height of the animal's hips and the length of its stride. When two dinosaurs have similar movement, then a ratio for dimensionless speed can be used. This type of equation can predict the speed and movement of similarly structured animals. A comparable equation is used for the animation of dinosaurs in movies such as Jurassic Park to make the dinosaurs and people look as if they were moving within the same scene.

Use the equation: Dimensionless speed $= \dfrac{speed}{\sqrt{lg}}$, where l = leg length, and g = gravitational acceleration to find the speed that a dinosaur moves who has a leg length of 2.8 meters. Let the dimensionless speed ratio be 0.4. Let gravitational acceleration be approximately 10 meters per second. Find the speed the dinosaur moves to the nearest tenth of a meter per second.

#43

| Math Application by Topic: | Line of Sight |
| Math Application by Concept: | Quadratic Equations |

The video discusses an equation which pilots use to determine the distance to the horizon by using the altitude of the plane as determined by the instrument panel. The equation is $D = 1.2\sqrt{A}$ where D is measured in miles and A is measured in feet.

a. Rewrite this equation to solve for the variable A.

b. Use your rewritten equation to approximate the altitude of a hot air balloon if the pilot can see the buildings of Springfield, Missouri while flying above Branson (40 miles away). Round to the nearest hundred feet.

Date_____ Name_____

Course_____ Section_____ Student Number_____

#44

Math Application by Topic: Electronics
Math Application by Concept: Quadratic Equations

Complex numbers may be applied to electrical circuits. Electrical engineers use the fact that resistance R to electrical flow, flow of the current I, and the voltage V, are related by the formula $V = RI$. (Voltage is measured in volts, resistance in ohms, and current in amperes.)

Find the resistance to electrical flow in a circuit that has a voltage $V = (80 + 20i)$ volts and current $I = (-6 + 2i)$ amps.

#45

Math Application by Topic: Population Growth
Math Application by Concept: Quadratic Equations

In 1829, former President James Madison made a prediction about the growth of the population of the United States. He felt the population growth could be modeled by the equation, $y = .029x^2 - 1.39x + 42$, where y was the population in millions of people x years from 1829.

Use the equation to determine the approximate year President Madison would have predicted that the U.S. population would reach 100,000,000 ($y = 100$).

#46

Math Application by Topic: Fighting Fires

Math Application by Concept: Quadratic Equations

As shown in the video, firefighters have gauges which are placed at the nozzle of the hose to measure the water flow velocity in feet per second. A tight straight stream is used for penetrating a burning structure, while a fog is for mopping-up at the end of a fire.

The equation: $V^2 = 147P$ shows the relationship between the water pressure (P) at the nozzle point and the velocity (V) of the water flow (in feet per second) as it is sprayed from the hose.

Determine the velocity (to the nearest foot per second) of the water discharged from a nozzle if the gauge indicates 52 pounds of pressure per square inch (psi).

Date_____ Name_____

Course_____ Section_____ Student Number_____

#47

Math Application by Topic: Atlatl
Math Application by Concept: Quadratic Equation

The distance an arrow travels is a function of time. The longer an arrow is in the air, the farther it flies. The compound bow shown in the video can shoot an arrow for a long distance. The equation, $S = -16t^2 + 192t$ describes the distance the arrow in the video would travel when shot straight into the air from a compound bow. S is the distance and t is the time involved.

a.   What is the distance the arrow would travel in 1.5 seconds?
b.   How long would it take the arrow to drop back down to the ground ($S = 0$)?

#48
Math Application by Topic: Optical Lab
Math Application by Concept: Quadratic Equations

Every optical lab can produce products in different amounts of time. There are many factors which can lead to this: skill of the employees, age of the equipment, grade of the materials, methods used, or employee motivation.

One equation which could be used to determine the number of glasses produced in a month by the optical lab in the video is based on the number of employees: $G = 2n^2 - 3n + 5$. Where G is the number of pairs of glasses and n is the number of employees working at the lab. For this equation assume all employees work at the same rate.

Use the equation to determine how many pairs of glasses can be produced by 24 employees working at the lab during a month.

Date_____ Name_____

Course_____ Section_____ Student Number_____

#49

Math Application by Topic: Investigating Accidents

Math Application by Concept: Exponents and Roots

The accident that was described in the video was on the street outside the studio. The investigating officer described a method for calculating how fast the vehicle was going when the breaks were first applied using the coefficient of friction of the road surface and the length of the skid marks. The formula, $S = \sqrt{30df}$ was used where S is the speed of vehicle in mph, d is the length of the skid, and f is a coefficient of friction.

Use the formula to determine how fast (to the nearest mile/hour) the driver was going when the breaks were applied. Because of the bad weather conditions, the wet road had a friction coefficient (f) of 0.55 The tire's skid marks measured 117 feet.

#50

| Math Application by Topic: | Population Growth |
| Math Application by Concept: | Exponents and Roots |

The video indicated that only the roughest of estimates exist for the world population. In fact, equation models have only been found to be good for the short-term estimates. Each time the U.S. census is counted, adjustments have to be made. One population model for the United States is based on an exponential equation. Using an initial value of 250 million people, the equation predicts a 0.155 percent increase in the population per year.

The population p at year t is computed as $p = 2.5 \times 10^8 (1.00155)^t$. Calculate the predicted population in 20 years ($t = 20$).

#51

Math Application by Topic: Earthquakes

Math Application by Concept: Additional Topics

The video showed footage of an earthquake in Japan. That earthquake which hit Kobe, Japan in 1992 recorded a magnitude of $M = 7.2$. The earthquake with the greatest magnitude ($M = 8.9$) ever recorded was also in Japan. Because the Richter scale is a logarithmic scale the difference between the two numbers is not very much, but the shock waves they measure are very different. The magnitude M of an earthquake is given by the formula: $M = \log_{10} T$, where T is the relative intensity of the shock wave.

Use the formula, $M = \log_{10} T$, in exponential form, to determine how many times greater the shock wave T was during the great Japanese earthquake ($M = 8.9$) than the Kobe earthquake of 1992 ($M = 7.2$) In solving round T to the nearest 10 million.

#52

Math Application by Topic: Earthquakes

Math Application by Concept: Additional Topics

The video indicated that there are approximately 6000 earthquakes worldwide per year, but 5500 of these earthquakes are either too small or remote to be felt. On a Richter Scale the magnitude M of an earthquake is given by the formula: $M = \log_{10} T$, where T is the relative intensity of the shock wave to the smallest shock wave measurable. What would be the magnitude of an earthquake that had a shock wave of $T = 100,000$?

Date_____ Name_____

Course_____ Section_____ Student Number_____

#53

Math Application by Topic: Aeronautics

Math Application by Concept: Sequences and Series

A pilot checks weather conditions with the National Weather Service before taking off and finds that the air temperature is dropping 4.7°F for every 1000 feet above the surface of the earth. (The higher the flight, the colder the air.)

a. If the air temperature is 38°F when the plane reaches 12,000 feet, write a sequence of numbers that gives the air temperature every 1000 feet as the plane climbs from 12,000 to 18,000 feet. Is this an arithmetic or geometric sequence?

b. Extend the sequence to determine approximately at what altitude the air temperature would fall below 0°F ?

#54

| Math Application by Topic: | Piano Pitch |
| Math Application by Concept: | Sequences and Series |

Dr. Pond discussed the harmonics of music as it applies to the length, mass and tension of piano strings. The sequence $1, \dfrac{1}{2}, \dfrac{1}{3}, \dfrac{1}{4}, \dfrac{1}{5}, \ldots$ is a natural harmonic sequence and can be mathematically associated with a vibrating musical string.

a. Find the next term in this sequence.
b. Find the 78th term in this sequence.
c. Write an expression for the term number n in the sequence.

A-108

#1

Math Application by Topic: Flying to Hawaii
Math Application by Concept: Basics
Answer: 11:10 AM

#2

Math Application by Topic: Beetle Infestation
Math Application by Concept: Basics
Answers:
a. -$62.75
b. -$45,493,750,000

#3

Math Application by Topic: Population Growth
Math Application by Concept: Basics
Answer: Loss of 585,000 people (-585,000)

#4

Math Application by Topic: Premature Infant Formula
Math Application by Concept: Basics
Answer: 4.3 grams

#5

Math Application by Topic: Aeronautics
Math Application by Concept: Basics
Answer: 634 feet

#6

Math Application by Topic: World War II Bombers
Math Application by Concept: Basics
Answers:
a. 66,725 feet per minute
b. Approximately 758 miles per hour which is slightly less than the speed of sound.

#7

Math Application by Topic: Air Pollution
Math Application by Concept: Linear Equations
Answers:
a. $C = 24T$
b. 300 trees

#8

Math Application by Topic: Investing and Debt
Math Application by Concept: Linear Equations
Answers:
a. $360.75
b. $55.71

#9

Math Application by Topic: Falling Objects
Math Application by Concept: Linear Equations
Answer: 30 feet per second

#10

Math Application by Topic: Acoustics
Math Application by Concept: Linear Equations
Answer: $T \cup 0.47$ second

#11

Math Application by Topic: Investing and Debt
Math Application by Concept: Linear Equations
Answers:
a. Difference during day = $10.2 million
b. Market value of stock at end of day = $3,876,000

#12

Math Application by Topic: Investigating Accidents
Math Application by Concept: Linear Equations
Answers:
a. 6.1 tons
b. 1.7 tons

#13

Math Application by Topic: Optical Lab
Math Application by Concept: Linear Equations
Answer: The glasses would not fit the customer's face correctly.
               4.5 is not between 5 and 8.

#14

Math Application by Topic: Constructing Buildings
Math Application by Concept: Linear Equations

Answer: The slope is: $\dfrac{3}{5}$

#15
Math Application by Topic: Map Evolution
Math Application by Concept: Linear Systems
Answer: 8 years

#16
Math Application by Topic: Air Pollution
Math Application by Concept: Linear Systems
Answer: $x = 20$
            $y = 14$

#17
Math Application by Topic: Metal Alloys
Math Application by Concept: Linear Systems
Answer: Approximately 238.64 pounds of the alloy with 82% copper
                Approximately 111.36 pounds of the alloy with 60% copper

#18
Math Application by Topic: Premature Infant Formula
Math Application by Concept: Linear Systems
Answer: 2 kl of 10% dextrose solution
             1 kl of 70% dextrose solution

#19
Math Application by Topic: Living Soil Crust
Math Application by Concept: Linear Systems
Answer: 10 gallons of the 30% solution
             15 gallons of the 60% solution

#20
Math Application by Topic: Electronics
Math Application by Concept: Linear Systems
Answer: $x = 4,\ y = 3, z = 1$

#21
Math Application by Topic: Constructing Buildings
Math Application by Concept: Basics
Answer: 11.6 feet

#22

Math Application by Topic: Line of Sight
Math Application by Concept: Exponents and Roots
Answer:

Mercury	5.8×10^7
Venus	1.081×10^8
Earth	1.5×10^8
Mars	2.2784×10^8
Jupiter	7.781×10^8
Saturn	1.427×10^9
Uranus	2.8703×10^9
Neptune	4.5×10^9
Pluto	5.913×10^9

#23

Math Application by Topic: Atlatl
Math Application by Concept: Exponents and Roots
Answer: The statement was correct. $I = 225mr^2$

#24

Math Application by Topic: Optical Lab
Math Application by Concept: Exponents and Roots

Answer: $f = \dfrac{ab}{a+b}$

#25

Math Application by Topic: Electronics
Math Application by Concept: Rational Equations

Answer: $\dfrac{R_1 R_2 R_3}{R_2 R_3 + R_1 R_3 + R_1 R_2}$

#26

Math Application by Topic: Flying to Hawaii
Math Application by Concept: Rational Equations
Answer: Average Speed ≈ 293 mph

#27

Math Application by Topic: Map Evolution
Math Application by Concept: Rational Equations
Answer: 200 miles

#28
Math Application by Topic: Beetle Infestation
Math Application by Concept: Rational Equations

Answer: $\frac{6}{5}$ hours or 1 hour 12 minutes.

#29
Math Application by Topic: Piano Pitch
Math Application by Concept: Rational Expressions
Answer: String length for the A note ≈ 36.3 inches

#30
Math Application by Topic: Lighting a Stage
Math Application by Concept: Rational Equations
Answer: 15 lumens

#31
Math Application by Topic: Flying to Hawaii
Math Application by Concept: Linear Equations-2 Variables
Answer:

Distance to Airport (miles)	Altitude (miles)
-65	5.7
-55	4.8
-45	3.9
-35	3.1
-25	2.2
-15	1.3
-5	0.4

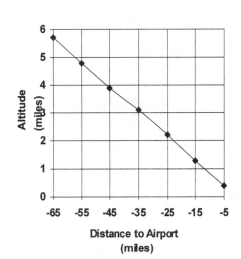

#32

Math Application by Topic: Lighting a Stage
Math Application by Concept: Linear Equations-2 Variables
Answer:

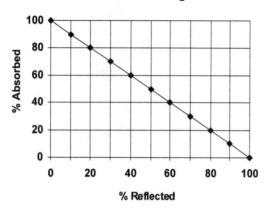

Relationship Between Reflected and Absorbed Light

#33

Math Application by Topic: Investing and Debt
Math Application by Concept: Linear Equations-2 Variables
Answers:
a. $188.00
b. $5937.50

#34

Math Application by Topic: Metal Alloys
Math Application by Concept: Linear Equations-2 Variables
Answers:
a.   Place graph from solution here. See next page…
b. $y = -3x + 300$, where x = % of zinc and y = cost in cents

---

#35

Math Application by Topic: World War II Bombers
Math Application by Concept: Linear Equations-2 Variables
Answers: a. $y = 22573x - 43785594$

         b. $y = 6300x - 12207400$

         c. $x \cup 1940.5$ Thus, about midway through 1940 the US production of airplanes overtook German production.

#36

Math Application by Topic:	Acoustics
Math Application by Concept:	Linear Equations-2 Variables
Answer:	504 feet

#37

Math Application by Topic:	Piano Pitch
Math Application by Concept:	Exponents and Roots

Answers:

a. Constant of variation, $k \cup 19$

b. The tension would have to be increased by a multiple of four to double the frequency.

#38

Math Application by Topic:	Dinosaur Motion
Math Application by Concept:	Rational Equations
Answer:	Dimensionless speed ≈ 2.2

#39

Math Application by Topic:	Fighting Fires
Math Application by Concept:	Exponents and Roots
Answer:	$P \approx 33.6$ psi (pounds per square inch)

#40

Math Application by Topic:	Pendulum
Math Application by Concept:	Exponents and Roots

Answers:

a. $T \cup 6.1$ seconds. Rounded to the nearest tenth.

b. $L = \dfrac{8T^2}{\pi^2}$

#41

Math Application by Topic:	Pendulum
Math Application by Concept:	Exponents and Roots
Answer:	0.81 foot. Rounded to the nearest hundredth.

#42

Math Application by Topic:	Dinosaur Motion
Math Application by Concept:	Exponents and Roots
Answer:	$S \approx 2.1$ meters per second

#43

Math Application by Topic: Line of Sight
Math Application by Concept: Quadratic Equations
Answers:

a. $A = \dfrac{D^2}{1.44}$

b. 1,100 feet

#44

Math Application by Topic: Electronics
Math Application by Concept: Quadratic Equations
Answer: $R = -11 - 7i$

#45

Math Application by Topic: Population Growth
Math Application by Concept: Quadratic Equations
Answer: 1904

#46

Math Application by Topic: Fighting Fires
Math Application by Concept: Quadratic Equations
Answer: $V = 87$ feet per second

#47

Math Application by Topic: Atlatl
Math Application by Concept: Quadratic Equation
Answers:

a. $S = 252$ feet

b. The arrow would fall back to earth in 12 seconds.

#48

Math Application by Topic: Optical Lab
Math Application by Concept: Quadratic Equations
Answer: $G = 1085$ pairs of glasses

#49

Math Application by Topic: Investigating Accidents
Math Application by Concept: Exponents and Roots
Answer: 44 miles per hour

#50

Math Application by Topic: Population Growth
Math Application by Concept: Exponents and Roots
Answer: $p \approx 25{,}786{,}500$ people

#51

Math Application by Topic: Earthquakes
Math Application by Concept: Additional Topics
Answer: The shock wave was approximately 53 times the intensity.

#52

Math Application by Topic: Earthquakes
Math Application by Concept: Additional Topics
Answer: $M = 5$

#53

Math Application by Topic: Aeronautics
Math Application by Concept: Sequences and Series
Answers:

a. {38°F, 33.3°F, 28.6°F, 23.9°F, 19.2°F, 14.5°F, 9.8°F}; Arithmetic sequence

b. Between 20,000 and 21,000 feet.

#54

Math Application by Topic: Piano Pitch
Math Application by Concept: Sequences and Series

Answers   a. $\dfrac{1}{6}$

        b. $a_{78=}\dfrac{1}{78}$

        c. $a_n = \dfrac{1}{n}$